algebraic geometry

LECTURE NOTES IN PURE AND APPLIED MATHEMATICS

1. *N. Jacobson,* Exceptional Lie Algebras
2. *L.-Å. Lindahl and F. Poulsen,* Thin Sets in Harmonic Analysis
3. *I. Satake,* Classification Theory of Semi-Simple Algebraic Groups
4. *F. Hirzebruch et al., Differentiable Manifolds and Quadratic Forms*
5. *I. Chavel,* Riemannian Symmetric Spaces of Rank One
6. *R. B. Burckel,* Characterization of C(X) Among Its Subalgebras
7. *B. R. McDonald et al.,* Ring Theory
8. *Y.-T. Siu,* Techniques of Extension on Analytic Objects
9. *S. R. Caradus et al.,* Calkin Algebras and Algebras of Operators on Banach Spaces
10. *E. O. Roxin et al.,* Differential Games and Control Theory
11. *M. Orzech and C. Small,* The Brauer Group of Commutative Rings
12. *S. Thomier,* Topology and Its Applications
13. *J. M. Lopez and K. A. Ross,* Sidon Sets
14. *W. W. Comfort and S. Negrepontis,* Continuous Pseudometrics
15. *K. McKennon and J. M. Robertson,* Locally Convex Spaces
16. *M. Carmeli and S. Malin,* Representations of the Rotation and Lorentz Groups
17. *G. B. Seligman,* Rational Methods in Lie Algebras
18. *D. G. de Figueiredo,* Functional Analysis
19. *L. Cesari et al.,* Nonlinear Functional Analysis and Differential Equations
20. *J. J. Schäffer,* Geometry of Spheres in Normed Spaces
21. *K. Yano and M. Kon,* Anti-Invariant Submanifolds
22. *W. V. Vasconcelos,* The Rings of Dimension Two
23. *R. E. Chandler,* Hausdorff Compactifications
24. *S. P. Franklin and B. V. S. Thomas,* Topology
25. *S. K. Jain,* Ring Theory
26. *B. R. McDonald and R. A. Morris,* Ring Theory II
27. *R. B. Mura and A. Rhemtulla,* Orderable Groups
28. *J. R. Graef,* Stability of Dynamical Systems
29. *H.-C. Wang,* Homogeneous Branch Algebras
30. *E. O. Roxin et al.,* Differential Games and Control Theory II
31. *R. D. Porter,* Introduction to Fibre Bundles
32. *M. Altman,* Contractors and Contractor Directions Theory and Applications
33. *J. S. Golan,* Decomposition and Dimension in Module Categories
34. *G. Fairweather,* Finite Element Galerkin Methods for Differential Equations
35. *J. D. Sally,* Numbers of Generators of Ideals in Local Rings
36. *S. S. Miller,* Complex Analysis
37. *R. Gordon,* Representation Theory of Algebras
38. *M. Goto and F. D. Grosshans,* Semisimple Lie Algebras
39. *A. I. Arruda et al.,* Mathematical Logic
40. *F. Van Oystaeyen,* Ring Theory
41. *F. Van Oystaeyen and A. Verschoren,* Reflectors and Localization
42. *M. Satyanarayana,* Positively Ordered Semigroups
43. *D. L Russell,* Mathematics of Finite-Dimensional Control Systems
44. *P.-T. Liu and E. Roxin,* Differential Games and Control Theory III
45. *A. Geramita and J. Seberry,* Orthogonal Designs
46. *J. Cigler, V. Losert, and P. Michor,* Banach Modules and Functors on Categories of Banach Spaces
47. *P.-T. Liu and J. G. Sutinen,* Control Theory in Mathematical Economics
48. *C. Byrnes,* Partial Differential Equations and Geometry
49. *G. Klambauer,* Problems and Propositions in Analysis
50. *J. Knopfmacher,* Analytic Arithmetic of Algebraic Function Fields
51. *F. Van Oystaeyen,* Ring Theory
52. *B. Kadem,* Binary Time Series
53. *J. Barros-Neto and R. A. Artino,* Hypoelliptic Boundary-Value Problems
54. *R. L. Sternberg et al.,* Nonlinear Partial Differential Equations in Engineering and Applied Science
55. *B. R. McDonald,* Ring Theory and Algebra III
56. *J. S. Golan,* Structure Sheaves Over a Noncommutative Ring
57. *T. V. Narayana et al.,* Combinatorics, Representation Theory and Statistical Methods in Groups

58. *T. A. Burton,* Modeling and Differential Equations in Biology
59. *K. H. Kim and F. W. Roush,* Introduction to Mathematical Consensus Theory
60. *J. Banas and K. Goebel,* Measures of Noncompactness in Banach Spaces
61. *O. A. Nielson,* Direct Integral Theory
62. *J. E. Smith et al.,* Ordered Groups
63. *J. Cronin,* Mathematics of Cell Electrophysiology
64. *J. W. Brewer,* Power Series Over Commutative Rings
65. *P. K. Kamthan and M. Gupta,* Sequence Spaces and Series
66. *T. G. McLaughlin,* Regressive Sets and the Theory of Isols
67. *T. L. Herdman et al.,* Integral and Functional Differential Equations
68. *R. Draper,* Commutative Algebra
69. *W. G. McKay and J. Patera,* Tables of Dimensions, Indices, and Branching Rules for Representations of Simple Lie Algebras
70. *R. L. Devaney and Z. H. Nitecki,* Classical Mechanics and Dynamical Systems
71. *J. Van Geel,* Places and Valuations in Noncommutative Ring Theory
72. *C. Faith,* Injective Modules and Injective Quotient Rings
73. *A. Fiacco,* Mathematical Programming with Data Perturbations I
74. *P. Schultz et al.,* Algebraic Structures and Applications
75. *L Bican et al.,* Rings, Modules, and Preradicals
76. *D. C. Kay and M. Breen,* Convexity and Related Combinatorial Geometry
77. *P. Fletcher and W. F. Lindgren,* Quasi-Uniform Spaces
78. *C.-C. Yang,* Factorization Theory of Meromorphic Functions
79. *O. Taussky,* Ternary Quadratic Forms and Norms
80. *S. P. Singh and J. H. Burry,* Nonlinear Analysis and Applications
81. *K. B. Hannsgen et al.,* Volterra and Functional Differential Equations
82. *N. L. Johnson et al.,* Finite Geometries
83. *G. I. Zapata,* Functional Analysis, Holomorphy, and Approximation Theory
84. *S. Greco and G. Valla,* Commutative Algebra
85. *A. V. Fiacco,* Mathematical Programming with Data Perturbations II
86. *J.-B. Hiriart-Urruty et al.,* Optimization
87. *A. Figa Talamanca and M. A. Picardello,* Harmonic Analysis on Free Groups
88. *M. Harada,* Factor Categories with Applications to Direct Decomposition of Modules
89. *V. I. Istrățescu,* Strict Convexity and Complex Strict Convexity
90. *V. Lakshmikantham,* Trends in Theory and Practice of Nonlinear Differential Equations
91. *H. L. Manocha and J. B. Srivastava,* Algebra and Its Applications
92. *D. V. Chudnovsky and G. V. Chudnovsky,* Classical and Quantum Models and Arithmetic Problems
93. *J. W. Longley,* Least Squares Computations Using Orthogonalization Methods
94. *L. P. de Alcantara,* Mathematical Logic and Formal Systems
95. *C. E. Aull,* Rings of Continuous Functions
96. *R. Chuaqui,* Analysis, Geometry, and Probability
97. *L. Fuchs and L. Salce,* Modules Over Valuation Domains
98. *P. Fischer and W. R. Smith,* Chaos, Fractals, and Dynamics
99. *W. B. Powell and C. Tsinakis,* Ordered Algebraic Structures
100. *G. M. Rassias and T. M. Rassias,* Differential Geometry, Calculus of Variations, and Their Applications
101. *R.-E. Hoffmann and K. H. Hofmann,* Continuous Lattices and Their Applications
102. *J. H. Lightbourne III and S. M. Rankin III,* Physical Mathematics and Nonlinear Partial Differential Equations
103. *C. A. Baker and L. M. Batten,* Finite Geometrics
104. *J. W. Brewer et al.,* Linear Systems Over Commutative Rings
105. *C. McCrory and T. Shifrin,* Geometry and Topology
106. *D. W. Kueke et al.,* Mathematical Logic and Theoretical Computer Science
107. *B.-L. Lin and S. Simons,* Nonlinear and Convex Analysis
108. *S. J. Lee,* Operator Methods for Optimal Control Problems
109. *V. Lakshmikantham,* Nonlinear Analysis and Applications
110. *S. F. McCormick,* Multigrid Methods
111. *M. C. Tangora,* Computers in Algebra
112. *D. V. Chudnovsky and G. V. Chudnovsky,* Search Theory
113. *D. V. Chudnovsky and R. D. Jenks,* Computer Algebra
114. *M. C. Tangora,* Computers in Geometry and Topology
115. *P. Nelson et al.,* Transport Theory, Invariant Imbedding, and Integral Equations
116. *P. Clément et al.,* Semigroup Theory and Applications

117. *J. Vinuesa,* Orthogonal Polynomials and Their Applications
118. *C. M. Dafermos et al.,* Differential Equations
119. *E. O. Roxin,* Modern Optimal Control
120. *J. C. Díaz,* Mathematics for Large Scale Computing
121. *P. S. Milojevič,* Nonlinear Functional Analysis
122. *C. Sadosky,* Analysis and Partial Differential Equations
123. *R. M. Shortt,* General Topology and Applications
124. *R. Wong,* Asymptotic and Computational Analysis
125. *D. V. Chudnovsky and R. D. Jenks,* Computers in Mathematics
126. *W. D. Wallis et al.,* Combinatorial Designs and Applications
127. *S. Elaydi,* Differential Equations
128. *G. Chen et al.,* Distributed Parameter Control Systems
129. *W. N. Everitt,* Inequalities
130. *H. G. Kaper and M. Garbey,* Asymptotic Analysis and the Numerical Solution of Partial Differential Equations
131. *O. Arino et al.,* Mathematical Population Dynamics
132. *S. Coen,* Geometry and Complex Variables
133. *J. A. Goldstein et al.,* Differential Equations with Applications in Biology, Physics, and Engineering
134. *S. J. Andima et al.,* General Topology and Applications
135. *P Clément et al.,* Semigroup Theory and Evolution Equations
136. *K. Jarosz,* Function Spaces
137. *J. M. Bayod et al.,* *p*-adic Functional Analysis
138. *G. A. Anastassiou,* Approximation Theory
139. *R. S. Rees,* Graphs, Matrices, and Designs
140. *G. Abrams et al.,* Methods in Module Theory
141. *G. L. Mullen and P. J.-S. Shiue,* Finite Fields, Coding Theory, and Advances in Communications and Computing
142. *M. C. Joshi and A. V. Balakrishnan,* Mathematical Theory of Control
143. *G. Komatsu and Y. Sakane,* Complex Geometry
144. *I. J. Bakelman,* Geometric Analysis and Nonlinear Partial Differential Equations
145. *T. Mabuchi and S. Mukai,* Einstein Metrics and Yang–Mills Connections
146. *L. Fuchs and R. Göbel,* Abelian Groups
147. *A. D. Pollington and W. Moran,* Number Theory with an Emphasis on the Markoff Spectrum
148. *G. Dore et al.,* Differential Equations in Banach Spaces
149. *T. West,* Continuum Theory and Dynamical Systems
150. *K. D. Bierstedt et al.,* Functional Analysis
151. *K. G. Fischer et al.,* Computational Algebra
152. *K. D. Elworthy et al.,* Differential Equations, Dynamical Systems, and Control Science
153. *P.-J. Cahen, et al.,* Commutative Ring Theory
154. *S. C. Cooper and W. J. Thron,* Continued Fractions and Orthogonal Functions
155. *P. Clément and G. Lumer,* Evolution Equations, Control Theory, and Biomathematics
156. *M. Gyllenberg and L. Persson,* Analysis, Algebra, and Computers in Mathematical Research
157. *W. O. Bray et al.,* Fourier Analysis
158. *J. Bergen and S. Montgomery,* Advances in Hopf Algebras
159. *A. R. Magid,* Rings, Extensions, and Cohomology
160. *N. H. Pavel,* Optimal Control of Differential Equations
161. *M. Ikawa,* Spectral and Scattering Theory
162. *X. Liu and D. Siegel,* Comparison Methods and Stability Theory
163. *J.-P. Zolésio,* Boundary Control and Variation
164. *M. Křížek et al.,* Finite Element Methods
165. *G. Da Prato and L. Tubaro,* Control of Partial Differential Equations
166. *E. Ballico,* Projective Geometry with Applications
167. *M. Costabel et al.,* Boundary Value Problems and Integral Equations in Nonsmooth Domains
168. *G. Ferreyra, G. R. Goldstein, and F. Neubrander,* Evolution Equations
169. *S. Huggett,* Twistor Theory
170. *H. Cook et al.,* Continua
171. *D. F. Anderson and D. E. Dobbs,* Zero-Dimensional Commutative Rings
172. *K. Jarosz,* Function Spaces
173. *V. Ancona et al.,* Complex Analysis and Geometry
174. *E. Casas,* Control of Partial Differential Equations and Applications
175. *N. Kalton et al.,* Interaction Between Functional Analysis, Harmonic Analysis, and Probability

176. *Z. Deng et al.*, Differential Equations and Control Theory
177. *P. Marcellini et al.* Partial Differential Equations and Applications
178. *A. Kartsatos*, Theory and Applications of Nonlinear Operators of Accretive and Monotone Type
179. *M. Maruyama*, Moduli of Vector Bundles
180. *A. Ursini and P. Aglianò*, Logic and Algebra
181. *X. H. Cao et al.*, Rings, Groups, and Algebras
182. *D. Arnold and R. M. Rangaswamy*, Abelian Groups and Modules
183. *S. R. Chakravarthy and A. S. Alfa*, Matrix-Analytic Methods in Stochastic Models
184. *J. E. Andersen et al.*, Geometry and Physics
185. *P.-J. Cahen et al.*, Commutative Ring Theory
186. *J. A. Goldstein et al..*, Stochastic Processes and Functional Analysis
187. *A. Sorbi*, Complexity, Logic, and Recursion Theory
188. *G. Da Prato and J.-P. Zolésio*, Partial Differential Equation Methods in Control and Shape Analysis
189. *D. D. Anderson*, Factorization in Integral Domains
190. *N. L. Johnson*, Mostly Finite Geometries
191. *D. Hinton and P. W. Schaefer*, Spectal Theory and Computational Methods of Sturm–Liouville Problems
192. *W. H. Schikhof et al.*, *p*-adic Functional Analysis
193. *S. Sertöz*, Algebraic Geometry

Additional Volumes in Preparation

algebraic geometry

proceedings of Bilkent Summer School

edited by

Sinan Sertöz
Bilkent University
Ankara, Turkey

MARCEL DEKKER

MARCEL DEKKER, INC. NEW YORK • BASEL • HONG KONG

Library of Congress Cataloging-in-Publication Data

Algebraic geometry : proceedings of of Bilkent summer school / edited by Sinan Sertöz.
p. cm. — (Lecture notes in pure and applied mathematics ; v. 193)
Papers from the Algebraic Geometry Summer School held Aug. 7-19, 1995, at Bilkent University, Ankara, Turkey.
Includes bibliographical references and index.
ISBN 0-8247-0123-2 (alk. paper)
1. Geometry, Algebraic—Congresses. I. Sertöz, Sinan. II. Algebraic Geometry Summer School (1995 : Bilkent University) III. Series.
QA564.A376 1997
516.3'5—dc21

97-22487
CIP

The publisher offers discounts on this book when ordered in bulk quantities. For more information, write to Special Sales/Professional Marketing at the address below.

This book is printed on acid-free paper.

MARCEL DEKKER, INC.
270 Madison Avenue, New York, New York 10016
http://www.dekker.com

Current printing (last digit):
10 9 8 7 6 5 4 3 2 1

PRINTED IN THE UNITED STATES OF AMERICA

I take great pleasure and honour
in dedicating this book to

Professor İhsan Doğramacı

founder of Bilkent University,
for his grand patronage of the arts and sciences

Preface

The amount of information that science is producing is growing at a tremendous rate. We can no longer be comfortable and content with our activities in one research problem in a special field of one particular area of human knowledge. The borders of neighboring fields are no longer visible. In fact more often than not we discover themes of a seemingly unrelated field in our own research. In the past we were tempted to find out how our subject relates to other fields. Today it is a necessity.

One major obstacle that hinders aspiring graduate students' enthusiasm is the apparent discrepancy between the wealth and breadth of information in textbooks and the depth of insight and specialization in published papers. Most of what they learn today will be out of date tomorrow. Besides, research requires quite different skills than those utilized by graduate students in passing their courses successfully. The skills required for acquiring knowledge involve an obedient acceptance of the powers of past masters. However, research requires, in addition to this, a questioning attitude toward what has been achieved so far. And how to strike a balance between these two extremes can be learned only from successful practitioners of the trade.

Recently a summer school on algebraic geometry was held at Bilkent University, Ankara. The school attracted sixty-one participants from twelve countries other than Turkey: Austria, England, France, Germany, Iran, Israel, Japan, Kyrgyzstan, Mexico, Poland, Russia, and the United States. In this truly international atmosphere fourteen speakers gave a total of sixty-five lectures in two weeks. This book constitutes a collection of the lectures given at the summer school. All the articles here are accounts of talks given at Bilkent with the exception of the article I wrote. Miles Reid, who gave the lectures on threefolds, was not able to provide a manuscript himself but was kind enough to read and referee my article for inclusion in these proceedings. Two lectures are not included in this volume. David Eisenbud's lecture notes will appear under the title "Résolutions libres des idéaux de courbes et d'ensembles des points"

(Diderot Editeurs, Paris). Alexander Reznikov's manuscript can be found in *Selecta Mathematica*.

The lectures in the summer school were organized to survey some fundamental ideas in the theory of curves, surfaces, and threefolds and then expose the students to applications of these ideas and techniques to a wide array of topics. Readers of the book will find that the particular spirit of the school is alive in this collection of articles.

All was not work during that summer. Besides academic connections, technical knowledge, and a wider perspective on the subject, the participants took home with them several anecdotes. On one occasion a tour of Ankara was planned. The tour company was asked to send the bus "to the backyard of the science faculty." While the bus was waiting at the backyard of the science faculty of *another* university in Ankara the participants were saved by Vice Rector Erol Sezer, who assigned a university bus to run the tour as scheduled. The Cappadocia tour, however, ran without a hitch. The opening of the school was filmed for TV and broadcast that same night during a live program in which Norbert Schappacher participated as a guest. İhsan Doğramacı, the founder of Bilkent University, President and Chairman of the Board of Trustees, invited the speakers to a dinner at his mansion where besides good food we were treated to good music. A talented young pianist from Azerbaijan gave a solo concert. This was of course after David Eisenbud sang, sometimes with the participation of Rick Miranda.

The summer school was made possible by the vision and aid of the Bilkent administration, which supported the school without reservation. The Scientific and Technical Research Council of Turkey, TÜBİTAK, generously provided travel funds for the lecturers. The British Council in Ankara kindly covered Miles Reid's airfare. Yapı Kredi Bank supplied stationery for the participants. Bilintur provided food services and touristic excursions. We also remember the graciousness of the Faculty of Music and Performing Arts of Bilkent University who arranged the opening concert of the school.

During the preparation and execution stages of the summer school project I received unending support and help from my colleagues, from graduate students, and from the secretarial staff of the faculty. To all I express my gratitude. In particular I thank my friend and colleague, Hurşit Önsiper, a member of the organizing committee, who gave me wise advice and freedom to act. And finally I thank my wife, Tülin, who organized the TV coverage of the school, for not only sharing my enthusiasm but for also helping me to keep it alive.

Sinan Sertöz

Contents

Contributors

Abd'Allah Al Amrani Université Louis Pasteur, Strasbourg, France

Jean-Paul Brasselet Institut de Mathématiques de Luminy, Marseille, France

Xavier Gómez-Mont Centro de Investigación en Matemáticas, A. C., Guanajuato, Gto., México

Michał Kwieciński Uniwersytet Jagielloński, Kraków, Poland

Rick Miranda Colorado State University, Fort Collins, Colorado

Donal O'Shea Mount Holyoke College, South Hadley, Massachusetts

Norbert Schappacher UFR de mathématique et d'informatique, Strasbourg, France

Sinan Sertöz Bilkent University, Ankara, Turkey

Serguei A. Stepanov Bilkent University, Ankara, Turkey

Andrei Tyurin Steklov Mathematical Institute, Moscow, Russia

Conference Program

Algebraic Geometry Summer School

Bilkent University, Ankara

Abd'Allah Al Amrani,
Weighted Projective Bundles and Chern Classes.

Jean-Paul Brasselet,
Geometry of Toric Varieties.

David Eisenbud,
Matrices of Linear Forms.

Xavier Gómez-Mont,
Curves and Their Jacobians.

Ryoichi Kobayashi,
Restructuring Value Distribution Theory.

Michał Kwieciński,
Around MacPherson's Graph Construction.

Rick Miranda,
Algebraic Surfaces.

Donal O'Shea,
Computing Invariants of Singularities of Hypersurfaces.

Murad Özaydın,
Hyperplane Arrangements.

Miles Reid,
Introduction to Threefolds.

Alexander Reznikov,
Three-manifolds Class Field Theory.

Norbert Schappacher,
Aspects of Complex Multiplication.

Serguei Stepanov,
Codes on Complete Intersections.

Andrei Tyurin,
Vector Bundles on Algebraic Curves and Surfaces.

Conference Participants

Abdymanapov, Usen; National Academy of Sciences, Kyrgyzstan
Aghassi, Mansour; University of Durham, England
Ağca, Serap; Boğaziçi University, Turkey
Akyar, Bedia; Dokuz Eylül Üniversitesi, Turkey
Al Amrani, Abd'Allah; Université Louis Pasteur, France
Altınok, Selma; University of Warwick, England
Arslan, Sefa Feza; Bilkent University, Turkey
Atasoy, Koray; Bilkent University, Turkey
Bennun, Hayati; METU, Turkey
Bıçak, Ali; Bilkent University, Turkey
Bilhan, Mehpare; METU, Turkey
Brasselet, Jean-Paul; IML, Marseille, France
Caibar, Mirel; University of Warwick, England
Caldararu, Andrei Horia; Cornell University, USA
Çakçak, Emrah; METU, Turkey
Çelikler, Fırat; Bilkent University, Turkey
Davarcıoğlu, Burhan; Niğde Üniversitesi, Turkey
Demir, Selçuk; Boğaziçi Üniversitesi, Turkey
Dönmez, Doğan; Çukurova Üniversitesi, Turkey
Eisenbud, David; Brandeis University, USA
Esen, Yakup; METU, Turkey
Gawlick, Thomas; Ruhr University, Germany
Gómez-Mont, Xavier; CIMAT, Mexico
Gönen, Alper; İTÜ, Turkey
Günaydın, Aytaç; Osmangazi Üniversitesi, Turkey
Güneri, Cem; METU, Turkey
Haghighi, Hassan; University of Tehran, Iran
Kara, Orhun; Bilkent University, Turkey
Kaygun, Atabey; Boğaziçi Üniversitesi, Turkey
Keyman, Ebru; University of Sussex, England
Kırdar, Mehmet; Bilkent University, Turkey
Kobayashi, Ryoichi; Nagoya University, Japan
Kraskiewicz, Witold; Nicolas Copernicus University, Torun, Poland
Kwieciński, Michał; Jagiellonian University, Krakow, Poland
Lorenz, Wotzlaw; Bremen University, Germany
Miranda, Rick; Colorado State University, USA

O'Shea, Donal; Mount Holyoke College, USA
Önsiper, Hurşit; METU, Turkey
Özaydın, Murad; University of Oklahoma, USA
Özbağcı, Burak; UC Irvine, USA
Özbudak, Ferruh; Bilkent University, Turkey
Özkara, Bülent; Zonguldak Karaelmas Üniversitesi, Turkey
Radtke, Achim; Bremen University, Germany
Reid, Miles; Warwick University, England
Reznikov, Alexander; Hebrew University, Israel
Satır, Ahmet; METU, Turkey
Schappacher, Norbert; Université Louis Pasteur, France
Sertöz, Sinan; Bilkent University, Turkey
Stepanov, Serguei; Bilkent University, Turkey
Şahin, Ayşe; North Dakota State University, USA
Tekinel, Can; METU, Turkey
Tekman, Okan; Bilkent Universitesi, Turkey
Tosun, Meral; Universite Paris VII, France
Tyurin, Andrei; Steklov Institute, Russia
Unterkircher, Andreas; University of Innsbruck, Austria
Uysal, Vahap Bülent; Bilkent University, Turkey
Ürtiş, Çetin; Bilkent University, Turkey
Yakar, Handan; Anadolu Üniversitesi, Turkey
Yalçın, Ergun; Wisconsin University, USA
Yücel, Gülnihal; Boğaziçi Üniversitesi, Turkey
Yüzügüllü, Tuba; Anadolu Üniversitesi, Turkey

algebraic geometry

Cohomological Study of Weighted Projective Spaces

Abd'Allah Al Amrani

Département de Mathématiques,
Université Louis Pasteur,
67084 Strasbourg, France

Introduction

Let $\pi : E \longrightarrow X$ be a complex vector bundle of constant rank and with a compact base X. Assume given an operation $\sigma : \mathbb{C}^* \times E \longrightarrow E$ of the multiplicative group $\mathbb{C}^*$ on E, compatible with the natural operation of $\mathbb{C}^*$. In the algebraic (resp. analytic) situation, σ is supposed to be defined by polynomial (resp. holomorphic) functions, so that E splits into a direct sum of vector bundles $E = E_0 \oplus \cdots \oplus E_n$ (over X), such that the operation σ is determined by the characters of $\mathbb{C}^*$ in the following way

$$\sigma(\lambda, (u_0, \cdots, u_n)) = (\chi_0(\lambda)\, u_0, \cdots, \chi_n(\lambda)\, u_n) \quad (u_i \in E_i)\ ,$$

where χ_i are characters of $\mathbb{C}^*$. These have the form $\chi_i(\lambda) = \lambda^{q_i}$ $(\lambda \in \mathbb{C}^*, q_i \in \mathbb{Z})$. Consider the case $q_0 > 0, \cdots, q_n > 0$, and denote by

$$\widetilde{\mathbb{P}}(E) = \mathbb{P}(\bigoplus_0^n E_i; q_0, \cdots, q_n) \quad \text{(with projection } \rho \text{ on } X\text{)}$$

The author thanks Feryal Alayont for her excellent typing.

the (topological) quotient of $E^*= E\backslash X$ modulo the operation σ. The problem we are interested in is the computation, à la GROTHENDIECK, of the integral cohomology $H^*(\widetilde{\mathbb{P}}(E))$ (étale cohomology in the algebraic case). More precisely we want to describe:
1) the additive structure of $H^*(\widetilde{\mathbb{P}}(E))$, as an $H^*(X)$-module (via ρ^*). The result is that $H^*(\widetilde{\mathbb{P}}(E))$ is a finitely generated free $H^*(X)$-module, and an explicit basis can be constructed.
2) the multiplicative structure of $H^*(\widetilde{\mathbb{P}}(E))$. Let us recall that in the standard case of the projective bundle $\mathbb{P}(E)$ (i.e. $n = 0$ or $q_0 = \cdots = q_n = 1$), the multiplicative structure of $H^*(\widetilde{\mathbb{P}}(E))$ is given by a unique relation

$$\xi_E^{r+1} = -\sum_1^{r+1} c_i\, \xi_E^{r-i+1} \qquad \text{(see III.3)}$$

which, at the same time, defines, in an algebraic way, the CHERN classes of the vector bundle E (after GROTHENDIECK [G1]). So the question is to find a generalization of the preceding relation for the weighted projective bundle $\widetilde{\mathbb{P}}(E)$, and to see how much it determines the multiplicative structure of $H^*(\widetilde{\mathbb{P}}(E))$. Instead of the usual CHERN classes $c_i(E)$ of E, we encounter here the following "characteristic" classes (given by their total class)

$$\tilde{c}(E) = (\psi^{l/q_0} c(E_0)) \cdots (\psi^{l/q_n} c(E_n))$$

where $l = \mathrm{lcm}\{q_0, \cdots, q_n\}$, ψ^i is the i-th ADAMS operation and $c(E_j)$ is the total CHERN class of E_j. These "twisted" classes have the same properties as CHERN classes, except the additivity one which is replaced by the following. Assume given $E' = \bigoplus_0^{n'} E'_i$ with weights $q'_0, \cdots, q'_{n'}$ and $E'' = \bigoplus_0^{n''} E''_i$ with weights $q''_0, \cdots, q''_{n''}$ (two copies of the situation above). Put $l' = \mathrm{lcm}\{q'_0, \cdots, q'_{n'}\}$, $l'' = \mathrm{lcm}\{q''_0, \cdots, q''_{n''}\}$ and $m = \mathrm{lcm}\{l', l''\}$. Weighting $E' \oplus E''$ by q'_i, q''_j in the obvious way, we obtain

$$\tilde{c}\,(E' \oplus E'') = (\psi^{m/l'} \tilde{c}(E'))\, (\psi^{m/l''} \tilde{c}(E''))\ .$$

We shall say here nothing more about the multiplication in $H^*(\widetilde{\mathbb{P}}(E))$.

This, and other more explicit results, will be the subject for a forthcoming work [A1].

To construct a basis of the free $H^*(X)$-module $H^*(\widetilde{\mathbb{P}}(E))$, we first study (in details) the case when the space X is a point, i.e. $\widetilde{\mathbb{P}}(E)$ is a weighted projective space $\widetilde{\mathbb{P}}^r$, and then use the LERAY-HIRSCH theorem.

The integral cohomology $H^*(\widetilde{\mathbb{P}}^r)$ has been first computed by KAWASAKI [K]. Then the author determined the étale cohomology of the scheme $\widetilde{\mathbb{P}}^r$. For this he first recomputed the integral cohomology of the space $\widetilde{\mathbb{P}}^r$ using algebraic methods, in such a way that these extend to the calculation of the étale case. This is done in Chapter I for integral cohomology and in Chapter II for étale cohomology. In these two chapters we tried to be as "self contained" as possible, having in mind beginners in the subject, like young researchers from Turkey (Tuba, Ali, Handan, Yakup, ... met in Ankara, Ağustos 1995).

In Chapter III we explain by a fundamental particular case, how the results announced above for $H^*(\widetilde{\mathbb{P}}(E))$ are obtained. No proofs are given. The general case (with proofs) will be treated in [A1].

The study of $H^*(\widetilde{\mathbb{P}}(E))$ was submitted to me by Jean-Pierre JOUANOLOU. Computations in Chapters I and II were achieved under his advisement. I would like to thank him warmly here, for his encouragement and for his help.

Contents

I. Integral Cohomology of Weighted Projective Spaces

Before computing integral cohomology $H^*(-, \mathbb{Z})$ of weighted projective spaces, we first recall definitions and some properties.

Through all this chapter (I), the ground field is the complex numbers $\mathbb{C}$, with multiplicative group $\mathbb{C}^*$.

Let n be an integer ≥ 0 and fix $q_0, \cdots, q_n$, $n+1$ integers ≥ 1. The group $\mathbb{C}^*$ acts on the space $(\mathbb{C}^{n+1})^* = \mathbb{C}^{n+1} \backslash (0)$ in the following way:

$$\lambda \cdot x = (\lambda^{q_0} x_0, \cdots, \lambda^{q_n} x_n) \tag{$*$}$$

where $\lambda \in \mathbb{C}^*$, $x = (x_0, \cdots, x_n) \in (\mathbb{C}^{n+1})^*$.

Notice that this action is free if, and only if, $q_0 = \cdots = q_n = 1$. We denote its quotient topological space by

$$\widetilde{\mathbb{P}}^n = \mathbb{P}_{\mathbb{C}}(q_0, \cdots, q_n) \, .$$

It is called *weighted* (or twisted) *projective space*, of type $(q_0, \cdots, q_n)$. If $q_0 = \cdots = q_n = 1$, then $\widetilde{\mathbb{P}}^n = \mathbb{P}^n$ is the usual complex projective space. For any integer $d \geq 1$, we have a natural homeomorphism $\mathbb{P}(dq_0, \cdots, dq_n) = \mathbb{P}(q_0, \cdots, q_n)$ (given by $\lambda \mapsto \lambda^d$ in $\mathbb{C}^*$).

1 First Properties of the Space $\widetilde{\mathbb{P}}^n$

(a) Local structure of $\widetilde{\mathbb{P}}^n$

For $i \in \{0, \cdots, n\}$, put

$$U_i = \{x = (x_0, \cdots, x_n) \in \widetilde{\mathbb{P}}^n \,|\, x_i \neq 0\}, \; Y_i = \widetilde{\mathbb{P}}^n \backslash U_i \, .$$

Then Y_i is nothing else but $\widetilde{\mathbb{P}}^{n-1} = \mathbb{P}(q_0, \cdots, \hat{q}_i, \cdots, q_n)$ considered as a closed subset of $\widetilde{\mathbb{P}}^n$ (defined by the equation $x_i = 0$)[1]. Let us now identify the open subset $U_i \subset \widetilde{\mathbb{P}}^n$.

For an integer $q \geq 1$, consider $\mu_q \subset \mathbb{C}^*$ the subgroup of q-th roots of unity. So μ_q acts on the space $\mathbb{C}^{n+1}$ by the formula (*) above. Denote the topological quotient-space by

$$\mathbb{C}^{n+1}/\mu_q = (\mathbb{C}^{n+1}/\mu_q)\,(q_0, \cdots, q_n) \, .$$

There is a homeomorphism $U_i \simeq (\mathbb{C}^n/\mu_{q_i})(q_0, \cdots, \hat{q}_i, \cdots, q_n)$ (sending $(x_0, \cdots, \hat{x}_i, \cdots, x_n) \in \mathbb{C}^n/\mu_{q_i}$ to $(x_0, \cdots, 1, \cdots, x_n) \in U_i$, in the obvious way!).

This shows that thc space $\widetilde{\mathbb{P}}^n$ is locally of the form $\mathbb{C}^n/\mu_q$ since the U_i's cover $\widetilde{\mathbb{P}}^n$. In particular, $\widetilde{\mathbb{P}}^n$ is an orbifold.

---

[1] $x = (x_0, \cdots, x_n)$ denotes a point in $\mathbb{C}^{n+1}$ or the class it represents in $\widetilde{\mathbb{P}}^n$. The symbol $\hat{a}$ means that the letter a is omitted.

(b) Maps between $\widetilde{\mathbb{P}}^n$ and $\mathbb{P}^n$

There are two maps, $\varphi : \mathbb{P}^n \longrightarrow \widetilde{\mathbb{P}}^n$ and $\psi : \widetilde{\mathbb{P}}^n \longrightarrow \mathbb{P}^n$, where $\widetilde{\mathbb{P}}^n = \mathbb{P}(q_0, \cdots, q_n)$ and $\mathbb{P}^n = \mathbb{P}(1, \cdots, 1)$ (the usual projective space over $\mathbb{C}$). The first one is defined by $\varphi(x_0, \cdots, x_n) = (x_0^{q_0}, \cdots, x_n^{q_n})$. For the second, put $m = \text{lcm}\{q_0, \cdots, q_n\}$ and $t_i = m/q_i$. Then ψ is defined by $\psi(x_0, \cdots, x_n) = (x_0^{t_0}, \cdots, x_n^{t_n})$.

Let us make precise the fibers of φ. Consider the group $G = (\mu_{q_0} \times \cdots \times \mu_{q_n})/\mu_d$ where $d = \gcd\{q_0, \cdots, q_n\}$, and μ_d is a subgroup of $\mu_{q_0} \times \cdots \times \mu_{q_n}$ via the diagonal. There is an operation of G on $\mathbb{P}^n$ given by $(\lambda_0, \cdots, \lambda_n) \cdot (x_0, \cdots, x_n) = (\lambda_0 x_0, \cdots, \lambda_n x_n)$, with $\lambda_i \in \mu_{q_i}$, $x = (x_0, \cdots, x_n) \in \mathbb{P}^n$. This operation is faithful, i.e. no non-trivial element of G fixes all points of $\mathbb{P}^n$.

The fibers of φ are the G-orbits, and φ induces a homeomorphism $\mathbb{P}^n/G \xrightarrow{\sim} \widetilde{\mathbb{P}}^n$. Note that G has order $|G| = (\prod_0^n q_i)/d$. We will see later that φ has a degree equal to $|G|$.

The map ψ allows to define a "canonical" line bundle $\widetilde{L}$ over $\widetilde{\mathbb{P}}^n$. Let L be the canonical line bundle over the projective space $\mathbb{P}^n$. Then put $\widetilde{L} := \psi^*(L)$, the pull-back of L by ψ. Recall that $\psi^*(L)$ is defined by

$$\psi^* L = \widetilde{\mathbb{P}}^n \times_{\mathbb{P}^n} L = \{(x, u) \in \widetilde{\mathbb{P}}^n \times L \mid \psi(x) = \pi(u)\} ,$$

where π is the projection of L on $\mathbb{P}^n$.

(c) Another way to describe $\widetilde{L}$ is the following. Let q be an integer ≥ 1. Consider $\widetilde{\mathbb{P}}^{n+1} = \mathbb{P}(q_0, \cdots, q_{n+1})$, the (complex) weighted projective space of type $(q_0, \cdots, q_{n+1})$, with $q_{n+1} = q$. Then the space $E = \widetilde{\mathbb{P}}^{n+1} \backslash (0, \cdots, 0, 1)$ projects on $\widetilde{\mathbb{P}}^n = \mathbb{P}(q_0, \cdots, q_n)$:

$$\rho : E \longrightarrow \widetilde{\mathbb{P}}^n, \ (x_0, \cdots, x_{n+1}) \mapsto (x_0, \cdots, x_n) .$$

If each q_i $(0 \leq i \leq n)$ divides q, then E is a (complex) line bundle over $\widetilde{\mathbb{P}}^n$. This is not difficult to see. The vector space structure on the fibers $\rho^{-1}(x) = E_x$ is the obvious one. To explain the local triviality, one uses the standard open subsets $U_i \subset \widetilde{\mathbb{P}}^n$ (see (a)). Let V_j be the corresponding open subsets in $\widetilde{\mathbb{P}}^{n+1}$. We have a commutative diagram $(0 \le i \le n)$:

$$\begin{array}{ccc} \rho^{-1}(U_i) = V_i \simeq \dfrac{\mathbb{C}^{n+1}}{\mu_{q_i}}(q_0, \cdots, \hat{q}_i, \cdots, q_{n+1}) & = & \left(\dfrac{\mathbb{C}^{n}}{\mu_{q_i}}(q_0, \cdots, \hat{q}_i, \cdots, q_n)\right) \times \mathbb{C} \\ \Big\downarrow \rho & \swarrow \text{1st proj.} & \\ U_i \simeq \dfrac{\mathbb{C}^n}{\mu_{q_i}}(q_0, \cdots, \hat{q}_i, \cdots, q_n) & & \end{array}$$

This shows that there exists an isomorphism $E|U_i \simeq U_i \times \mathbb{C}$ over U_i (linear on the fibers).

In case $q = m = \mathrm{lcm}\{q_0, \cdots, q_n\}$ the vector bundle $\widetilde{L} = \psi^* L$ is isomorphic to E (over $\mathbb{P}^n$).

<u>**Proof.**</u> Write $L = \mathbb{P}^{n+1} \backslash (0, \cdots, 0, 1)$. Then the map

$$E \to \psi^* L = \widetilde{\mathbb{P}}^n \times_{\mathbb{P}^n} L : (x_0, \cdots, x_{n+1}) \mapsto ((x_0, \cdots, x_n), (x_0^{t_0}, \cdots, x_n^{t_n}, x_{n+1}))$$

(cf. (b) for notation) is an isomorphism of vector bundles (over $\mathbb{P}^n$).

More generally, if q is a multiple of m, and $s = q/m$, then one has $\psi^*(L^{\otimes s}) \simeq E$, since $L^{\otimes s} = \mathbb{P}^{n+1}(1, \cdots, 1, s) \backslash (0, \cdots, 0, 1)$.

(d) Consider the real sphere

$$S^{2n+1} = \{(x_0, \cdots, x_n) \in \mathbb{C}^{n+1} \mid \sum_0^n |x_i|^2 = 1\} \subset (\mathbb{C}^{n+1})^* .$$

The operation $(*)$ of $\mathbb{C}^*$ on $(\mathbb{C}^{n+1})^*$ induces, by restriction, an operation of S^1 on S^{2n+1}. Denote the corresponding quotient-space

$$S^{2n+1}/S^1 = (S^{2n+1}/S^1)(q_0, \cdots, q_n) .$$

The natural map $S^{2n+1}/S^1 \longrightarrow \widetilde{\mathbb{P}}^n$ is a homeomorphism. In particular a weighted projective space is compact Hausdorff.

<u>**Proof.**</u> The mentioned map is induced by the closed inclusion $S^{2n+1} \subset (\mathbb{C}^{n+1})^*$. So it is continuous and closed. To see injectivity, let $x, y \in S^{2n+1}$, $\lambda \in \mathbb{C}^*$ such that: $y_i = \lambda^{q_i} x_i$ $(0 \le i \le n)$. Since

$\sum|x_i|^2 = 1$ and $\sum|\lambda|^{2q_i}|x_i|^2 = 1$, one has $|\lambda| = 1$. For surjectivity, given $(y_0,\cdots,y_n)$ in $(\mathbb{C}^{n+1})^*$, it is enough to find $\lambda \in \mathbb{C}^*$ such that $\sum|\lambda|^{2q_i}|y_i|^2 = 1$ (since then $(x_0,\cdots,x_n) \in S^{2n+1}$ maps to $(y_0,\cdots,y_n) \in \widetilde{\mathbb{P}}^n$, with $x_i = \lambda^{q_i}y_i$). But any equation of the form

$$a_0t^{q_0} + \cdots + a_nt^{q_n} = 1 \quad (\text{all } a_i \geq 0,\ \text{some } a_j \neq 0,\ t \in \mathbb{R})$$

has a (unique) solution $t_0 \in \mathbb{R}, t_0 > 0$. In case $a_i = |y_i|^2$, take $\lambda = \sqrt{t_0}$.

Now $\widetilde{\mathbb{P}}^n \simeq S^{2n+1}/S^1$ is Hausdorff since S^1 is a compact Hausdorff group acting on a Hausdorff space!

2 Further Properties of the Space $\widetilde{\mathbb{P}}^n$

First we establish a special cellular decomposition of $\widetilde{\mathbb{P}}^n$, where the cells have the form B^m/μ_q (B^m: real ball, μ_q: q-th roots of unity).

(a) A cellular decomposition of $\widetilde{\mathbb{P}}^n$

Here we identify $\widetilde{\mathbb{P}}^n = \mathbb{P}(q_0,\cdots,q_n)$ with S^{2n+1}/S^1 (after 1.(d)). Consider the real ball

$$B^{2n} = \{(x_0,\cdots,x_{n-1}) \in \mathbb{C}^n | \sum_0^{n-1} |x_i|^2 \leq 1\}\,.$$

The action $\mathbb{C}^*\times(\mathbb{C}^n)^* \longrightarrow (\mathbb{C}^n)^*$ which defines $\widetilde{\mathbb{P}}^{n-1} = \mathbb{P}(q_0,\cdots,q_{n-1})$ induces an action of the q-th roots of unity μ_q on B^{2n}, where $q = q_n$ (see (*) at the beginning). The boundary $S^{2n-1} \subset B^{2n}$ is stable under this action. So we obtain a canonical closed inclusion

$$\alpha : S^{2n-1}/\mu_q \longrightarrow B^{2n}/\mu_q\,,$$

and a map $\rho : B^{2n}/\mu_q \longrightarrow S^{2n+1}/S^1$, well-defined by:

$$\rho(x_0,\cdots,x_{n-1}) = (x_0,\cdots,x_n) \text{ with } x_n = \left(1 - \sum_0^{n-1}|x_i|^2\right)^{\frac{1}{2}}$$

Let $\beta : S^{2n-1}/S^1 \longrightarrow S^{2n+1}/S^1$ be the canonical closed inclusion $\widetilde{\mathbb{P}}^{n-1} \subset \widetilde{\mathbb{P}}^n$ (defined by $x_n = 0$).

We can now state the main result in this paragraph.

Proposition *There is a commutative diagram*

$$\begin{array}{ccc} S^{2n-1}/\mu_q & \overset{\alpha}{\hookrightarrow} & B^{2n}/\mu_q \\ {\scriptstyle can.}\downarrow\gamma & & \downarrow\rho \\ S^{2n-1}/S^1 & \underset{\beta}{\hookrightarrow} & S^{2n+1}/S^1 \end{array}$$

where ρ maps homeomorphically the open subset, complementary to $Im(\alpha)$ on the open subset, complementary to $Im(\beta)$.

This means that $\widetilde{\mathbb{P}}^n$ is obtained from $\widetilde{\mathbb{P}}^{n-1}$ by adjoining a "cell" of the form B^{2n}/μ_q. Such a structure could be called VCW-complex or orbi-CW complex (think of V-manifold, orbifold).

<u>**Proof.**</u> Commutativity of the diagram is clear. Let us see that the map ρ induces a bijection

$$\rho' : (B^{2n}\backslash S^{2n-1})/\mu_q \xrightarrow{\sim} (S^{2n+1}\backslash S^{2n-1})/S^1 .$$

(i) ρ' <u>is onto</u>: let $x = (x_0, .., x_n) \in S^{2n+1}$ with $x_n \neq 0$. Choose λ such that $\lambda^q = x_n/|x_n|$, and put $y_i = \lambda^{-q_i} x_i$ $(0 \leq i \leq n-1)$. Then $y = (y_0, \cdots, y_{n-1}) \in B^{2n}$, $y \notin S^{2n-1}$ and $\rho'(y) = x$.
(ii) ρ' <u>is one-one</u>: assume $\rho'(x) = \rho'(y)$. So there exists $\lambda \in S^1$ such that $y_i = \lambda^{q_i} x_i$ $(0 \leq i \leq n-1)$ and

$$\left(1 - \sum_0^{n-1} |y_i|^2\right)^{\frac{1}{2}} = \lambda^q \left(1 - \sum_0^{n-1} |x_i|^2\right)^{\frac{1}{2}} \neq 0 .$$

Therefore $|y_i| = |x_i|$, and so $\lambda^q = 1$. Hence $x = y$ in $(B^{2n}\backslash S^{2n-1})/\mu_q$.

Corollary *The diagram*

$$S^{2n-1}/\mu_q \xrightarrow[can.]{\gamma} S^{2n-1}/S^1 \overset{\beta}{\hookrightarrow} S^{2n+1}/S^1$$

is a cofibration, that is S^{2n+1}/S^1 is the mapping cone $C(\gamma)$ of γ and β is the canonical injection of S^{2n-1}/S^1 into $C(\gamma)$.

It is enough to see that B^{2n}/μ_q is the cone of $S^{2n-1}/\mu_q =: X$, with canonical inclusion α. This is given by the homeomorphism

$$CX = (X\times I)/X\times\{0\} \to B^{2n}/\mu_q : ((x_0,\cdots,x_{n-1}),t) \mapsto (tx_0,\cdots,tx_{n-1})$$

where $I=[0,1]$.

REMARK. From the preceding we deduce (1.(a))

$$\mathbb{C}^n/\mu_q = (\mathbb{C}/\mu_q)(q_0,..,q_{n-1}) = \widetilde{\mathbb{P}}^n\backslash\widetilde{\mathbb{P}}^{n-1} = CX\backslash X$$

where $X = S^{2n-1}/\mu_q$, $CX =$ cone of $X, q = q_n$. The point $0\in\mathbb{C}^n/\mu_q$ corresponds to the vertex of CX.

(b) Degree of the map φ

We show that the map $\varphi : \mathbb{P}^n \longrightarrow \widetilde{\mathbb{P}}^n$ has a degree, and that the degree is equal to $(\prod_0^n q_i)/d$ (cf. 1.(b)).

Because of what we saw in 1.(b), our situation is a particular case of the following.

Theorem *Let X be an orientable connected manifold. Assume a finite group G acts faithfully on X, and that G preserves the orientation. Then*
(i) $H_c^r(X/G,\mathbb{Z}) = \mathbb{Z}$ ($r = \dim X$, $H_c^{\cdot} =$ cohomology with compact supports).
(ii) the canonical projection $\pi : X \longrightarrow X/G$ has a degree equal to $|G|$.

This result must be “well-known”, but we do not know a reference for a proof.

Proof. (i) We may assume $G\neq\{1\}$. Let $g\in G$, $g\neq 1$. Then $X^q = \{x\in X|\, gx = x\}$ is a closed submanifold of X, with dimension $\leq r-1$. The open subset $U = X\backslash\cup_{h\neq 1}X^h$ is non-empty, G-invariant, and the action of G on U is free. Assume for a while that we have proved $\dim(X^g)\leq r-2$. Put $Y = \cup_{h\neq 1}X^h$. The Mayer-Vietoris exact sequence ([B] II.13.(1)) shows that $H_c^i(Y\mathbb{Z}) = 0$ for $i\geq r-1$.

Hence we obtain the following commutative diagram (cohomology groups have coefficients in $\mathbb{Z}$).

$$\begin{array}{ccc} H_c^r(U) & \xrightarrow{\sim} & H_c^r(X) = \mathbb{Z} \\ {\scriptstyle (\pi_U)^*}\uparrow & & \uparrow{\scriptstyle \pi^*} \\ H_c^r(U/G) & \xrightarrow{\sim} & H_c^r(X/G) \to 0 \\ \parallel & & \\ \mathbb{Z} & & \end{array} \qquad (U \text{ is connected; } \pi, \pi_u \text{ are proper})$$

So $H_c^r(X/G)$ has the form $\mathbb{Z}/a\mathbb{Z}$ $(a \geq 0)$. But $\pi^* \otimes \mathbb{Q}$ is an isomorphism ([B] II (19.1)), whence $a = 0$. This proves (i).

It remains to show dim $(X^g) \leq r - 2$. We can assume $X^g \neq \emptyset$. Consider a point $x \in X^g$. The subgroup (g), spanned by g, acts on the tangent space $T_x(X)$ (by automorphisms), and one has $T_x(X^g) = T_x(X)^g$. Let us assume that dim $(X^g) = r - 1$. It follows that $T_x(X)/T_x(X)^g = \mathbb{R}$ and that (g) acts on $\mathbb{R}$ (by homotheties), preserving orientation. Now g has finite order, so it acts on $\mathbb{R}$ as a homothety the coefficient of which is 1. Hence $T_x(X)^g = T_x(X)$ (The action of (g) on $T_x(X)$ is semi-simple, by Maschke theorem). This gives a contradiction!

(ii) The commutative diagram above shows it is enough to see that $\rho = \pi_U : U \longrightarrow U/G$ has a degree equal to $|G|$. Since G acts freely on U, there exists a non-empty open subset $V \subset U$ such that $V = \cup V_g$ $(g \in G)$, where the V_g's are connected disjoint open subsets with the property:

$$\rho_g - \rho_{V_g} : V_g \longrightarrow V/G \text{ is a homeomorphism for all } g \in G.$$

One obtains a commutative diagram (coefficients are in $\mathbb{Z}$) of cohomology groups

$$
\begin{array}{ccccc}
H_c^r(V) & & \longrightarrow & & \\
\uparrow \rho_V^* & \| & & & \\
 & \bigoplus_g H_c^r(V_g) & \longrightarrow & H_c^r(U) & \\
 & \uparrow \oplus\rho_g^* & & \uparrow \rho^* & \\
 & H_c^r(V/G) & \xrightarrow{\sim} & H_c^r(U/G) , &
\end{array}
$$

the square of which is

$$
\begin{array}{ccc}
\bigoplus_g \mathbb{Z} & \longrightarrow & \mathbb{Z}, \\
\uparrow \Delta & & \uparrow \mu \\
\mathbb{Z} & \xrightarrow[=]{} & \mathbb{Z}
\end{array}
\quad \text{with } \Delta(z) = (z, \cdots, z),\ \sigma(z_1, \cdots, z_s) = z_1 + \cdots + z_s ,
$$

where $s = |G|$. This implies that μ is multiplication by s, as desired.

(c) We shall need the following topological property of $\widetilde{\mathbb{P}}^n$.

In the space $\widetilde{\mathbb{P}}^n$ every point has a fundamental system of contractible open neighborhoods.

Since $\widetilde{\mathbb{P}}^n$ is homeomorphic to $\mathbb{P}^n/G$ (see 1.(b)), our assertion is a corollary of the more general following fact.

Proposition *Let X be a manifold on which a finite group G acts. Then every point in the quotient-space X/G has a fundamental system of contractible open neighborhoods.*

Proof. Take $\bar{x} \in X/G$ and an open subset $\bar{U} \subset X/G$ with $\bar{x} \in \bar{U}$. Let U be the inverse image of $\bar{U}$ in X and x a point in the fiber over $\bar{x}$. Then there exists ([HM] 1.6. SATZ page 7) a neighborhood V of x, $V \subset U$, and a coordinate system

$$f = (f_1, \cdots, f_r) : V \longrightarrow \mathbb{R}^r$$

such that $f(x) = 0$, $f(V)$ is an open ball with center 0 and the following are satisfied.

(i) the stabilizing subgroup G_x acts linearly on $\mathbb{R}^r$, V is G_x-invariant and f is G_x-equivariant;
(ii) there exists a G_x-invariant linear subspace L in $\mathbb{R}^r$ so that $T = f^{-1}(f(V) \cap L)$ is a "slice" of U at the point x,
i.e. (1) $x \in T$ and T is stable under the action of G_x,
(2) for every $g \in G$, if $(gT) \cap T \neq \emptyset$ then $g \in G_x$,
(3) for every local section $s : E \longrightarrow G$ of the canonical projection $G \longrightarrow G/G_x$, where E contains the identity element, the map

$$E \times T \longrightarrow U : (e, t) \mapsto s(e)t$$

is a diffeomorphism on an open subset of U.

Now $(f(V) \cap L)/G_x$ is contractible, so T/G_x is also contractible. On the other hand, the projection $\pi : U \longrightarrow \bar{U} = U/G$ induces a map $T/G_x \longrightarrow U/G$ which is a homeomorphism on $\pi(T)$ ([HM] 1.6. FOLGERUNG page 8). But $\pi(T)$ is an open subset in $\bar{U}$ containing $\bar{x}$ (in (ii)(3) above, take $E = \{\bar{1}\}$ and $s(\bar{1}) = 1$), and $\pi(T)$ is contractible.

(d) Consider the open subset $\widetilde{\mathbb{P}}^n = \mathbb{P}(q_0, \cdots, q_n)$

$$\mathbb{C}^n/\mu_q = (\mathbb{C}^n/\mu_q)\,(q_0, \cdots, q_{n-1})$$

where $q = q_n$ (see 1.(b)). We want to show that the natural map

$$H_0^*(\mathbb{C}^n/\mu_q, \mathbb{Z}) \longrightarrow H_c^*(\mathbb{C}^n/\mu_q, \mathbb{Z})$$

is an isomorphism (H_0^*= local cohomology at the point $0 = (0, \cdots, 0) \in \mathbb{C}^n/\mu_q$, H_c^*=cohomology with compact supports).

The remark in (a) above shows that this result is a particular case of the following one.

Proposition *Take a compact topological space X and put*

$$Y = CX \backslash X, \text{ where } CX = \text{ cone of } X .$$

Call 0 the vertex of CX. Let $\mathfrak{F}$ be a constant abelian sheaf over CX. Then the natural homomorphism

$$H_0^*(U, \mathfrak{F}|U) \longrightarrow H_c^*(U, \mathfrak{F}|U)$$

is bijective.

Proof. Denote by $i : U \longrightarrow CX$ the natural inclusion of U in CX. For any abelian sheaf $\mathcal{A}$ over CX, one has a morphism between exact sequences

$$\begin{array}{ccccccc} 0 \longrightarrow & \Gamma_0(CX,\mathcal{A}) & \longrightarrow & \Gamma(CX,\mathcal{A}) & \longrightarrow & \Gamma(CX\backslash 0,\mathcal{A}) \\ & \downarrow & & \downarrow \Vert & & \downarrow \\ 0 \longrightarrow & \Gamma(CX, i_!(\mathcal{A}|\mathcal{U})) & \longrightarrow & \Gamma(CX,\mathcal{A}) & \longrightarrow & \Gamma(X,\mathcal{A}|\mathcal{X}) \end{array}$$

where $i_!$ means extension by zero. Apply this morphism to an injective resolution of $\mathfrak{F}$ and take cohomology. One obtains the following exact sequence morphism:

$$\begin{array}{ccccccccc} \cdots \to & H_0^j(CX,\mathfrak{F}) & \to & H^j(CX,\mathfrak{F}) & \to & H^j(CX\backslash 0,\mathfrak{F}) & \to & H_0^{j+1}(CX,\mathfrak{F}) & \to \cdots \\ & \downarrow & & \downarrow \Vert & & \downarrow \sim & & \downarrow & \\ \cdots \to & H_c^j(U,\mathfrak{F}|U) & \to & H^j(CX,\mathfrak{F}) & \to & H^j(X,\mathfrak{F}|X) & \to & H_0^{j+1}(U,\mathfrak{F}|U) & \to \cdots \end{array}$$

(the inclusion $X \subset CX\backslash 0$ is a homotopy equivalence). By the five lemma it follows that we have isomorphisms

$$H_0^j(CX,\mathfrak{F}) \xrightarrow{\sim} H_c^j(U,\mathfrak{F}|U) \quad (j \geq 0) .$$

But the homomorphisms $H_0^j(CX,\mathfrak{F}) \longrightarrow H_0^j(U,\mathfrak{F}|U)$ are bijective (U is open in CX and $0 \in U$), and there are commutative diagrams

$$\begin{array}{ccc} H_0^j(U,\mathfrak{F}|U) & \longrightarrow & H_c^j(U,\mathfrak{F}|U) \\ & \nwarrow \sim \quad \sim \nearrow & \\ & H_0^j(CX,\mathfrak{F}) & \end{array}$$

which ends the proof.

3 Grothendieck Theorem

Here we state some notation, and recall a theorem by GROTHENDIECK we shall use later.

Let X be a topological space on which a group G acts. Denote by $\pi : X \longrightarrow Y = X/G$ the canonical inclusion. Let $\mathfrak{F}$ be an abelian G-sheaf over X. Put

$$\Gamma_X^G(\mathfrak{F}) = \Gamma(X, \mathfrak{F})^G \quad \text{(invariant sections)}$$

and write $\pi_*^G(\mathfrak{F})$ for the sheaf over Y

$$V \mapsto \Gamma(V, \pi_*\mathfrak{F})^G \ , \ V \text{ open subset in } Y \ ,$$

(G acts identically on Y, $\pi_*\mathfrak{F}$ is a G-sheaf). So one has two left exact functors

$$\mathfrak{F} \mapsto \Gamma_X^G(\mathfrak{F}) \ , \ \mathfrak{F} \mapsto (\pi_*\mathfrak{F})^G \ ,$$

and can consider their derived functors

$$H^i(X; G, \mathfrak{F}) := R^i\Gamma_X^G(\mathfrak{F}) \ , \ \mathfrak{H}^i(G, \mathfrak{F}) := R^i\Pi_*^G(\mathfrak{F}) \quad (i \geq 0) \ .$$

Alexander GROTHENDIECK established the following results [G2].
(a) **Theorem** ([G2]5.2.1) *There are two spectral sequences*

$$\begin{aligned} {}'E_2^{i,j} &= H^i(Y, \mathfrak{H}^j(G, \mathfrak{F})) \Longrightarrow H^{i+j}(X; G, \mathfrak{F}) \ , \\ {}''E_2^{i,j} &= H^i(G, H^j(X, \mathfrak{F})) \Longrightarrow H^{i+j}(X; G, \mathfrak{F}) \ . \end{aligned}$$

Let us call (D) the following condition:

> for every $x \in X$, the stabilizing subgroup G_x of x is finite, and there exists a neighborhood V_x of x such that $(gV_x) \cap V_x = \emptyset$ for all $g \in G \backslash G_x$.

For instance (D) is verified if X is Hausdorff and G is finite.

(b) **Theorem** ([G2] 5.3.1) *If the condition (D) is satisfied, then there are natural isomorphisms*

$$\mathfrak{H}^i(G, \mathfrak{F})_y = H^i(G_x, \mathfrak{F}_x)$$

for all $y \in Y$ and all $x \in X$ with $\pi(x) = y$.

(c) Corollary ([G2] Corollaire 1 to 5.3.1) *Assume (D) is true. If, for every $x \in X$, multiplication by $|G_x|$ is a group automorphism of $\mathfrak{F}_x$, then there is a spectral sequence*

$$''E_2^{i,j} = H^i(G, H^j(X, \mathfrak{F})) \Longrightarrow H^{i+j}(Y, \pi_*^G \mathfrak{F})$$

(by Theorem (b), $\mathfrak{H}^i(G, \mathfrak{F}) = 0$ for $i > 0$; apply then Theorem (a)).

The following theorem is a consequence of Corollary (c) ([G2] Corollaire to 5.2.3).

(d) GROTHENDIECK Theorem: *Assume G finite and X Hausdorff. If multiplication by the order of G is an automorphism of $\mathfrak{F}$, then*

$$H^j(Y, \pi_*^G \mathfrak{F}) = H^j(X, \mathfrak{F})^G \quad (j \geq 0) .$$

4 The Groups $H^i(\widetilde{\mathbb{P}}^n, \mathbb{Z})$

Consider an integer $a \geq 0$, and put $A = \mathbb{Z}/a\mathbb{Z}$. If X is a topological space, we write A_X, or simply A, for the constant abelian sheaf defined by A over X. For instance, cohomology groups of X with values in A_X are denoted by $H^i(X, A)$ $(i \geq 0)$.

Recall that $\widetilde{\mathbb{P}}^n$ is the weighted projective space $\mathbb{P}(q_0, \cdots, q_n)$ over the complex numbers $\mathbb{C}$, defined in the beginning. The main result in this paragraph is the following.

(a)Theorem *We have*

$$H^i(\widetilde{\mathbb{P}}^n, \mathbb{Z}) = \begin{cases} \mathbb{Z} & \text{if } i = 2r,\ 0 \leq r \leq n , \\ 0 & \text{if } i \text{ is odd or } i > 2n. \end{cases}$$

To prove this problem we first compute the groups $H^i(\widetilde{\mathbb{P}}^n, \mathbb{Z}/a\mathbb{Z})$ where $a > 0$, and then deduce the groups $H^i(\widetilde{\mathbb{P}}^n, \mathbb{Z})$ by use of the universal-coefficient formula. Denote $\mathbb{Z}_a = \mathbb{Z}/a\mathbb{Z}$.

Lemma 1 For any integer $a > 0$, one has

$$H^i(\widetilde{\mathbb{P}}^n, \mathbb{Z}_a) = \begin{cases} \mathbb{Z}_a & \text{if } i = 2r,\ 0 \le r \le n\ , \\ 0 & \text{if } i \ne 2r,\ 0 \le r \le n. \end{cases}$$

<u>**Proof.**</u> We may assume $a = p^\alpha$ with p prime number $\ne 1$ and $\alpha > 0$ (if not, decompose $\mathbb{Z}_a$!). Because $\mathbb{P}(dq_0, \cdots, dq_n) = \mathbb{P}(q_0, \cdots, q_n)$ we may also assume gcd $\{q_0, \cdots, q_n\} = 1$. Now, let us proceed by induction on n. The case $n = 0$ is clear since $\widetilde{\mathbb{P}}^0$ is a point. So let $n \ge 1$. There exists j, $0 \le j \le n$, such that p does not divide q_j. Consider then

$$\widetilde{\mathbb{P}}^{n-1} = \mathbb{P}(q_0, .., \hat{q}_j, \cdots, q_n), \mathbb{C}^n/\mu_q = (\mathbb{C}^n/\mu_{q_j})(q_0, \cdots, \hat{q}_j, \cdots, q_n)(\text{cf. } 1.(\text{a})).$$

Put $q = q_j$. Since $\mathbb{C}^n/\mu_q$ is the complement of the closed subset $\widetilde{\mathbb{P}}^{n-1} \subset \widetilde{\mathbb{P}}^n$, there is a cohomology exact sequence, with compact supports, and coefficients in $\mathbb{Z}_{p^\alpha}$,

$$\cdots \to H^i_c(\mathbb{C}^n/\mu_q) \to H^i(\widetilde{\mathbb{P}}^n) \to H^i(\widetilde{\mathbb{P}}^{n-1}) \to H^{i+1}_c(\mathbb{C}^n/\mu_q) \to \cdots$$

Hence it is enough to show the following, and to use induction.

Lemma 2 We have

$$H^i_c(\mathbb{C}^n/\mu_q, \mathbb{Z}_{p^\alpha}) = \begin{cases} \mathbb{Z}_{p^\alpha} & \text{if } i = 2n, \\ 0 & \text{if } i \ne 2n. \end{cases}$$

where $(p, q) = 1$.

<u>**Proof.**</u> Lemma 2 is well-known in case $q = 1$. So it suffices to show:

$$H^i_c(\mathbb{C}^n/\mu_q, \mathbb{Z}_{p^\alpha}) = [H^i_c(\mathbb{C}^n, \mathbb{Z}_{p^\alpha})]^{\mu_q} \quad (i \ge 0) \qquad (1)$$

(since μ_q acts identically on $H^i_c(\mathbb{C}^n, \mathbb{Z}_{p^\alpha})$). Consider the open inclusion in the usual projective space

$$\mathbb{C}^n \hookrightarrow \mathbb{P}^n : (x_0, \cdots, \hat{x}_j, \cdots, x_n) \mapsto (x_0, \cdots, 1, \cdots, x_n) \text{ (1 is at the } j\text{-th place)},$$

and the action of μ_q on $\mathbb{P}^n$ is defined by

$$\lambda \cdot (x_0, \cdots, x_n) = (\lambda^{q_0} x_0, \cdots, \lambda^{q_n} x_n)\ .$$

The inclusion above is then μ_q-equivariant and there is a commutative diagram

$$\begin{array}{ccc} \mathbb{C}^n & \xhookrightarrow[\text{open}]{} & \mathbb{P}^n \\ \text{can.}\downarrow & & \downarrow\text{can.} \\ \mathbb{C}^n/\mu_q & \xhookrightarrow[\text{open}]{} & \mathbb{P}^n/\mu_q \ . \end{array}$$

Therefore (1) is a particular case of the following proposition.

Proposition *Let X be a compact Hausdorff topological space on which a finite group G acts. Let U be a G-stable open subset of X with canonical projection $\pi : U \longrightarrow U/G$. Suppose $\mathfrak{F}$ is an abelian G-sheaf over U such that the multiplication by the order of G is an automorphism of $\mathfrak{F}$. Then*

$$H_c^i(U/G, \pi_*^G\mathfrak{F}) = [H_c^i(U, \mathfrak{F})]^G \quad (i \geq 0)$$

(for notation see 3.)

Proof. There is a commutative diagram, with obvious notation,

$$\begin{array}{ccc} U & \xhookrightarrow{\ i\ } & X \\ \pi\downarrow & & \downarrow\sigma \\ U/G & \xhookrightarrow[\bar{i}]{} & X/G \ . \end{array}$$

We have

$$H_c^k(U, \mathfrak{F}) = H^k(X, i_!\mathfrak{F}), \ H_c^k(U/G, \pi_*^G\mathfrak{F}) = H^k(X/G, \bar{i}_!\pi_*^G\mathfrak{F}),$$

where $i_!$, $\bar{i}_!$ are extension by 0. Multiplication by $|G|$ is an automorphism of $i_!\mathfrak{F}$ ($i_!$ is exact). Therefore, Grothendieck theorem shows that we have only to verify (3.(d))

$$\bar{i}_!\pi_*^G\mathfrak{F} = \sigma_*^G i_!\mathfrak{F} \ .$$

But one has

$$\bar{i}^*\sigma_*^G i_!\mathfrak{F} = \pi_*^G i^* i_!\mathfrak{F} = \pi_*^G\mathfrak{F} \ ,$$

and, because $\bar{i}_!$ is left adjoint for $\bar{i}^*$, one obtains a natural morphism

$$\bar{i}_! \pi_*^G \mathfrak{F} \longrightarrow \sigma_*^G i_! \mathfrak{F} ,$$

which is an isomorphism over the open subset U/G of X/G. To see that is an isomorphism over the whole X/G, it is enough to check

$$(\sigma_*^G i_! \mathfrak{F})_y = 0 \text{ for all } y \in X/G,\ y \notin U/G .$$

But $\sigma_*^G i_! \mathfrak{F}$ is a subsheaf of $\sigma_* i_! \mathfrak{F}$ and

$$(\sigma_* i_! \mathfrak{F})_y = H^0(\sigma^{-1}(y), (i_! \mathfrak{F})|\sigma^{-1}(y)) \quad (\sigma \text{ is proper!}) ,$$

and $(i_! \mathfrak{F})|\sigma^{-1}(y) = 0$ $(\sigma^{-1}(y) \subset X \backslash U)$.

The proof of the Proposition is finished. Hence Lemma 1 and Lemma 2 are proved.

(b) Proof of the Theorem

To simplify notation, we put $H^i(A) =. H^i(\widetilde{\mathbb{P}}^n, A)$, where $A = \mathbb{Z}$ or $\mathbb{Z}_a$ $(a \geq 1)$. The exact sequence $0 \to \mathbb{Z} \xrightarrow{.a} \mathbb{Z} \to \mathbb{Z}_a \to 0$ induces a cohomology exact sequence

$$\cdots \longrightarrow H^i(\mathbb{Z}) \xrightarrow{.a} H^i(\mathbb{Z}) \longrightarrow H^i(\mathbb{Z}_a) \longrightarrow H^{i+1}(\mathbb{Z}) \xrightarrow{.a} \cdots .$$

which in turn gives the following exact sequences

$$0 \longrightarrow H^i(\mathbb{Z}) \otimes \mathbb{Z}_a \longrightarrow H^i(\mathbb{Z}_a) \longrightarrow \mathrm{Tor}(H^{i+1}(\mathbb{Z}), \mathbb{Z}_a) \longrightarrow 0 .$$

By [D] 3.8., these sequences split. This proves the universal coefficient formula:

$$H^i(\mathbb{Z}_a) = (H^i(\mathbb{Z}) \otimes \mathbb{Z}_a) \oplus \mathrm{Tor}(H^{i+1}(\mathbb{Z}), \mathbb{Z}_a) . \qquad (2)$$

The groups $H^i(\mathbb{Z})$ are finitely generated (2.(c) and [B] II.16.5). Fix $i \geq 0$ and write $H^i(\mathbb{Z})$ as

$$\mathbb{Z}^\beta \oplus \mathbb{Z}_{\beta_1} \oplus \cdots \oplus \mathbb{Z}_{\beta_s} \quad (\beta \geq 0, \beta_k \geq 1).$$

(i) If i is odd or $i > 2n$, then Lemma 1 and (2) imply

$$\mathbb{Z}_a \otimes H^i(\mathbb{Z}) = 0 \text{ for all } a > 0 ,$$

and so: $\beta = 0, \beta_1 = \cdots = \beta_s = 1$, i.e. $H^i(\mathbb{Z}) = 0$.
(ii) If $i = 2r,\ 0 \le r \le n$, then Lemma 1, (2) and (i) imply

$$\mathbb{Z}_a = \mathbb{Z}_a \otimes H^i(\mathbb{Z}) \text{ for all } a > 0 .$$

Whence $\beta = 1$ (choose a prime with all the β_k's). So

$$\mathbb{Z}_a = \mathbb{Z}_a \oplus \mathbb{Z}_{(a,\beta_1)} \oplus \cdots \oplus \mathbb{Z}_{(a,\beta_s)} ,$$

which means $(a, \beta_1) = 1, \cdots, (a, \beta_s) = 1$, for all $a > 0$. This gives $\beta_1 = 1, \cdots, \beta_s = 1$, i.e. $H^i(\mathbb{Z}) = \mathbb{Z}$. The theorem is proved.

5 The Homomorphism φ^*

Our purpose here is to compute the homomorphism $\varphi^* : H^*(\widetilde{\mathbb{P}}^n, \mathbb{Z}) \longrightarrow H^*(\mathbb{P}^n, \mathbb{Z})$ induced by the map $\varphi : \mathbb{P}^n \longrightarrow \widetilde{\mathbb{P}}^n$ (see 1.(b)). We need first to define some notations.
Take $k \in \{0, \cdots, n\}$ and consider

$$I = \{i_0, \cdots, i_k\} \text{ with } 0 \le i_0 < \cdots < i_k \le n .$$

Then put

$$l_I = l_I(q_{i_0}, \cdots, q_{i_k}) = q_{i_0} q_{i_1} \cdots q_{i_k} / (q_{i_0}, \cdots, q_{i_k}) ,$$

where $(a_0, \cdots, a_k) = \gcd\{a_0, \cdots, a_k\}$. The following integers will play an important role in the sequel.

$$l_k = l_k(q_0, \cdots, q_n) = \operatorname{lcm}\{l_I | I \subset \{0, \cdots, n\}, |I| = k + 1\} .$$

For instance $l_0 = 1, l_1 = \operatorname{lcm}\{q_0, \cdots, q_n\}$ and $l_n = q_0 \cdots q_n / \gcd\{q_0, \cdots, q_n\}$. Notice that $l_n = \deg(\varphi)$ (2.(b)). To compute the integer l_k, consider $E = \{p \text{ prime } | p \text{ divides some } q_i,\ 0 \le i \le n\}$. Then

$$l_k = \prod_{p \in E} p^{\alpha(p,k)} ,$$

where $\alpha(p, k) = v_p(l_k)$ is determined as follows. Put $\alpha_i = v_p(q_i)$ and order the α_i's: $\alpha_{i_0} \le \alpha_{i_1} \le \cdots \le \alpha_{i_n}$ ($p \in E$ is fixed). Then

$$\alpha(p, k) = \alpha_{i_n} + \cdots + \alpha_{i_{n-k+1}} \quad (0 \le k \le n) .$$

Let $\xi = c_1(L^*)$ be the usual generator of $H^2(\mathbb{P}^n, \mathbb{Z})$ (L is the canonical bundle over $\mathbb{P}^n$). It is well-known that $\{1, \xi, \cdots, \xi^n\}$ is a $\mathbb{Z}$-basis of $H^*(\mathbb{P}^n, \mathbb{Z})$ [BO].

Theorem *For each k, $0 \leq k \leq n$, there exists a unique $\xi_k \in H^{2k}(\widetilde{\mathbb{P}}^n, \mathbb{Z})$ such that $\varphi^*(\xi_k) = l_k \xi^k$, and $\{1, \xi, \cdots, \xi^n\}$ is a $\mathbb{Z}$-basis of the free abelian group $H^*(\widetilde{\mathbb{P}}^n, \mathbb{Z})$ (cf. 4.(a)). In other words there are commutative diagrams*

$$\begin{array}{ccc} H^{2k}(\widetilde{\mathbb{P}}^n, \mathbb{Z}) & \xrightarrow{\varphi^*} & H^{2k}(\mathbb{P}^n, \mathbb{Z}) \qquad (0 \leq k \leq n) \\ \| & & \| \\ \mathbb{Z} & \xrightarrow[\cdot l_k]{} & \mathbb{Z} \end{array}$$

<u>**Proof.**</u> Let $k \in \{0, \cdots, n\}$, fixed in all the proof. For $I = \{i_0, \cdots, i_k\}$, with $0 \leq i_0 < \cdots < i_k \leq n$, we put

$$\mathbb{P}_I^k = \mathbb{P}_{\mathbb{C}}(q_{i_0}, \cdots, q_{i_k}) ,$$

and denote by u_I the closed inclusion $\mathbb{P}_I^k \subset \widetilde{\mathbb{P}}^n$ defined by $x_i = 0$ for all $i \notin I$. Let m_I be the integer defined by the diagram

$$\begin{array}{ccc} H^{2k}(\widetilde{\mathbb{P}}^n, \mathbb{Z}) & \xrightarrow{u_I^*} & H^{2k}(\widetilde{\mathbb{P}}^k, \mathbb{Z}) \\ \| & & \| \\ \mathbb{Z} & \xrightarrow[\cdot m_I]{} & \mathbb{Z} \end{array}$$

Lemma 1: *The integers, elements of the set*

$$M = \{m_I | I \subset \{0, \cdots, n\},\ |I| = k+1\} ,$$

have gcd equal to 1.

<u>**Proof of Lemma 1**</u>. We use induction on n. If $n = 0$, then $M = \{1\}$, and the lemma is clear. Assume then $n \geq 1$ and the result true in case $n-1$. If $k = n$, then $M = \{1\}$. So we can suppose $k \leq n-1$. Now, for $0 \leq i \leq n$, consider $\widetilde{\mathbb{P}}_i^{n-1} = \mathbb{P}_{\mathbb{C}}(q_0, \cdots, \hat{q}_i, \cdots, q_n)$, $u_i : \widetilde{\mathbb{P}}_i^{n-1} \subset \widetilde{\mathbb{P}}^n$ the inclusion defined by $x_i = 0$. Denote by m_i the integer defined by the diagram

$$\begin{array}{ccc} H^{2k}(\widetilde{\mathbb{P}}^n, \mathbb{Z}) & \xrightarrow{u_i^*} & H^{2k}(\widetilde{\mathbb{P}}_i^{n-1}, \mathbb{Z}) \\ \| & & \| \\ \mathbb{Z} & \xrightarrow[\cdot m_I]{} & \mathbb{Z} \end{array}$$

Take $I = \{i_0, \cdots, i_k\}$, with $0 \leq i_0 < \cdots < i_k \leq n$, and i, $0 \leq i \leq n$, such that $i \notin I$. Then the map u_I factorizes through $\widetilde{\mathbb{P}}_i^{n-1}$:

$$\begin{array}{ccc} \widetilde{\mathbb{P}}_I^k & \xrightarrow{u_I} & \widetilde{\mathbb{P}}^n \\ u_{I,i} \nwarrow & & \nearrow u_i \\ & \widetilde{\mathbb{P}}_i^{n-1} & \end{array}$$

Hence $u_I^* = u_{I,i}^* \circ u_i^*$, which shows that $m_I = m_{I,i} m_i$. Therefore M is the set of integers m such that:

> there exist $I \subset \{0, \cdots, m\}$, $|I| = k+1$, and i, $0 \leq i \leq$, $i \notin I$ verifying $m = m_i m_{I,i}$.

(Recall $k \leq n-1$). But, by induction hypothesis, for each i, $0 \leq i \leq n$, the elements of the set

$$M_i = \{M_{I,i} | I \subset \{0, \cdots, n\},\ i \notin I,\ |I| = k+1\}$$

have gcd equal to 1. So we need only to show the next lemma.

Lemma 2: *One has $gcd\{m_0, \cdots, m_n\} = 1$.*

Proof of Lemma 2. Since $\mathbb{P}(dq_0, \cdots, dq_n) = \mathbb{P}(q_0, \cdots, q_n)$, we may assume $\gcd\{q_0, \cdots, q_n\} = 1$. The inclusion $u_i : \widetilde{\mathbb{P}}_i^{n-1} \subset \widetilde{\mathbb{P}}^n$ has $\mathbb{C}^n/\mu_q$ as complementary open set (1.(a)), showing that there is an exact sequence (with coefficients in $\mathbb{Z}_a = \mathbb{Z}/a\mathbb{Z}$, $a \geq 0$):

$$\begin{array}{ccccccc} H^{2k}(\widetilde{\mathbb{P}}^n) & \xrightarrow{u_i^*} & H^{2k}(\widetilde{\mathbb{P}}_i^{n-1}) & \longrightarrow & H_c^{2k+1}(\mathbb{C}^n/\mu_q) & \longrightarrow & 0 \\ \| & & \| & & & & \\ \mathbb{Z}_a & \xrightarrow[\cdot m_i]{} & \mathbb{Z}_a & & & & \end{array}$$

Assume a is a prime number which does not divide q_i. Then, by 4.Lemma 2, $H_c^{2k+1}(\mathbb{C}^n/\mu_{q_i}, \mathbb{Z}_a) = 0$, which means that multiplication by m_i is an automorphism of $\mathbb{Z}_a$. Hence a does not divide m_i. This implies $\gcd\{m_0, \cdots, m_n\} = 1$.

To prove the theorem, take $I = \{i_0, \cdots, i_k\}$ with $0 \leq i_0 < \cdots < i_k \leq n$ and consider the commutative diagram

$$\begin{array}{ccc} \mathbb{P}^n & \xrightarrow{\varphi} & \widetilde{\mathbb{P}}^n \\ \uparrow & & \uparrow u_I \\ \mathbb{P}^k & \xrightarrow[\varphi_I]{} & \widetilde{\mathbb{P}}^k_I \end{array} \quad , \text{ where } \varphi_I(x_{i_0}, \cdots, x_{i_k}) = (x_{i_0}^{q_{i_0}}, \cdots, x_{i_k}^{q_{i_k}}) \ .$$

Thus induces a cohomology commutative diagram (coefficients in $\mathbb{Z}$)

$$\begin{array}{ccc} H^{2k}(\widetilde{\mathbb{P}}^n) & \xrightarrow{\varphi^*} & H^{2k}(\mathbb{P}^n) \\ u_I^* \downarrow & & \downarrow \\ H^{2k}(\widetilde{\mathbb{P}}^k_I) & \xrightarrow[\varphi_I^*]{} & H^{2k}(\mathbb{P}^k) \end{array} \quad = \quad \begin{array}{ccc} \mathbb{Z} & \xrightarrow{\cdot a} & \mathbb{Z} \\ \cdot m_i \downarrow & & \downarrow \| \\ \mathbb{Z} & \xrightarrow[\cdot a_I]{} & \mathbb{Z} \end{array}$$

Whence $a = m_I a_I$, for all $I \subset \{0, \cdots, n\}$ with $|I| = k+1$. Lemma 1 implies then $\operatorname{lcm}\{a_i | I \subset \{0, \cdots, n\},\ |I| = k+1\}$. But $a_I = \deg(\varphi_I) = q_{i_0} \cdots q_{i_k} / (q_{i_0}, \cdots, q_{i_k})$ (2.(b)). The proof of the theorem is finished.

Consider the closed inclusion $\widetilde{\mathbb{P}}^{n-1} = \mathbb{P}(q_0, \cdots, q_{n-1}) \subset \widetilde{\mathbb{P}}^n$ $(n \geq 1)$ and its complementary

$$\mathbb{C}^n/\mu_q = (\mathbb{C}^n/\mu_q)(q_0, \cdots, q_{n-1}), \text{ where } q = q_n \quad \text{(cf. 1.(a))} \ .$$

For $0 \leq k \leq n-1$, consider the integer (see beginning of this §5.)

$$m_k = l_k(q_0, \cdots, q_n)/l_k(q_0, \cdots, q_{n-1}) \ .$$

Corollary *The cohomology groups, with compact supports, of $\mathbb{C}^n/\mu_q$ are:*

$$H^i_c(\mathbb{C}^n/\mu_q, \mathbb{Z}) = \begin{cases} \mathbb{Z} & \text{if } i = 2n \ , \\ \mathbb{Z}/m_k\mathbb{Z} & \text{if } i = 2k+1,\ 1 \leq k \leq n-1 \ , \\ 0 & \text{if not.} \end{cases}$$

Proof. There are exact sequences (coefficients in $\mathbb{Z}$)

$$0 \to H_c^{2i}(\mathbb{C}^n/\mu_q) \to H^{2i}(\widetilde{\mathbb{P}}^n) \to H^{2i}(\widetilde{\mathbb{P}}^{n-1}) \to H_c^{2i+1}(\mathbb{C}^n/\mu_q) \to 0 .$$

On the other hand, putting $f = \varphi(q_0, \cdots, q_n) : \mathbb{P} \longrightarrow \widetilde{\mathbb{P}}^n$, $g = \varphi(q_0, \cdots, q_{n-1}) : \mathbb{P}^{n-1} \longrightarrow \widetilde{\mathbb{P}}^{n-1}$ (1.(b)), we have commutative squares

$$\begin{array}{ccc} H^{2i}(\widetilde{\mathbb{P}}^n) & \longrightarrow & H^{2i}(\widetilde{\mathbb{P}}^{n-1}) \\ f^* \downarrow & & \downarrow g^* \\ H^{2i}(\mathbb{P}^n) & \longrightarrow & H^{2i}(\mathbb{P}^{n-1}) \end{array}$$

Then apply the theorem above.

6 The Ring $H^*(\widetilde{\mathbb{P}}^n, \mathbb{Z})$

We are going to make precise the multiplicative structure of the cohomology $H^*(\widetilde{\mathbb{P}}^n, \mathbb{Z})$. Since $\varphi^* : H^*(\widetilde{\mathbb{P}}^n, \mathbb{Z}) \longrightarrow H^*(\mathbb{P}^n, \mathbb{Z})$ is a ring homomorphism, it follows from Theorem in 5.

$$\xi_i \xi_j = \begin{cases} e_{ij}\xi_{i+j} & \text{if } i+j \leq n , \\ 0 & \text{if not} , \end{cases}$$

where $e_{ij} = l_i l_j / l_{i+j}$, $1 \leq i, j \leq n$.

In the polynomial ring $\mathbb{Z}[T_1, \cdots, T_n]$, let $\mathfrak{a}$ be the ideal generated by the elements

$$T_i T_j \ (i+j > n \text{ and } T_i T_j - e_{ij} T_{i+j} \ (i+j \leq n) .$$

We obtain a ring isomorphism

$$H^*(\widetilde{\mathbb{P}}^n, \mathbb{Z}) \simeq \mathbb{Z}[T_1, \cdots, T_n]/\mathfrak{a}$$

where ξ_i corresponds to the class of T_i.

7 The Generator ξ_1

Let us determine the homomorphism $\psi^* : H^*(\mathbb{P}^n, \mathbb{Z}) \longrightarrow H^*(\widetilde{\mathbb{P}}^n, \mathbb{Z})$ (1.(b)). Write $\psi = \psi(q)$, $\varphi = \varphi(q)$, $q = (q_0, \cdots, q_n)$, and make the identification $\mathbb{P}^n = \mathbb{P}^n(m, \cdots, m)$ where $m = l_1 = \text{lcm}\{q_0, \cdots, q_n\}$. Then it holds $\psi(q) \circ \varphi(q) = \varphi(\underline{m})$. This implies $\psi^*(\xi) = \xi_1$ (Theorem in 5.), which determines completely ψ^*.

Since $\xi = c_1(L^*)$, it follows that $\xi_1 = c_1(\widetilde{L}^*)$, where $\widetilde{L} = \psi^*(L)$ is the canonical line bundle over $\widetilde{\mathbb{P}}^n$ (1.(b)). Denote $\widetilde{\xi} = c_1(\widetilde{L}^*)$ $(= \xi_1)$. From 6., we obtain

$$\xi^k = (l_1^k / l_k)\xi_k,\ 0 \leq k \leq n\ .$$

II. Étale Cohomology of Weighted Projective Spaces

We fix through all this chapter (II) a field k. Schemes, morphisms of schemes, will always mean k-schemes, k-morphisms of k-schemes ... For basic facts about schemes one may consult [M1], [H].

First of all, we want to define weighted projective spaces as quotients of $A^{n+1}\backslash\{0\}$ by the multiplicative group G_n, in the category schemes (A^{n+1} is the affine space, over k, and $G_m = GL(1)$).

1 Quotients of Schemes by Groups

A) Let X be a scheme and G be a group scheme acting on X [M2].

(a) DEFINITION. *A geometric quotient of X by G is a couple (Y, π), where Y is a scheme, $\pi : X \longrightarrow Y$ a morphism of schemes, such that (in the category of locally ringed spaces) the following sequence be exact*

$$(1) \quad G \times_k X \underset{p}{\overset{\sigma}{\rightrightarrows}} X \xrightarrow{\pi} Y ,$$

where σ stands for the action of G on X, and p is the second projection. This means that the next conditions hold.
$(QG)_1$ *(i)* $\pi \circ \sigma = \pi \circ p$,
(ii) if x, $x' \in X$ and $\pi(x) = \pi(x')$, then there exists $z \in G \times_k X$ such that $\sigma(z) = x$ and $p(z) = x'$,
(iii) π is onto;
$(QG)_2$ *the topology of Y is the quotient topology defined by X and π;*
$(QG)_3$ *if $f = \pi \circ \sigma = \pi \circ p$, then the sequence of sheaves (given by (1))*

$$(2) \quad 0 \longrightarrow O_Y \longrightarrow \pi_* O_X \rightrightarrows f_* O_{G \times X}$$

is exact. [2]

The kernel of the double arrow in the sequence (2) is, by definition, the subsheaf of invariants of $\pi_* O_X$ under G and is denoted $(\pi_* O_X)^G$. The condition $(QG)_3$ says that the structural sheaf of the scheme Y is $(\pi_* O_X)^G$.

The next proposition justifies the preceding definition.

Proposition ([M2]0, §2, 0.1)
A geometric quotient (Y, π) of X by G is a (categorical) quotient of X by G, i.e. for any commutative diagram (in the category of schemes)

---

[2] In this definition, the condition $(QG)_2$ is weaker than the corresponding one in the first edition of Geometric Invariant Theory, by D. Mumford (Chap.0, §1, 0.6).

$$\begin{array}{ccc} G \times_k X & \xrightarrow{\sigma} & X \\ {\scriptstyle p}\downarrow & & \downarrow{\scriptstyle h} \\ X & \xrightarrow[h]{} & Z \end{array} ,$$

there is a unique morphism $\bar{h} : Y \longrightarrow Z$ *with* $\bar{h} \circ \pi = h$*:*

$$X \xrightarrow[\pi]{} Y \xrightarrow[\bar{h}]{} Z \quad (\text{composite } h) .$$

Hence a geometric quotient is unique up to isomorphism.

(b) Existence of a geometric quotient in the affine case

Assume the scheme X is affine:

$$X = \text{Spec}\,(A)\,, \quad \text{where } A \text{ is a } k\text{-algebra.}$$

Then the operation $\sigma : G \times_k X \longrightarrow X$ induces a k-algebra morphism

$$f : A \longrightarrow R \otimes_k A\,, \quad \text{where } R = \Gamma(G, O_G)\,.$$

An element $a \in A$ is invariant under G if $f(a) = 1 \otimes a$. The set A^G of invariant elements is a sub-k-algebra of A, and the inclusion $A^G \subset A$ defines a morphism of schemes $\pi : X \longrightarrow Y = \text{Spec}\,(A^G)$.

Theorem ([M2] 1, §2, 1.1 and 1.3)
If G is a linearly reductive algebraic group (over k), and if the orbits of geometric points of X are closed, then (Y, π) is a geometric quotient of X by G. Moreover, if X is noetherian (resp. algebraic), then Y is noetherian (resp. algebraic).

Recall that *linearly reductive* means linear and that every representation $G \longrightarrow GL(n)$ is completely reducible (cf. [M2] 1, §1).

B) Let now X be a scheme and G be a finite abstract group, acting on X by automorphisms. That is, we have a group homomorphism $G \longrightarrow \text{Aut}_{\,k}(X)$. This is equivalent to the datum of an action $G_k \times X \longrightarrow X$, where G_k is the constant group scheme associated to G. A *geometric quotient* of X by G is a geometric quotient of X by G_k (in the sense of A.(a)).

Assume X is affine: $X = \text{Spec }(A)$ (A k-algebra). Then $\text{Aut }_k(X) = \text{Aut }_k(A)$, and let $A^G \subset A$ denote the subalgebra of invariants (Notice that $A^G = A^{G_k}$, A^{G_k} being defined in A.(b)). This induces a morphism $\pi : X \longrightarrow Y = \text{Spec }(A^G)$.

From [G3] Exp. V, 1.1, 1.2, 1.5, 1.8, we have the following theorem about the existence of a geometric quotient of a scheme by the group G.

Theorem

i) (Y, π) *is a geometric quotient of* X *by* G. *If* X *is algebraic, then so is* Y.

ii) If X' *is a scheme on which* G *acts, then a geometric quotient of* X' *by* G *exists if, and only if,* X' *is covered by affine open subsets which are stable under* G.

2 Weighted Projective Spaces as Geometric Quotients

Remember that a ground field k has been fixed.

Let $q_0, \cdots, q_n$ be positive integers. Set

$$X = A^{n+1} \backslash \{0\} = \text{Spec }(R) \setminus \text{Spec }(k) \ ,$$

where $R = k[T_0, \cdots, T_n]$ is the polynomial k-algebra with $n+1$ indeterminates. Denote G the multiplicative group over k

$$G_m = \text{Spec }(k, [T, T^{-1}]) \ .$$

(a) Action of G on X

Consider the localizations

$$R_i = R_{T_i} \ , \ R_{ij} = R_{T_iT_j} \quad (0 \le i, j \le n) \ ,$$

and the associated affine schemes

$$X_i = \text{Spec } (R_i) \ , \ X_{ij} = \text{Spec } (R_{ij}) \ .$$

There are canonical k-algebra isomorphisms $R_{ij} \simeq R_{ji}$, inducing isomorphisms $X_{ji} \simeq X_{ij}$. We shall view X as the scheme obtained by gluing the X_i's along these isomorphisms between the open subsets $X_{ij} \subset X_i$.

There are k-algebra homomorphisms

$$R_i \longrightarrow k[T, T^{-1}] \otimes_k R_i \ , \ R_{ij} \longrightarrow k[T, T^{-1}] \otimes_k R_{ij}$$

defined by $T_s \longrightarrow T^{q_s} \otimes T_s$ $(0 \leq s \leq n)$. This gives actions of the group G on X_i and X_{ij}, let us say σ_i and σ_{ij}, such that the open immersions $X_{ij} \subset X_i$ are G-equivariant.

The operations σ_i's $(0 \leq i \leq n)$ glue together along the natural isomorphisms $\sigma_{ij} \simeq \sigma_{ji}$, into an operation $\sigma : G \times_k X \longrightarrow X$ of G on X. To make precise that this action depends on the integers $q_0, \cdots, q_n$, we shall write $\sigma = \sigma(q_0, \cdots, q_n)$.

(b) Quotient of X by G

For each i, $0 \leq i \leq n$, put $Y_i = \text{Spec } (R_i^G)$ and $\pi_i : X_i \longrightarrow Y_i$ the morphism induced by the inclusion $R_i^G \subset R_i$.

Lemma *The couple (Y_i, π_i) is a geometric quotient of X_i by G.*

Proof. Since G is linearly reductive, and because of Theorem 1.(b), we need only to show that the orbits of geometric points of X_i are closed. So take a geometric point $\epsilon : \text{Spec } (K) \longrightarrow X_i$, where K is an algebraically closed extension of k. Let $u : R_i \longrightarrow K$ be the corresponding k-homomorphism. We have to prove the image of the morphism

$$(\sigma_i \circ (1 \times \epsilon), p) : G \times_k \text{Spec } (K) \longrightarrow X_i \times_k \text{Spec } (K)$$

is closed (p is the second projection $G \times_k \text{Spec } (K) \longrightarrow \text{Spec } (K)$). But this corresponds to the k-homomorpishm

$$h : K[T_0, \cdots, T_n, T_i^{-1}] = R_i \otimes_k K \longrightarrow k[T, T^{-1}] \otimes_k K = K[T, T^{-1}] \ ,$$

defined by $h(T_s) = u(T_s)T^{q_s}$ $(0 \leq s \leq n)$. This h is integral, indeed:

$$T^{q_i} = u(T_i^{-1})(u(T_i)\,T^{q_i}) = u(T_i^{-1})h(T_i) = h(u(T_i^{-1})\,T_i)$$

and

$$T^{-q_i} = h(u(T_i)\,T_i^{-1})\ .$$

Therefore $(\sigma_i \circ (1 \times \epsilon), p)$ is closed, and so the orbit of ϵ is closed as required.

Similarly to the geometric quotients $\pi_i : X_i \longrightarrow Y_i$, there are geometric quotients $\pi_{ij} : X_{ij} \longrightarrow Y_{ij}$ $(o \leq i, j \leq n)$ in such a why that there exist open immersions $Y_{ij} \subset Y_i$ and isomorphisms $Y_{ij} \simeq Y_{ji}$ (use $(QG)_2$ and Proposition in 1.(a)). This gives the following commutative diagrams

$$\begin{array}{ccccccc} X_i & \supset & X_{ij} & \simeq & X_{ji} & \subset & X_j \\ \pi_i\downarrow & & \pi_{ij}\downarrow & & \downarrow\pi_{ji} & & \downarrow\pi_j \\ Y_i & \supset & Y_{ij} & \simeq & Y_{ji} & \subset & Y_j \end{array}$$

Gluing the π_i's along the isomorphisms $\pi_{ij} \simeq \pi_{ji}$, we obtain a morphism of schemes $\pi : X \longrightarrow Y$. So (Y, π) is a geometric quotient of X by G (since this property is local on Y; see Definition 1.(a) above).

DEFINITION. *The geometric quotient (Y, π) we just constructed is called* weighted (or twisted) projective space *of type $(q_0, \cdots, q_n)$ and is denoted*

$$\widetilde{\mathbb{P}}^n_k = \mathbb{P}_k(q_0, .., q_n)$$

(k is any field, $q_0, \cdots, q_n$ are any positive integers).

(c) The k-algebra R_i^G

By construction, the scheme $\widetilde{\mathbb{P}}^n$ is covered by open subsets

$$Y_i = Y_i(q_0, \cdots, q_n) = \text{Spec }(R_i^G) \quad (0 \leq i \leq n)\ .$$

Let us determine the algebra of invariants R_i^G. The action $\sigma_i : G \times_k X_i \longrightarrow X_i$ is defined by the k-algebra homomorphism

$$f_i : R_i = k[T_0, \cdots, T_n, T_i^{-1}] \to k[T, T^{-1}] \otimes_k R_i = k[T, T^{-1}, T_0, \cdots, T_n, T_i^{-1}],$$

$f_i(T_s) = T^{q_s} T_s \quad (0 \le s \le n)$.

An element of R_i of the form

$$a_{\underline{\alpha}} = T_0^{\alpha_0} \cdots T_n^{\alpha_n} \text{ where } \alpha_s \in \mathbb{N}\ (s \neq i),\ \alpha_i \in \mathbb{Z}\ ,$$

is invariant under G if, and only if,

$$T^\beta a_{\underline{\alpha}} = f_i(a_{\underline{\alpha}}) = a_{\underline{\alpha}}\ , \quad \text{with } \beta = \sum_0^n \alpha_s q_s\ ,$$

which is equivalent to saying $\beta = 0$. Thus the k-algebra of invariants R_i^G is generated by the elements $a_{\underline{\alpha}}$ above which verify $\sum_0^n \alpha_s q_s = 0$.

(d) REMARKS
i) Our construction of $\widetilde{\mathbb{P}}^n = \mathbb{P}_k(q_0, \cdots, q_n)$ works for all integers $q_0, \cdots, q_n \in \mathbb{Z}\backslash\{0\}$. However we keep the hypothesis $q_i > 0$ $(0 \le i \le n)$.
ii) Consider the $\mathbb{N}$-graduation on the k-algebra $R = k[T_0, \cdots, T_n]$ given by

$$\deg\,(T_s) = q_s\ (0 \le s \le n)\ ,\ \deg\,(\lambda) = 0\ (\lambda \in k)\ .$$

We shall write $R = R(q_0, \cdots, q_n)$ to specify this graduation. This induces a $\mathbb{Z}$-graduation on $R_i = R_{T_i}$, the 0-degree component of which is $R_{i,0} = R_i^G$ (after (c) above). Since the scheme $\text{Proj}(R(q_0, \cdots, q_n))$ ([GD] II, §2) is constructed by gluing the affine schemes $\text{Spec}(R_i, 0)$ $(0 \le i \le n)$, we obtain a canonical isomorphism

$$\mathbb{P}_k^n(q_0, \cdots, q_n) = \text{Proj}\ (R(q_0, \cdots, q_n))\ .$$

In particular, it follows that $\mathbb{P}_k^n(1, \cdots, 1) = \mathbb{P}_k^n$ is the usual projective space over k.
iii) Let $q_0', \cdots, q_m'$ be integers > 0.
(α)Assume $m \ge n$ and q_i' divides q_i $(0 \le i \le n)$. Let $V = \cup_{i=0}^n Y_i'$, where $Y_i' = Y_i(q_0', \cdots, q_n')$ is the affine open subset of $\mathbb{P}(q_0', \cdots, q_m')$ defined above (see (b) and (c)). It is clear that there is a morphism $V \longrightarrow \widetilde{\mathbb{P}}^n = \mathbb{P}(q_0, \cdots, q_n)$, induced by the k-algebra homomorphism

$$R(q_0, .., q_n) \longrightarrow R(q_0', \cdots, q_n')\ ,\quad T_s \mapsto T_s^{\alpha_s}\ (0 \le s \le n)\ ,$$

where $\alpha_s = q_s/q'_s$ (glue the corresponding morphisms $Y'_i \longrightarrow Y_i$, $0 \le i \le n$).
(β) Assume now $m \le n$ and q'_i divides q_i ($0 \le i \le n$). Then the k-algebra homomorphism

$$\begin{array}{rcl} k[T_0, \cdots, T_n] & \longrightarrow & k[T_0, \cdots, T_m] \\ T_s & \mapsto & T_s^{\alpha_s}\ (0 \le s \le m),\ \alpha_s = q_s/q'_s \\ T_s & \mapsto & 0\ (m+1 \le s \le n) \end{array}$$

induces a morphism $\mathbb{P}(q'_0, \cdots, q'_m) \longrightarrow \mathbb{P}(q_0, \cdots, q_n)$, which is a closed immersion if $q'_i = q_i$ ($0 \le i \le m$).
iv) For any integer $d > 0$, there is a natural isomorphism $\mathbb{P}(dq_0, \cdots, dq_n) = \mathbb{P}(q_0, \cdots, q_n)$. Indeed, consider the endomorphism u of G defined by $T \mapsto T^d$. Then $\sigma' = \sigma \circ (u \times \mathrm{id}_X)$, where σ and σ' are the actions of G defining the quotients $\mathbb{P} = \mathbb{P}(q_i)$ and $\mathbb{P}' = \mathbb{P}(dq_i)$ (see 2.(a)). This induces a morphism $\mathbb{P}' \longrightarrow \mathbb{P}$, which is an isomorphism since locally σ and σ' have the same invariants (by 2.(c)).

3 Some Properties of the Scheme $\widetilde{\mathbb{P}}^n_k$

(a) The scheme $\widetilde{\mathbb{P}}^n = \mathbb{P}_k(q_0, .., q_n)$ is integral and algebraic over k

It is integral since it is the quotient of $X = A_k^{n+1} \backslash \{0\}$ (by $G = G_n$) (after 2. and Proposition in 1.A.), and X is integral. Now it is algebraic since $\widetilde{\mathbb{P}}^n = \cup_{i=0}^n Y_i$, where $Y_i = Y_i(q_0, \cdots, q_n) \subset \widetilde{\mathbb{P}}^n$ are open immersions, and the Y_i's are algebraic (cf. 2.(c)).

(b) Proposition *The k-scheme $\widetilde{\mathbb{P}}^n$ is projective. More precisely there is a closed immersion $\widetilde{\mathbb{P}}^n \subset \mathbb{P}^m$ where the integer m is defined as follows. Let s be the smallest integer which is a multiple of $l = \mathrm{lcm}\{q_0, \cdots, q_n\}$ and greater or equal to the sum $\sum_0^n (l - q_i)$. Let f be the complex function*

$$f(z) = [z^{s+1}(1 - z^{q_0} \cdots (1 - z^{q_n})]^{-1} .$$

Then $m = \mathrm{res}\ _0(f) - 1$ (res = residue).

Proof. Consider the graded algebra $R = R(q_0, \cdots, q_n)$ defined in 2.(d).ii. For any integer $r \geq 1$, $R^{(r)}$ denotes the subalgebra generated by homogeneous elements of degree multiple of r. There is a graduation on $R^{(r)}$ in such a way that $(R^{(r)})_\alpha = R_{r\alpha}$. The integer s is defined in the statement above, the algebra $R^{(s)}$ is generated by its elements of degree 1 ([GD] II.2.1.6, ii) and v); in the proof of ii) the integer n_1 is equal to $\sum_0^n (l - q_i)$ in our situation). The monomials $T_0^{\alpha_0} \cdots T_n^{\alpha_n}$, of degree $s = \alpha_0 q_1 + \cdots + \alpha_n q_n$ in $R = k[T_0, \cdots, T_n]$ (i.e. of degree 1 in $R^{(s)}$) form a set the cardinality of which is $\mathrm{res}_0(f) = m + 1$. Let $a_0, \cdots, a_m$ be homogeneous elements of degree 1, generating the k-algebra $R^{(s)}$. So there is a graded k-algebra epimorphism

$$S = k[t_0, \cdots, t_m] \longrightarrow R^{(s)} : t_i \mapsto a_i \; (0 \leq i \leq m) ,$$

where S is the usual graded polynomial algebra. This induces a closed immersion ([GD] II.2.9.2.i).)

$$\text{Proj } (R^{(s)}) \longrightarrow \text{Proj } (S) = \mathbb{P}_k^m .$$

But $\text{Proj}(R^{(s)}) = \text{Proj}(R)$ (loc. cit. 2.4.7 i).) and $\text{Proj}(R) = \widetilde{\mathbb{P}}_k^m$ (2.(d).ii).

(c) Local structure of $\widetilde{\mathbb{P}}^n$

Let q be an integer ≥ 1, and let $U_q = U_q(k)$ be the subgroup of the q-th roots of unity, in the multiplicative group $k^* = k \backslash \{0\}$. Denote by d the order of U_q (if k is algebraically closed and char (k) if 0 or is prime with q, then $d = q$). The constant algebraic group (over k), defined by the abstract group U_q, is the group of the d-th roots of unity:

$$\mu_d = \text{Spec } (k[T]/(T^d - 1)) \; (\text{since } T^d - 1 = \prod_{\lambda \in U_q} (T - \lambda)) .$$

In particular, an action of U_q (by automorhisms) on a scheme S is equivalent to an action of μ_d on S.

We have $\widetilde{\mathbb{P}}^n = \cup_0^n Y_i$, where $Y_i = Y_i(q_0, \cdots, q_n)$ is an open subset (see 2.(c)), complementary to the closed immersion

$$\mathbb{P}^{n-1}(q_0, \cdots, \hat{q}_i, \cdots, q_n) \subset \widetilde{\mathbb{P}}^n = \mathbb{P}^n(q_0, \cdots, q_n)$$

defined by the projection

$$k[T_0,\cdots,T_n] \longrightarrow k[T_0,\cdots,\hat{T}_i,\cdots,T_n] : T_j \mapsto (1-\delta_{ij})T_j \ .$$

If the order of U_{q_i} is q_i, then Y_i is a geometric quotient of the affine space $A^n = \text{Spec}\ (k[T_0,\cdots,\hat{T}_i,\cdots,T_n])$ by U_{q_i}. In this case we shall write

$$Y_i = A^n/U_{q_i} = (A^n/U_{q_i})(q_0,\cdots,\hat{q}_i,\cdots,q_n) \ .$$

Proof. To simplify the notation, we assume $i=0$, and put $q=q_0$. The action of U_q on $A^n = \text{Spec}\ (k[T_1,\cdots,T_n])$ is defined by the group homomorphism

$$U_q \longrightarrow \text{Aut}_k(k[T1,\cdots,T_n]) : \lambda \mapsto u_\lambda \ , \ u\lambda(T_s) = \lambda^{q_s}T_s \ .$$

Let B be the associated subalgebra of invariants. Since $Y_0 = \text{Spec}(R_0^G)$ (2.(c)), it is enough to verify that the two algebras B and R_0^G are isomorphic. But B is generated by the monomials $T_1^{\alpha_1}\cdots T_n^{\alpha_n}$ with $\sum_{s=1}^{n}\alpha_s q_s \equiv 0 \bmod |U_q|$. Using 2.(c), we have the following homomorphism

$$\begin{array}{rcccccl} B & \rightarrow & R_0^G & : & T_1^{\alpha_1}\cdots T_n^{\alpha_n} & \mapsto & T_0^{\alpha_0}\cdots T_n^{\alpha_n} \text{ with } \alpha_0 = -(\sum_1^n \alpha_s q_s)/q_0 \ , \\ R_0^G & \rightarrow & B & : & T_0^{\alpha_0}\cdots T_n^{\alpha_n} & \mapsto & T_1^{\alpha_1}\cdots T_n^{\alpha_n}(\text{where } \sum_0^n \alpha_s q_s = 0) \ , \end{array}$$

which are inverse each to the other.

(d) Morphisms between $\widetilde{\mathbb{P}}^n$ and $\mathbb{P}^n$

There are two morphisms

$$\begin{array}{rcl} \varphi = \varphi(q_0,\cdots,q_n) & : & \mathbb{P}^n \longrightarrow \widetilde{\mathbb{P}}^n = \mathbb{P}(q_0,\cdots,q_n) \\ \psi = \psi(q_0,\cdots,q_n) & : & \widetilde{\mathbb{P}}^n \longrightarrow \mathbb{P}^n = \mathbb{P}(1,\cdots,1) \ , \end{array}$$

defined as follows. The first one is induced by the graded k-algebra homomorphism (2.(d) ii.)

$$R(q_0,\cdots,q_n) \longrightarrow R(1,\cdots,1) : T_i \mapsto T_i^{q_i} \ (0 \le i \le n) \ .$$

To define the second one, let $l = \operatorname{lcm}\{q_0, \cdots, q_n\}$ and put $s_i = l/q_i$. Then the graded k-algebra homomorphism

$$R(l, \cdots, l) \longrightarrow R(q_0, \cdots, q_n) : T_i \mapsto T_i^{s_i} \ (0 \leq i \leq n)$$

induces a morphism $\widetilde{\mathbb{P}}^n \longrightarrow \mathbb{P}^n(l, \cdots, l)$ which, composed with the natural isomorphism $\mathbb{P}^n(l, \cdots, l) \xrightarrow{\sim} \mathbb{P}^n$ (2.(d) iv.), gives ψ.

Proposition *Let $q = (q_0, \cdots, q_n)$ and $U_q = U_{q_0} \times \cdots \times U_{q_n}$. Then the group U_q acts (by automorphisms) on $\mathbb{P}^n$ in such a way that a geometric quotient $(\mathbb{P}^n/U_q, \rho)$ exist with a commutative diagram*

$$\begin{array}{ccc} \mathbb{P}^n/U_q & \xrightarrow{\bar{\varphi}} & \widetilde{\mathbb{P}}^n \\ \rho \uparrow & \nearrow_{\varphi} & \\ \mathbb{P}^n & & \end{array}$$

If $|U_{q_i}| = q_i$ for every i, then $\bar{\varphi}$ is an isomorphism.

Proof. i) Action of U_q on $\mathbb{P}^n$: for $\lambda = (\lambda_0, \cdots, \lambda_n) \in U_q$, let $h(\lambda)$ be the automorphism of $\mathbb{P}^n$ defined by $T_i \longrightarrow \lambda_i T_i$ in the usual graded algebra $R(1) = k[T_0, \cdots, T_n]$ (2.(d) ii). We obtain a group homomorphism $h : U_q \longrightarrow \operatorname{Aut}_k(\mathbb{P}^n)$, that is an action of U_q on $\mathbb{P}^n$. The standard affine open subsets $Y_i(1)$ $(0 \leq i \leq n)$ are stable under U_q. Therefore there exists a geometric quotient $(\mathbb{P}^n/U_q, \rho)$ (cf. Theorem in 1.B).

ii) Let us recall some notation: $R = k[T_0, \cdots, T_n]$, $R_i = R_{T_i}$; $R(q) = R(q_0, \cdots, q_n)$ means that R is graded by $\deg(T_i) = q_i$; $Y_i(q) = \operatorname{Spec}(R_i^G(q))$ where G is the multiplicative group (see 2.(c) and 2.(d) ii). Now we construct the morphism $\bar{\varphi}$.

Let $B_i \subset R_i^G(1) = k[\frac{T_0}{T_i}, \cdots, \frac{T_n}{T_i}]$ denote the subalgebra of invariants under U_q. So we have (Theorem in 1.B)

$$\rho(Y_i(1)) = Y_i(1)/U_q = \operatorname{Spec}(B_i) .$$

On the other hand, there are commutative diagrams of k-algebras

$$\begin{array}{ccc} R_i^G(q) & \xrightarrow{\ u\ } & R_i^G(1) \\ {\scriptstyle v}\uparrow & \nearrow & \text{inclusion} \\ B_i & & \end{array}$$

such that
$$\begin{array}{rcl} u(T_0^{\alpha_0}\cdots T_n^{\alpha_n}) & = & T_0^{\alpha_0 q_0}\cdots T_n^{\alpha_n q_n} \ , \\ v(T_0^{\alpha_0}\cdots T_n^{\alpha_n}) & = & T_0^{\alpha_0 q_0}\cdots T_n^{\alpha_n q_n} \ : \end{array}$$

recall that $T_0^{\alpha_0}\cdots T_n^{\alpha_n} \in R_i^G(q)$ if and only if $\sum_0^n \alpha_s q_s = 0$ and that such "monomials" generate $R_i^G(q)$. Note also that u is induced by the homomorphism $R(q) \longrightarrow R(1)$ defining φ. Now, if $|U_{q_i}| = q_i$ $(0 \leq i \leq n)$, then v is an isomorphism. Indeed, in that case, an element of $R_i^G(1)$ of the form

$$a_{\underline{\alpha}} = T_0^{\alpha_0}\cdots T_n^{\alpha_n} \ , \quad \text{where } \alpha_s \in \mathbb{N}(s \neq i),\ \alpha_i \in \mathbb{Z}, \text{ and } \sum_0^n \alpha_s = 0 \ ,$$

is invariant under U_q if, and only if,

$$\lambda_0^{\alpha_0}\cdots\lambda_n^{\alpha_n} a_{\underline{\alpha}} = a_{\underline{\alpha}} \text{ for all } \lambda = (\lambda_0, \cdots, \lambda_n) \in U_q \ ,$$

that is if, and only if, q_s divides α_s for all s.

The diagrams above define commutative diagrams

$$\begin{array}{ccc} Y_i(1) & \longrightarrow & Y_i(q) \\ & \searrow & \uparrow \\ & & Y_i(1)/U_q \end{array}$$

which, when glued together, give the required diagram.

(e) Canonical sheaves over $\widetilde{\mathbb{P}}^n$

The canonical sheaves $O_X(s)$ over $X = \widetilde{\mathbb{P}}^n$ are defined by the graded $R(q)$-modules $R(q)[s]$. By definition of the morphism $\psi : X = \widetilde{\mathbb{P}}^n \longrightarrow Y = \mathbb{P}^n$, we see that $\psi^* O_Y(r) = O_X(lr)$, where $l = \text{lcm}\{q_0, \cdots, q_n\}$. In particular the sheaf $O_X(l)$ is invertible, and is considered as the canonical invertible sheaf over $\widetilde{\mathbb{P}}^n$. Modulo a normalization of the degrees $q_0, \cdots, q_n$, it is shown in [A2] that $O_X(s)$ is invertible if, and only if, l divides s; moreover Pic $(\widetilde{\mathbb{P}}^n) = \mathbb{Z}$ with generator $O_X(l)$.

4 Degree of the Morphism φ

In this section, we assume the ground field k is algebraically closed and the integers $q_0, .., q_n$ are prime to the characteristic exponent of k (equal to 1 if char $(k) = 0$ and equal to char (k) if not). Let a be an integer ≥ 1, prime to the characteristic exponent of k, and put $F = \mathbb{Z}/a\mathbb{Z}$. We want to compute the étale cohomology homomorphism $\varphi^* : H^{2n}(\widetilde{\mathbb{P}}^n, F_{\widetilde{\mathbb{P}}}) \longrightarrow H^{2n}(\mathbb{P}, F_{\mathbb{P}})$, induced by $\varphi : \mathbb{P} = \mathbb{P}^n_k \longrightarrow \widetilde{\mathbb{P}} = \mathbb{P}^n_k(q_0, \cdots, q_n)$.

(a) First let us see that $\widetilde{\mathbb{P}}$ has the form $\mathbb{P}/\bar{H}$, where $\bar{H}$ is a finite abstract group acting faithfully on $\mathbb{P}$.
i) A finite abstract group H, acting on a scheme S (by automorhisms), acts faithfully on S if the homomorphism $H \longrightarrow$ Aut (S), defining the action of H on S, is injective. Consider an action of H on S and put $H_0 =$ Ker $(H \longrightarrow$ Aut $(S))$. Then $\bar{H} = H/H_0$ acts on S faithfully and, for every (geometric) quotient $(S/H, \rho)$ there is a (geometric) quotient $(S/\bar{H}, \bar{\rho})$ with a commutative diagram (see Theorem in 1.B)

$$\begin{array}{ccc} & S & \\ {}^{\rho}\swarrow & & \searrow{}^{\bar{\rho}} \\ S/H & \xrightarrow{\sim} & S/\bar{H} \end{array} .$$

ii) In the situation of Proposition (3.(d)), it is immediate that $H_0 = U_d$ with $d = \gcd\{q_0, \cdots, q_n\}$ $(S = \mathbb{P}^n, H = U_q)$. Since $|U_{q_i}| = q_i$ (by hypotheses above), we have

$$\widetilde{\mathbb{P}}^n = \mathbb{P}^n/\bar{H} \text{ with } \bar{H} = U_q/U_d ,$$

and $|\bar{H}| = q_0 \cdots q_n/d$. Therefore the homomorphism $\varphi^* : H^{2n}(\widetilde{\mathbb{P}}^n, F_{\widetilde{\mathbb{P}}}) \longrightarrow H^{2n}(\mathbb{P}, F_{\mathbb{P}})$ we want to determine has the form

$$\bar{\rho}^* : H^{2n}(\mathbb{P}^n, F_{\widetilde{\mathbb{P}}}) \longrightarrow H^{2n}(\mathbb{P}, F_{\mathbb{P}}) \text{ where } \bar{\rho} : \mathbb{P} \longrightarrow \mathbb{P}/\bar{H} .$$

The following theorem shows that this is nothing else but multiplication by the integer $|\bar{H}| = q_0 \cdots q_n/d$ in the group F.

(b) Theorem *Let S be an integral, smooth scheme, of finite type over k and of dimension n. Let H be a finite group acting faithfully on S such that a geometric quotient $(S/H, \rho)$ of S by H exist. Then we have*

$$(i)\ H_c^{2n}(S/H, F_{S/H}) = F\ ,$$
$$(ii)\ (H_c^{2n}(S/H, F_{S/H}) \xrightarrow{\rho^*} H_c^{2n}(S, F_s)) = (F \xrightarrow{\cdot|H|} F)\ ,$$

where $H_c^{\cdot}$ stands for cohomology groups with compact supports, [DV].

<u>**Proof.**</u> We may assume $H \neq \{1\}$. So let $g \in H$, $g \neq 1$, and denote by S^g the closed subscheme of invariants of S by the subgroup generated by g (By hypothesis a geometric quotient of S by H exist, hence (Theorem 1.B) S is covered by affine open subsets $V_\alpha = \text{Spec }(A_\alpha)$ which are stable under H. Let $\mathfrak{a}_\alpha^g$ be the ideal of A_α generated by $gx - x$, $x \in A_\alpha$; then $S^g \cap V_\alpha = \text{Spec }(A_\alpha/\mathfrak{a}_\alpha^g)$.) The action of H on S is faithful, so $S^g \neq S$ (S is reduced). Hence $\dim\,(S^g) \leq n-1$, and so $\dim\,(T) \leq n-1$ with $T = \cup_{g\neq 1} S^g$. On the other hand, the open subset $V = S\backslash T$ is non-empty, stable under H, and the action of H on it has no inertia. Thus we have a principal H-covering $\rho|_V : V \longrightarrow V/H$. Now the closed subscheme T of S is invariant by H and covered by stable affine open subsets ($T = \cup_\alpha \text{Spec }(B_\alpha)$, $B_\alpha = A_\alpha / \cap_{g\neq 1} \mathfrak{a}_\alpha^g$). Therefore a geometric quotient $(T/H, \gamma)$ exists so that T/H is a closed subscheme of S/H, complementary to V/H. This shows that we have an exact sequence morphism (coefficients are in $F = \mathbb{Z}/a\mathbb{Z}$)

$$\begin{array}{ccccccc}
0 = H_c^{2n-1}(T) & \to & H_c^{2n}(V) & \to & H_c^{2n}(S) & \to & H_c^{2n}(T) = 0 \\
\uparrow \gamma^* & & \uparrow \rho|_V^* & & \uparrow \rho^* & & \uparrow \gamma^* \\
0 = H_c^{2n-1}(T/H) & \to & H_c^{2n}(V/H) & \to & H_c^{2n}(S/H) & \to & H_c^{2n}(T/H) = 0
\end{array}$$

($\rho^*, \gamma^*, \cdots$ are well-defined since $\rho, \gamma, \cdots$ are proper!). It follows that:
1) $H_c^{2n}(S/H) = H_c^{2n}(V/H) = F$ (V/H is irreducible since V is),
2) to see (ii), it is sufficient to establish

$$(H_c^{2n}(V/H) \xrightarrow{\rho|_V^*} H_c^{2n}(V)) = (F \xrightarrow{\cdot|H|} F)\ .$$

Set $\tau = \rho|_V$ and $\bar{V} = V/h$. There is a commutative triangle

$$\begin{array}{ccc} H^{2n}_c(V) & \xrightarrow{\tau_*} & H^{2n}_c(\bar{V}) \\ & {\scriptstyle \mathrm{Tr}_V}\searrow\tilde{} \quad \tilde{}\swarrow{\scriptstyle \mathrm{Tr}_{\bar{V}}} & \\ & F & \end{array} \qquad (\mathrm{Tr} = \text{trace homomorphism}).$$

Hence it is enough to see that

$$\tau^*\tau_*(x) = |H|x, \text{ for all } x \in H^{2n}_c(V)\,.$$

But this is the case if $\tau^*\tau_*(1) = |H|1$ in $H^0_c(V)$ because of the projection formula $\tau^*\tau_*(x) = x\tau^*\tau_*(1)$. Since $\tau : V \longrightarrow V/H$ is a principal H-covering, there is a commutative diagram

$$\begin{array}{ccc} H \times V = V \times_{\bar{V}} V & \xrightarrow{p_1} & V \\ \searrow \quad {\scriptstyle p_2}\downarrow & & \downarrow{\scriptstyle \tau} \\ V & \xrightarrow[\tau]{} & \bar{V} \end{array} \qquad (H \times V = \coprod_H V)$$

Therefore $p_{1*}p_2^*(1) = |H| \cdot 1$ in $H^0_c(V)$. The proper change of basis theorem ([AR1] 5.1) says that $\tau^*\tau_*(1) = p_{1*}p_2^*(1)$, and this ends the proof.

5 Cohomology of a Cone

(a) Let q be an integer ≥ 1 and assume that $|U_q| = q$ (3.(c)). Then the quotient scheme $A^{n+1}/U_q = (A^{n+1}/U_q)(q_0, \cdots, q_n)$, defined in 3.(c), is the affine projecting cone of a projective variety (over k) the vertex of which is the image of $(0, \cdots, 0) \in A^{n+1}$.

Indeed the proof in 3.(c) shows that $A^{n+1}/U_q = \mathrm{Spec}(B)$, where B is the subalgebra of $R = k[T_0, \cdots, T_n]$ generated by the monomials

$$a_{\underline{\alpha}} = T_0^{\alpha_0} \cdots T_n^{\alpha_n} \text{ such that } q \text{ divides } \sum_0^n \alpha_s q_s\,.$$

Setting, for these elements, deg $(a_{\underline{\alpha}}) := (\sum_0^n \alpha_s q_s)/q$, we define an $\mathbb{N}$-graduation on B, that is an action of the line $D_k = \mathrm{Spec}(k[T])$

on A^{n+1}/U_q. Hence A^{n+1}/U_q is an affine cone with vertex the point $(0, \cdots, 0)$. On the other hand, there is an integer $m \geq 1$ such that a closed immersion Proj $(B) \subset \mathbb{P}_K^N$ exist. This is because the k-algebra B of invariants is finitely generated (Theorem in 1.B), and, for each integer $r \geq 1$, the number of monomials $T_0^{\alpha_0} \cdots T_n^{\alpha_n}$ with $\sum_0^n \alpha_s q_s = rq$ is finite. Continuation of the proof is then similar to the one in 3.(b).

The following proposition shows that, if k is algebraically closed, then $H^0(A^{n+1}/U_q, F) = F$, $, H^i(A^{n+1}/U_q, F) = 0$ $(i \neq 0)$, and $H_0^i(A^{n+1}/U_q, F) \longrightarrow H_c^i(A^{n+1}/U_q, F)$ is an isomorphism for all i.

(b) Cohomology of a cone ([DE] 2.1.2, 2.1.3)[3]

We assume the ground field k algebraically closed.

Proposition *Let C be the affine projecting cone of a projective variety $V \subset \mathbb{P}_k^r$, and let 0 be the vertex of C. Let F be a torsion abelian group, prime to the characteristic exponent of k. Then we have:*
(i) $H^0(C, F) = F$, $H^i(C, F) = 0$ $(i \neq 0)$,
(ii) $H_0^i(C, F) \xrightarrow{\sim} H_c^i(C, F)$ $(i \geq 0)$.

Before the proof, we make a definition and establish a lemma.

Let S, T be two schemes (over k), A a torsion ring prime to the characteristic exponent of k, K an object in $D^+(S, A)$ and L an object in $D+(T, A)$ (derived category of A_T-Module complexes which are bounded below). A morphism $\varphi : (S, K) \longrightarrow (T, L)$ consists (by definition) of two morphisms $(S \xrightarrow{\varphi} T \ , \ \psi^* L \longrightarrow K)$. It induces $\psi^* : H^*(T, L) \longrightarrow H^*(S, K)$.

DEFINITION. *Two morphisms $\varphi_0, \varphi_1 : (S, K) \longrightarrow (T, L)$ are said to be homotopic if there exist a connected scheme Γ, of finite type over k, two points 0 and 1 in Γ and a morphism $\Phi : (S \times \Gamma, p_1^* K) \longrightarrow (T, L)$, where p_1 is the projection $S \times \Gamma \longrightarrow S$, with the following property. On the fibers over 0 and 1 of the projection $p_2 : S \times \Gamma \longrightarrow \Gamma$, Φ induces φ_0 and φ_1.*

[3] This paragraph is entirely borrowed from [DE] for our convenience.

Lemma *If $\varphi_0, \varphi_1 : (S, K) \longrightarrow (T, L)$ are homotopic, then $\varphi_0^* = \varphi_1^*$.*

Proof of Lemma. Keeping the same notation as in the definition above, there exist a sequence of points $0 = x_0, x_1, \cdots, x_m = 1$ in Γ, connected smooth curves Γ_i and morphisms $\gamma_i : \Gamma_i \longrightarrow \Gamma$ such that x_i and x_{i+1} belong to he image of γ_i (Γ_i is obtained by normalization of one-dimensional connected subscheme of Γ containing 0 and 1). Thus we may assume that Γ is a smooth curve. Now apply the smooth change of basis theorem ([AR2] 1.2) to the commutative square

$$\begin{array}{ccc} S \times \Gamma & \xrightarrow{p_1} & S \\ \beta \downarrow & & \downarrow f \\ \Gamma & \xrightarrow[g]{} & \mathrm{Spec}(k) \end{array}$$

Whence $g^* R^i f_* K \xrightarrow{\sim} R^i p_{2*}(p_1^* K)$, which gives the following commutative diagrams (with $x = 0, 1$)

$$\begin{array}{ccccc} H^i(T, L) & \xrightarrow{\varphi_x^*} & H^i(S, K) & \xrightarrow{\sim} & H^0(\Gamma, g^* H^i(S, K)) . \\ \Phi^* \downarrow & & \uparrow \text{fiber over } x & \nearrow \sim & \\ H^i(S \times \Gamma, p_1^* K) & \longrightarrow & H^0(\Gamma, R^i p_{2*}(p_1^* K)) & & \end{array}$$

But this means that φ_x^* is independent of x.

Proof of Proposition. (i) The projective variety $V \subset \mathbb{P}_k^r$ is defined by equations $f_1 = \cdots = f_s = 0$, where the f_i's are homogeneous polynomials in $r + 1$ variables $x_0, \cdots, x_r$. The affine projecting cone $C \subset A_k^{r+1}$ of V is given by the same equations. The morphism $C \times A_k^1 \longrightarrow C$, defined by $((x_0, \cdots, x_r), t) \mapsto (tx_0, \cdots, tx_r)$, induces the identity of C at $t = 1$ and the constant morphism at $t = 0$ with value 0 (vertex of C). So the Lemma implies $H^i(C, F) = H^i(\{0\}, F)$.
(ii) The projective projecting cone $\hat{C} \subset \mathbb{P}_k^{r+1}$ of V is defined by the same equations $f_1 = \cdots = f_s = 0$ as in (i) except that the homogenous polynomials f_i are considered with $r+2$ variables $x_0, \cdots, x_{r+1}$. The morphism $(\hat{C} \backslash \{0\}) \times A_k^1 \longrightarrow \hat{C} \backslash \{0\}$, given by $(((x_0, \cdots, x_{r+1}, t) \mapsto (x_0, \cdots, x_r, tx_{r+1})$, shows that $V \subset \hat{C} \backslash \{0\}$ is a deformation re-

tract of $\hat{C}\backslash\{0\}$. Whence $H^i(\hat{C}\backslash\{0\}) \tilde{\longrightarrow} H^i(V)$ by Lemma above (coefficients are in F). The 5 lemma and the following commutative diagram give the required result.

$$\begin{array}{ccccccccc}
\cdots \longrightarrow & H^{i-1}(\hat{C}\backslash\{0\}) & \longrightarrow & H_0^i(\hat{C}) & \longrightarrow & H^i(\hat{C}) & \longrightarrow & H^i(\hat{C}\backslash\{0\}) & \longrightarrow \cdots \\
& \downarrow\wr & & \downarrow & & \downarrow = & & \downarrow\wr & \\
\cdots \longrightarrow & H^{i-1}(V) & \longrightarrow & H_c^i(C) & \longrightarrow & H^i(\hat{C}) & \longrightarrow & H^i(V) & \longrightarrow \cdots
\end{array}$$

$(H_0^i(\hat{C}) = H_0^i(C))$.

6 Étale Cohomology of $\widetilde{\mathbb{P}}^n$

In all this section, the ground field k is algebraically closed, and the integers $q_0, \cdots, q_n$ are prime to its characteristic exponent. The weighted projective space $\mathbb{P}_k(q_0, \cdots, q_n)$ is denoted by $\widetilde{\mathbb{P}}^n$.

Let a be an integer ≥ 1 and l a prime number, both prime to the characteristic exponent of k. Let $F = \mathbb{Z}/a\mathbb{Z}$ or $F = \mathbb{Z}_l$ (ring of l-adic integers) and S be a scheme. We write $H^i(S, F) = H^i(S, F_S)$ for the i-th étale ($F = \mathbb{Z}/a\mathbb{Z}$) or l-adic ($F = \mathbb{Z}_l$) cohomology group of the scheme S, with coefficients in the constant sheaf F_S defined by F. By definition one has

$$H^i(S, \mathbb{Z}_l) = \varprojlim_{\alpha \geq 1} H^i(S, \mathbb{Z}/l^\alpha \mathbb{Z}) .$$

A good reference to learn about étale cohomology (and other cohomologies of algebraic varieties) is DANILOV [DA].

(a) The groups $H^i(\widetilde{\mathbb{P}}^n, F)$

Theorem *Let $\widetilde{\mathbb{P}}^n = \mathbb{P}_k(q_0, \cdots, q_n)$ be the weighted projective space over k, of type $(q_0, \cdots, q_n)$. Then*
i) $H^i(\widetilde{\mathbb{P}}^n, \mathbb{Z}/a\mathbb{Z}) = \mathbb{Z}/a\mathbb{Z}$ if $i = 2r$ $(0 \leq r \leq n)$, $= 0$ if not;
ii) $H^i(\widetilde{\mathbb{P}}^n, \mathbb{Z}_l) = \mathbb{Z}_l$ if $i = 2r$ $(0 \leq r \leq n)$, $= 0$ if not.

Proof. Obviously (i) implies (ii). The proof of (i) is quite similar to the proof of Lemma 1 in I.4.(a) (in the Proposition in I.4.(a) take a proper scheme over k instead of a compact space).

(b) The homomorphism φ^*

Theorem *Let $\varphi : \mathbb{P}^n \longrightarrow \widetilde{\mathbb{P}}^n$ be the morphism defined in 3.(d), and $l_i = l_i(q_0, \cdots, q_n)$ the integers considered in I.5. For $0 \leq i \leq n$, we have:*

$$(H^{2i}(\widetilde{\mathbb{P}}^n, \mathbb{Z}/a\mathbb{Z}) \xrightarrow{\varphi^*} H^{2i}(\mathbb{P}^n, \mathbb{Z}/a\mathbb{Z})) = (\mathbb{Z}/a\mathbb{Z} \xrightarrow{\cdot l_i} \mathbb{Z}/a\mathbb{Z}) .$$

Proof. Because of 4.(a).ii, the proof is the same as that of the Theorem in I.5.

(c) Corollary *Let q be an integer ≥ 1, prime to the characteristic exponent of k. Consider the quotient scheme $A^{n+1}/\mu_q = (A^{n+1}/U_q)(q_0, \cdots, q_n)$ (cf. 3.(c)). Its cohomology with compact supports is as follows.*

$$\begin{aligned} H^u_c(A^{n+1}/\mu_q, \mathbb{Z}/a\mathbb{Z}) &= \mathbb{Z}/a\mathbb{Z} \text{ if } i = 2(n+1) , \\ &= \mathbb{Z}/(a, m_r)\mathbb{Z} \text{ if } i = 2r \text{ or } 2r+1,\ 0 \leq r \leq n, \\ &= 0 \text{ if not,} \end{aligned}$$

where $(a, m_r) = gcd\{a, m_r\}$, $m_r = l_r(q_0, \cdots, q_n, q)/l_r(q_0, \cdots, q_n)$ (I.5).

Proof. By 3.(c) and setting $\widetilde{\mathbb{P}}^{n+1} = \mathbb{P}(q_0, \cdots, q_n, q)$, A^{n+1}/μ_q is the complementary open subset of the closed immersion $\widetilde{\mathbb{P}}^n \subset \widetilde{\mathbb{P}}^{n+1}$. Thus we have the exact sequences (with coefficients in $\mathbb{Z}/a\mathbb{Z}$)

$$0 \to H^{2i}_c(A^{n+1}/\mu_q) \to H^{2i}_c(\widetilde{\mathbb{P}}^{n+1}) \to H^{2j}(\widetilde{\mathbb{P}}^n) \to H^{2j+1}_c(A^{n+1}/\mu_q) \to 0.$$

But the preceding theorem shows that the homomorphism $H^{2j}(\widetilde{\mathbb{P}}^{n+1}) \longrightarrow H^{2j}(\widetilde{\mathbb{P}}^n)$ is equal to multiplication by m_j in $\mathbb{Z}/a\mathbb{Z}$ $(0 \leq i \leq n)$.

(d) The ring $H^*(\widetilde{\mathbb{P}}^n, \mathbb{Z}_l)$

In the same manner and with the same notation as I.6, we have a ring isomorphism

$$H^*(\widetilde{\mathbb{P}}^n, \mathbb{Z}_l) \simeq \mathbb{Z}_l[T_1, \cdots, T_n]/\mathfrak{a} .$$

III. Cohomology of Weighted Projective Bundles

1 Definition of a Weighted Projective Bundle

The ground field in this chapter is the complex numbers $\mathbb{C}$. Let $q_0, \cdots, q_n$ be integers ≥ 1 and let $E = \bigoplus_{0}^{n} E_i$ be a direct sum of vector bundles E_i with constant rank ≥ 1, over a compact space X. The projection $E \longrightarrow X$ is denoted by π. The multiplicative group $\mathbb{C}^*$ acts on E as follows:

$$\sigma : \mathbb{C}^* \times E \longrightarrow E, \ \sigma(\lambda, (u_0, \cdots, u_n)) = (\lambda^{q_0} u_0, \cdots, \lambda^{q_n} u_n)$$

where $u_i \in E_i$. A motivation of considering such an action is given in Introduction.

Let $E^* = E \backslash X$ where $X \subset E$ is the null section of E. The topological quotient $E^*/\mathbb{C}^*$ of E^* modulo the operation σ will be denoted by

$$\widetilde{\mathbb{P}}(E) = \mathbb{P}(\bigoplus_{0}^{n} E_i \, ; q_0, \cdots, q_n) \, .$$

There is a projection $\rho : \widetilde{\mathbb{P}}(E) \longrightarrow X$, induced by $\pi : E \longrightarrow X$, the fibers of which are weighted projective spaces. More precisely, if $rk(E_i) = r_i$, $r = (r_0 + \cdots + r_n) - 1$, then any fiber of ρ is equal to the complex weighted projective space (cf. I.)

$$\widetilde{\mathbb{P}}^r = \mathbb{P}_{\mathbb{C}}(q_0, \cdots, q_n) \text{ with } q_i = (q_i, \cdots, q_i) \in \mathbb{N}^{r_i} \, .$$

The bundle $\rho : \widetilde{\mathbb{P}}(E) \longrightarrow X$ is called *weighted projective bundle.* If $n = 0$ or $q_0 = \cdots = q_n = 1$, then $\widetilde{\mathbb{P}}(E) = \mathbb{P}(E)$ is the usual projective bundle associated to the vector bundle E.

2 Canonical Line Bundle over $\widetilde{\mathbb{P}}(E)$

Let $q_{n+1} = \text{lcm}\{q_0, \cdots, q_n\}$ and $E_{n+1} = X \times \mathbb{C}$ (trivial line vector bundle over X). Consider then the weighted projective bundle $\widetilde{\mathbb{P}}(F)$ associated to $(F = \bigoplus_0^{n+1} E_i \; ; q_0, \cdots, q_{n+1})$. There is a closed immersion:

$$X = \mathbb{P}(E_{n+1}; q_{n+1}) \subset \mathbb{P}(\bigoplus_0^{n+1} E_i; q_0, \cdots, q_{n+1}) = \mathbb{P}(F) \,,$$

defined by the canonical inclusion $E_{n+1} \subset \bigoplus_0^{n+1} E_i$. Denote $U = \widetilde{\mathbb{P}}(F) \backslash X$ its complementary open subset. Thus we have a projection $U \longrightarrow \widetilde{\mathbb{P}}(E)$ (given by $(u_0, \cdots, u_{n+1}) \mapsto (u_0, \cdots, u_n)$, $u_i \in E_i$), and this is a rank 1 vector bundle over $\widetilde{\mathbb{P}}(E)$ (the reason is that q_{n+1} is divisible by each q_i, $0 \leq i \leq n$; compare with I.1.(c)). We write $\widetilde{L}_E$ for this bundle and call it *the canonical line bundle* over $\widetilde{\mathbb{P}}(E)$.

More generally if, instead of q_{n+1}, we take any common multiple s of the integers q_i, we obtain a line bundle over $\widetilde{\mathbb{P}}(E)$ isomorphic to $(\widetilde{L})_E^{\otimes k}$ with $k = s/q_{n+1}$.

The restriction of $(\widetilde{L})_E$ to the fiber $\widetilde{\mathbb{P}}^r$ of $\widetilde{\mathbb{P}}(E)$ is the canonical line bundle defined in I.1.(b).

When Y is a topological space, then $H^*(Y)$ will stand for the integral cohomology $H^*(Y, \mathbb{Z})$. The class of $\widetilde{L}_E^*$ in the cohomology group $H^2(\widetilde{\mathbb{P}}(E))$ is denoted by $\widetilde{\xi}_E$.

There is a structure of $H^*(X)$-module on $H^*(\widetilde{\mathbb{P}}(E)) = \bigoplus_{i \geq 0} H^i(\widetilde{\mathbb{P}}(E))$ via the homomorphism $\rho^* : H^*(X) \longrightarrow H^*(\widetilde{\mathbb{P}}(E))$, induced by the projection $\rho : \widetilde{\mathbb{P}}(E) \longrightarrow X$.

3 Cohomology of $\mathbb{P}(E)$

(a) Let us suppose that $q_0 = \cdots = q_n = 1$. So $\widetilde{\mathbb{P}}(E) = \mathbb{P}(E)$, and $\widetilde{\xi}_E = \xi_E$ is the class of the canonical line bundle over $\mathbb{P}(E)$. It is well-known that the $H^*(X)$-module $H^*(\mathbb{P}(E))$ is free of rank $r+1$ with $\{1, \xi_E, \cdots, \xi_E^r\}$ as a basis ([BO],[J]). Moreover, the multiplicative structure of $H^*(\mathbb{P}(E))$ is determined by the unique relation ([G1],[BO],[J])

$$\xi_E^{r+1} + c_1\xi_E^r + \cdots + c_r\xi_E + c_{r+1} = 0 \qquad (c)$$

where $c_i = c_i(E)$ is the i-th Chern class of E.

(b) Now in the general case ($q_i \geq 1$), the elements $1, \widetilde{\xi}_E, \cdots, \widetilde{\xi}_E^r$ do not generate the $H^*(X)$-module $H^*(\widetilde{\mathbb{P}}(E))$. Indeed if X is one point, then $\widetilde{\mathbb{P}}(E)$ is reduced to its fiber $\widetilde{\mathbb{P}}^r$, and we know from I.7. that $1, \widetilde{\xi}_E, \cdots, \widetilde{\xi}_E^r$ do not generate $H^*(\widetilde{\mathbb{P}}^r)$.

4 The $H^*(X)$-module $H^*(\widetilde{\mathbb{P}}(E))$

(a) The result is that $H^*(\widetilde{\mathbb{P}}(E))$ is a $H^*(X)$-free module of rank $r+1$. This is done by constructing elements $w_0, \cdots, w_r$, $w_i \in H^{2i}(\widetilde{\mathbb{P}}(E))$, the restriction of which to the fiber $\widetilde{\mathbb{P}}^r$ gives the basis of $H^*(\widetilde{\mathbb{P}}^r)$ defined in Theorem I.5. This will prove that $\{w_0, \cdots, w_r\}$ is a basis for the module $H^*(\widetilde{\mathbb{P}}(E))$ by LERAY-HIRSCH theorem.

(b) We explain now how to construct the elements w_i in a particular case which is, in fact, fundamental for the general case.

So let us assume that the vector bundles $E_0, \cdots, E_n$ are line bundles and that the integers $q_0, \cdots, q_n$ are such that q_i is divisible by q_{i-1} for all $i = 1, \cdots, n$. We make a change of notation, putting $L_i = E_i$. So $E = \bigoplus_0^n L_i$, and $rk(E) = n+1$. The fiber of $\rho : \widetilde{\mathbb{P}}(E) \longrightarrow X$ is $\widetilde{\mathbb{P}}^n = \mathbb{P}_{\mathbb{C}}(q_0, \cdots, q_n)$.

(i) Since we are going to proceed by induction on n, let us first remark that there are closed immersions:

$$X = \widetilde{\mathbb{P}}(F_0) \subset \widetilde{\mathbb{P}}(F_1) \subset \cdots \subset \widetilde{\mathbb{P}}(F_n) = \widetilde{\mathbb{P}}(E) ,$$

where $F_i = L_0 \oplus \cdots \oplus L_i$ and $\widetilde{\mathbb{P}}(F_i) = \mathbb{P}(F_i; q_0, \cdots, q_i)$. These inclusions are induced by the canonical $F_i \subset F_{i+1}$. For simplification of notation, set $M_i = \widetilde{L}_{F_i}$, the canonical line bundle over $\mathbb{P}_i = \widetilde{\mathbb{P}}(F_i)$. An important fact is

$$M_i | P_{i-1} = M_{i-1}^{\otimes q_i / q_{i-1}} \quad (1 \leq i \leq n) .$$

(ii) Consider the open subset $V = \widetilde{\mathbb{P}}(E) \backslash \widetilde{\mathbb{P}}(L_n)$. This is a line bundle over $\mathbb{P}_{n-1}$ (because q_i divides q_n, $0 \leq i \leq n-1$), and

$$V = M_{n-1}^{\otimes q_n / q_{n-1}} \times p^*(L_m) , \text{ where } p : \mathbb{P}_{n-1} \longrightarrow X \text{ is the projection.}$$

In particular $H^*_c(V)$ is a free $H^*(\mathbb{P}_{n-1})$-module of rank 1, generated by the Thom class $\tau \in H^2_c(V)$ of V ([BO]).

On the other hand, there is a cohomology exact sequence

$$\cdots \to H^i_c(V) \to H^i(\widetilde{\mathbb{P}}(E)) \to H^i(\widetilde{\mathbb{P}}(L_n)) \to H^{i+1}_c(V) \to \cdots$$

The image of τ in $H^2(\widetilde{\mathbb{P}}(E))$ is $\widetilde{\xi}_E + c_1(L_m)$ $(\widetilde{\mathbb{P}}(L_m) = X)$.
(iii) Construction of the elements $w_0, \cdots, w_n \in H^*(\widetilde{\mathbb{P}}(E))$: by induction on n. Take $w_0 = 1 \in H^0(\widetilde{\mathbb{P}}(E))$. Assume that elements $w'_0 = 1, w'_1, \cdots, w'_{n-1}$ are defined in $H^*(\mathbb{P}_{n-1}) : w'_i \in H^{2i}(\mathbb{P}_{n-1})$, $\mathbb{P}_{n-1} = \widetilde{\mathbb{P}}(F_{n-1})$. Consider then $w'_i \cdot \tau$ in the $H^*(\mathbb{P}_{n-1})$-module $H^*(V)$ and denote by w_{i+1} its image in $H^*(\widetilde{\mathbb{P}}(E)) = H^*(\mathbb{P}_n)$.

Hence the elements $w_0, \cdots, w_n \in H^*(\widetilde{\mathbb{P}}(E))$ are well-defined. Their restriction to the fiber $\widetilde{\mathbb{P}}^n$ of $\widetilde{\mathbb{P}}(E)$ gives the basis $\xi_0 = 1, \xi_1, \cdots, \xi_n$ of $H^*(\widetilde{\mathbb{P}}^n)$ constructed in I.5. Thus, by LERAY-HIRSCH theorem, the $H^*(X)$-module $H^*(\widetilde{\mathbb{P}}(E))$ is free with $\{w_0, \cdots, w_n\}$ as a basis. Note that $w_1 = \widetilde{\xi}_E + c_1(L_n)$.

5 About the Multiplicative Structure of $H^*(\widetilde{\mathbb{P}}(E))$

The hypothesis made in 4.(b) is still valid in this paragraph. Recall that $P_i = \widetilde{\mathbb{P}}(F_i)$, $F_i = L_0 \oplus \cdots \oplus L_i$. Let $\{w^i_o, \cdots, w^i_{i-1}\}$ be the basis of the $H^*(X)$-module $H^*(P_i)$ constructed above (4.(b) iii).

(a) Consider the homomorphism $h_n : H^*_c(V) \longrightarrow H^*(\mathbb{P}_n), \mathbb{P}_n = \widetilde{\mathbb{P}}(E)$ (see 4.(b) ii). Since $h_n(\tau) = w^n_1$, and $H^*_c(V)$ is a $H^*(\mathbb{P}_{n-1})$-module, we shall write

$$a \cdot w^n_1 := h_n(a\tau) \text{ for } a \in H^*(\mathbb{P}_{n-1}) .$$

Thus, by construction of the w^j_i's, we obtain

$$\begin{aligned} w^n_0 = 1, w^n_1 &= \widetilde{\xi}_{F_n} + c_1(L_n), \\ w^n_2 = w^{n-1}_1 \cdot w^n_1, \cdots, \; w^n_n &= w^{n-1}_{n-1} \cdot w^n_1 \\ w^{n-1}_0 = 1, w^{n-1}_1 &= \widetilde{\xi}_{F_{n-1}} + c_1(L_{n-1}), \\ w^{n-1}_2 = w^{n-2}_1 \cdot w^{n-1}_1, \; \cdots \; , w^{n-1}_{n-1} &= w^{n-2}_{n-2} \cdot w^{n-1}_1 \\ &\cdots \end{aligned}$$

So, by induction, it holds that

$$w^n_i = w^{n-i+1}_1 \cdot w^{n-i+2}_1 \cdots w^n_1 \quad (1 \leq i \leq n) .$$

(similarly to h_n, there are homomorphisms $h_i : H^*_c(V_i) \longrightarrow H^*(\mathbb{P}_i)$, with $V_i = \widetilde{\mathbb{P}}(F_i) \backslash \widetilde{\mathbb{P}}(L_i)$ allowing to define $w^{i-1}_1 \cdot w^i_1$.) Whence

$$w^n_i = (\widetilde{\xi}_{F_s} + c_1(L_s)) \cdot (\widetilde{\xi}_{F_{s+1}} + c_1(L_{s+1})) \cdots (\widetilde{\xi}_{F_n} + c_1(L_n))$$

where $s = n - i + 1, 1 \leq i \leq n$.

Because $\widetilde{\xi}_{F_0} + c_1(L_0) = 0$ in $H^2(\widetilde{\mathbb{P}}(F_0))$ $(F_0 = L_0, \widetilde{\mathbb{P}}(F_0) = X)$, there is the following relation:

$$(\widetilde{\xi}_{F_0} + c_1(L_0)) \cdot (\widetilde{\xi}_{F_1} + c_1(L_1)) \cdots (\widetilde{\xi}_{F_n} + c_n(L_n)) = 0 \qquad (c')$$

If $q_0 = \cdots = q_n = 1$, then this relation coincides with the relation

(c) mentioned in 3.(a), which defines the Chern classes of $E = L_0 \oplus \cdots \oplus L_n$, that is:

$$(\xi_E + c_1(L_0))(\xi_E + c_1(L_1)) \cdots (\xi_E + c_1(L_n)) = 0 ,$$

the first member of which is the usual product in $H^*(\mathbb{P}(E))$. In fact, when $q_0 = \cdots = q_n = 1$, one has

$$w_i^n = (\xi_E + c_1(L_s))(\xi_E + c_1(L_{s+1})) \cdots (\xi_E + c_1(L_n))$$

$((s = n - i + 1)$, the second member being the usual product in $H^*(\mathbb{P}(E)))$, and -of course- these $w_0^n, w_1^n, \cdots, w_n^n$ form a basis of the $H^*(X)$-module $H^*(\mathbb{P}(E))$, but this is not the usual basis $\xi_E^0, \xi_E^1, \cdots, \xi_E^n$!

(b) So the relation (c') gives a generalization of (c). We want to write it in terms of the class $\widetilde{\xi}_E$ of $\widetilde{L}_E^*$ (dual of the canonical line bundle over $\widetilde{\mathbb{P}}(E)$), using the usual product in $H^*(\widetilde{\mathbb{P}}(E))$. This leads to the following relation in $H^*(\widetilde{\mathbb{P}}(E))$

$$(\widetilde{\xi}_E + c_1(L_0^{\otimes t_0}))(\widetilde{\xi}_E + c_1(L_1^{\otimes t_1})) \cdots (\widetilde{\xi}_E + c_1(L_n^{\otimes t_n})) = 0 \qquad (c'')$$

with $t_i = q_n/q_i$.

More generally we have

$$(\widetilde{\xi}_E + c_1(L_s^{\otimes t_s}))(\widetilde{\xi}_E + c_1(L_{s+1}^{\otimes t_{s+1}})) \cdots (\widetilde{\xi}_E + c_1(L_n^{\otimes t_n})) = t_s t_{s+1} \cdots t_n w_i^n$$

where $s = n - i + 1, 1 \leq i \leq n$.

6 The Classes $\tilde{c}_i(E)$

We continue to suppose that the vector bundles $E_i = l_i$ are of rank 1, but no hypothesis is made about the integers $q_0, \cdots, q_n$. Let $l = \text{lcm}\{q_0, \cdots, q_n\}$. Then the relation (c'') is a particular case of the following

$$(\widetilde{\xi}_E + c_1(L_0^{\otimes t_0}))(\widetilde{\xi}_E + c_1(L_1^{\otimes t_1})) \cdots (\widetilde{\xi}_E + c_1(L_n^{\otimes t_n})) = 0 \qquad (\tilde{c})$$

where $t_i = l/q_i$. It is this relation we consider as a relevant generalization of (c) in the case of $\widetilde{\mathbb{P}}(E)$. This will be justified in [A1]. Let $\tilde{c}_i(E)$ be the coefficient of $\widetilde{\xi}_E^{n-i+1}$ in the left member of $(\tilde{c})$, and $\tilde{c}(E) = \tilde{c}_0(E) + \cdots + \tilde{c}_n(E)$. Then

$$\tilde{c}(E) = (\psi^{l/q_0}c(L_0)) \cdots (\psi^{l/q_n}c(L_n)) ,$$

where ψ^k is the k-th Adams operation and $c(F)$ is the total Chern class of F.

Now, for general vector bundles E_i $(E = E_0 \oplus \cdots \oplus E_n)$, we put

$$\tilde{c}(E) := (\psi^{l/q_0}c(E_0)) \cdots (\psi^{l/q_n}c(E_n)) ,$$

which defines classes $\tilde{c}_i(E) \in H^{2i}(X)$ $(0 \le i \le r = rk(E))$ such that $\tilde{c}(E) = \sum_0^n \tilde{c}_i(E)$.

These classes verify the same operations as the $c_i(E)$ except the additivity one. If $(E' = \bigoplus_0^{n'} E_i'; q_0', \cdots, q_{n'}')$ and $(E'' = \bigoplus_0^{n''} E_i''; q_0'', \cdots, q_{n''}'')$ are two examples of our datum $(E = \bigoplus_0^n E_i; q_0, \cdots, q_n)$, then it holds that

$$\tilde{c}(E' \oplus E'') = (\psi^{m/l'}\tilde{c}(E'))(\psi^{m/l''}\tilde{c}(E'')) ,$$

where $l' = \text{lcm}\{q_0', \cdots, q_{n'}'\}, l'' = \text{lcm}\{q_0'', \cdots, q_{n''}''\}, m = \text{lcm}\{l', l''\}$, and $E' \oplus E'' = E_0' \oplus \cdots \oplus E_{n'}' \oplus E_0'' \oplus \cdots \oplus E_{n''}''$ is weighted by the integers $q_0', \cdots, q_n', q_0'', \cdots, q_{n''}''$.

References

[A1] Al Amrani, A., *Fibrés projectifs tordus et classes de Chern (àparaître).*

[A2] Al Amrani, A., *Classes d'idéaux et groupe de Picard des fibrés projectifs tordus*, K-Theory 2 (1989) 559-578.

[AR1] Artin, M., *Théorème de changement de base par un morpisme propre*, Séminaire de Géometrie Algébrique du Bois-Marie S.G.A.4, Exp XII-XIII Lect. Notes in Math. 305 (1973) (Springer-Verlag).

[AR2] Artin, M., *Théorème de changement de base par un morphisme lisse et applications*, S.G.A. 4, tome 4, Exp. XVI, Lect. Notes in Math. 305 (1973) (Springer-Verlag).

[BO] Bott, R., *Lectures on $K(X)$*, (Benjamin, 1969).

[B] Bredon; G.E., *Sheaf Theory* (McGraw-Hill, 1967).

[DA] Danilov, V.I., *Cohomology of algebraic varieties*, in Algebraic Geometry II, Encyclopedia of Mathematical Sciences Vol. 35 (Springer, 1996).

[DE] Deligne, P., *Formule de Picard-Lefschetz*, Séminaire de Géométrie Alg'ebrique de Bois-Marie S.G.A.7 II, Exp. XV, Lect. Notes in Math. 340 (1973) (Springer-Verlag).

[D] Dold, A., *Universelle Koeffizienten*, Math. Zeitschr. 80 (1962) 63-88.

[G] Grothendieck, A., *La théorie des classes de Chern*, Bull. Soc. math. France 86 (1958) 137-154.

[G2] Grothendieck, A., *Sur quelques points d'algèbre homologique*, Tôhoku Math. J. (2nd series) 9 (1957) 119-221.

[G3] Grothendieck, A., *Revêtements étales et groupe fondamental*, Séminaire de Géometrie Algébrique du Bois-Marie S.G.A.1, Lect. Notes in Math. 224 (1971) (Springer-Verlag).

[GD] Grothendieck, A.-Dieudonné, J., *Éléments de géometrie algébrique*, Chapitre II, Publ. Math. I.H.E.S. 8 (1961).

[H] Hartshorne, R., *Algebraic Geometry* (Springer-Verlag, 1977).

[HM] Hirzebruch,F.-Mayer, K.H., *$O(n)$-Mannigfaltigkeiten, exotische Splären und Singularitäten*, Lect. Notes in Math. 57 (1968) (Springer-Verlag).

[J] Jouanolou,J.-P., *Cohomologie de quleques schémas classiques et Théorie cohomologique des classes de Chern*, Séminaire de Géometrie Algébrique du Bois-Marie S.G.A.5, Exp. VII, Lect. Notes in Math. 589 (1977) (Sprinder-Verlag).

[K] Kawasaki, T., *Cohomology of twisted projective spaces and lens complexes*, Math. Ann. 206 (1973) 243-248.

[M1] Mumford, D., *Red book of varieties and schemes*, Lect. Notes. in Math. 1358 (1988) (Springer-Verlag).

[M2] Mumford, D.-Fogarty, J., *Geometric Invariant Theory*, 2nd edition (Springer-Verlag, 1982).

Geometry of Toric Varieties

Jean-Paul Brasselet

Institut de Mathématiques de Luminy
Campus de Luminy, Case 930
F-13288 Marseille Cedex 9
France

e-mail: `jpb@iml.univ-mrs.fr`

These notes present a course given during the Algebraic Geometry Summer School in Bilkent University (August 1995). This elementary introduction does not pretend to originality but to give examples and motivation for the study of toric varieties. These varieties play a prominent part giving explicit relations between combinatorial geometry and algebraic geometry. They provide also an important field of examples and models. The preface of [5] explains very well the interest of these objects "*Toric varieties provide a ... way to see many examples and phenomena in algebraic geometry... For example, they are rational, and, although they may be singular, the singularities are rational. Nevertheless, toric varieties have provided a remarkably fertile testing ground for general theories.*"

Basic references for toric varieties are [4], [5] and [7]. These references give proofs of the results and complete descriptions. They were used (a lot !) for writing these notes and the reader can consult them for useful complementary references and details. In particular, many points are omitted here, for lack of time during the course. For example, the reader will look at references for the notions and points of view such that character groups, one parameter subgroups, line bundles, Betti numbers etc...

I want to thank particularly Professor Sinan Sertöz and University of Bilkent for their hospitality during the Summer School.

1 From combinatorial geometry to toric varieties.

The procedure of the construction of (affine) toric varieties associates to a cone σ in $\mathbb{R}^n$ successively : the dual cone $\check{\sigma}$, a monoid S_σ, a finitely generated $\mathbb{C}$-algebra R_σ and finally the algebraic variety X_σ. In the following, we describe the steps of this procedure :

$$\sigma \quad \mapsto \quad \check{\sigma} \quad \mapsto \quad S_\sigma \quad \mapsto \quad R_\sigma \quad \mapsto \quad X_\sigma$$

and recall some useful definitions and results of algebraic geometry.

Let $A = \{x_1, \ldots, x_r\}$ be a finite set of vectors in $\mathbf{R}^n$, the set

$$\{x \in \mathbf{R}^n \ : x = \lambda_1 x_1 + \cdots + \lambda_r x_r, \quad \lambda_i \in \mathbf{R}, \quad \lambda_i \geq 0\}$$

is called a polyhedral cone.

The *dimension* of σ, denoted $\dim \sigma$, is the dimension of the smallest linear space containing σ.

If $A = \emptyset$ then $\sigma = \{0\}$ is the zero cone.

In the following, N will denote a fixed lattice $N \cong \mathbf{Z}^n \subset \mathbf{R}^n$.

Definition. A cone σ is a *lattice* (or rational) cone if all the vectors $x_i \in A$ defining σ belong to N.

Examples 1. In $\mathbb{R}^2$ with canonical basis (e_1, e_2), we have the following cones :

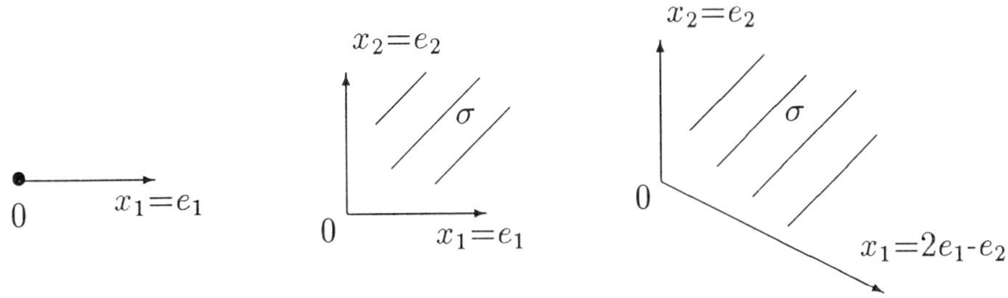

Definition. A cone is *strongly convex* if it does not contain any straight line going through the origin (i.e. $\sigma \cap (-\sigma) = \{0\}$).

One important step of the procedure of the construction of toric varieties is to define the dual cone associated to a cone. Let $(\mathbf{R}^n)^*$ be the dual space of $\mathbf{R}^n$ and $\langle\ ,\ \rangle$ the dual pairing. To each cone we associate the dual cone $\check{\sigma}$

$$\check{\sigma} = \{u \in (\mathbf{R}^n)^* \ : \ \langle u, v\rangle \geq 0 \quad \forall v \in \sigma\}$$

Examples 2. The canonical dual basis of $(\mathbf{R}^2)^*$ is denoted (e_1^*, e_2^*)

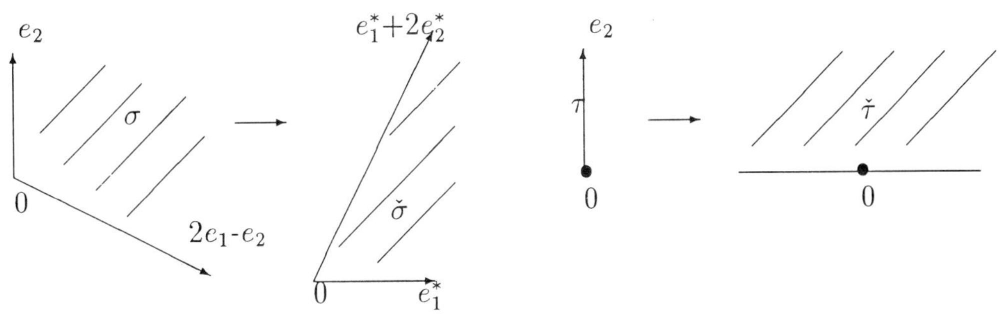

Giving a lattice N in $\mathbf{R}^n$, we define the dual lattice $M = \mathrm{Hom}_{\mathbf{Z}}(N; \mathbb{Z}) \cong \mathbb{Z}^n$ in $(\mathbf{R}^n)^*$ and we have the property :

Property. If σ is a lattice cone, then $\check{\sigma}$ is also a lattice cone (relatively to M).

Remark that if σ is a strongly convex cone, then $\check{\sigma}$ is not necessarily a strongly convex cone (see τ in examples 2).

Definition. Let σ be a cone and let $\lambda \in \check{\sigma} \cap M$, then $\tau = \sigma \cap \lambda^{\perp} =$

$\{v \in \sigma \ : \ \langle \lambda, v \rangle = 0\}$ is called a *face* of σ. We will write $\tau < \sigma$.

This definition coincides with the intuitive one. (Exercise).

A one dimensional face is called an *edge.*

Remark. If $\tau < \sigma$ then $\check{\sigma} \subset \check{\tau}$. (Easy exercise).

Definition. A *fan* Δ is a finite union of cones such that :

(i) every cone of Δ is a strongly convex, polyhedral, lattice cone,

(ii) every face of a cone of Δ is a cone of Δ,

(iii) if σ and σ' are cones of Δ, then $\sigma \cap \sigma'$ is a common face of σ and σ'.

Examples 3.

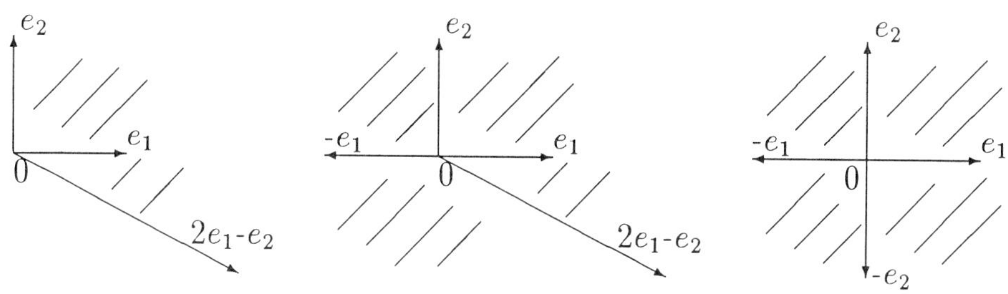

The aim of this course is to explain how we can associate an algebraic variety to each fan and to explain how properties of this algebraic variety can be read on the fan (in a combinatorial way). The first step is to associate to every cone σ a finitely generated monoid S_σ.

Definition. A *monoid* is a non empty set S with an associative and commutative law $+ : S \times S \to S$, with a zero element and satisfying the simplification law, i.e. :

$$s + t = s' + t \ \Rightarrow \ s = s' \text{ for } s, s' \text{ and } t \in S$$

Lemma. If σ is a cone, then $\sigma \cap N$ is a monoid.

Proof : If $x, y \in \sigma \cap N$, then $x + y \in \sigma \cap N$ and the rest is easily verified.

Definition A monoid S is *finitely generated* if there exists $a_1, \ldots, a_k \in S$ such that

$$\forall s \in S,\ s = \lambda_1 a_1 + \cdots + \lambda_k a_k \text{ with } \lambda_i \in \mathbf{Z} \text{ and } \lambda_i \geq 0$$

Lemma. (Gordon's Lemma). If σ is a polyhedral lattice cone, then $\sigma \cap N$ is a finitely generated monoid.

Proof : Let $A = \{x_1, \cdots, x_r\}$ be the set of vectors defining the cone σ. Each x_i is an element of $\sigma \cap N$. The set $K = \{\sum t_i x_i,\quad 0 \leq t_i \leq 1\}$ is compact and N is discrete, therefore $K \cap N$ is a finite set. We show that it generates $\sigma \cap N$. In fact, every $x \in \sigma \cap N$ can be written $x = \sum(n_i + r_i)x_i$ where $n_i \in \mathbf{Z}_{\geq 0}$ and $0 \leq r_i \leq 1$. Each x_i and the sum $\sum r_i x_i$ belong to $K \cap N$, so we obtain the result.

We will use this lemma for the polyhedral lattice cone $\check{\sigma}$ and will denote by S_σ the monoid $\check{\sigma} \cap M$.

Example 4. In $\mathbb{R}^2$, consider the cone reduced to the origin :

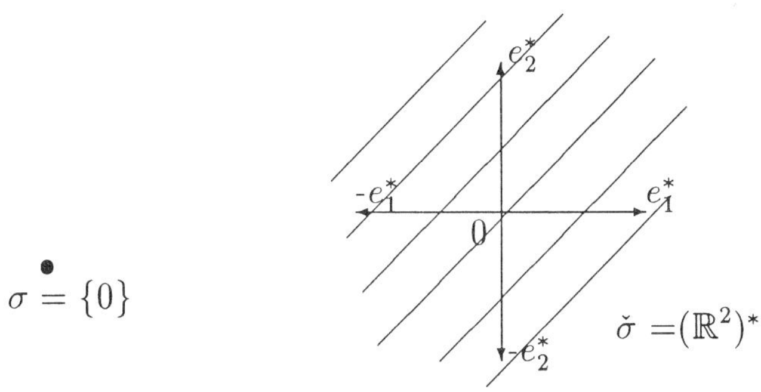

$S_\sigma = \check{\sigma} \cap M$ is generated by $(e_1^*, -e_1^*, e_2^*, -e_2^*)$.

Example 5. In $\mathbb{R}^2$, consider the following cone

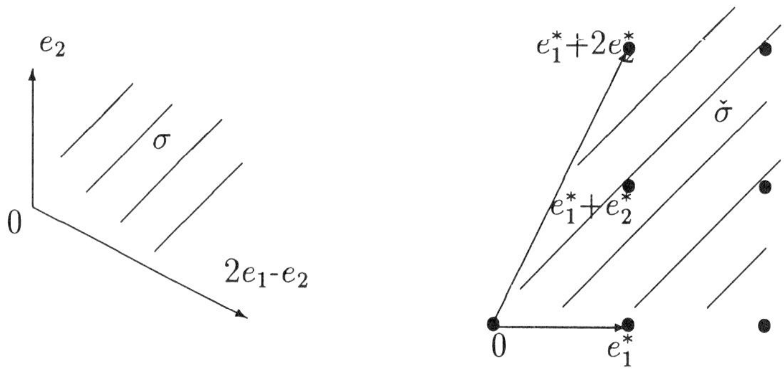

$S_\sigma = \check{\sigma} \cap M$, marked •, is generated by $(e_1^*, e_1^* + e_2^*, e_1^* + 2e_2^*)$.

Laurent polynomials

Denote by $\mathbf{C}[z, z^{-1}] = \mathbf{C}[z_1, \ldots, z_n, z_1^{-1}, \ldots, z_n^{-1}]$ the ring of Laurent polynomials. One of the important result in the definition of toric varieties, and key of the second step, is that the mapping

$$\begin{array}{rcl} \theta\ : \mathbf{Z}^n & \rightarrow & \mathbf{C}[z, z^{-1}] \\ a = (\alpha_1, \ldots, \alpha_n) & \mapsto & z^a = z_1^{\alpha_1} \cdots z_n^{\alpha_n} \end{array}$$

is an isomorphism between the additive group $\mathbf{Z}^n$ and the multiplicative group of monic Laurent monomials. Monic means that the coefficient of the monomial is 1. This isomorphism is easy to prove and let as an exercise.

Definition. The support of a Laurent polynomial $f = \sum_{\text{finite}} \lambda_a z^a$ is defined by

$$\text{supp}(f) \ = \ \{a \in \mathbf{Z}^n \ : \ \lambda_a \neq 0\}$$

Proposition. For a lattice cone σ, the ring

$$R_\sigma = \{f \in \mathbf{C}[z, z^{-1}] \ : \text{supp}(f) \subset \check{\sigma} \cap M\}$$

is a finitely generated monomial algebra (i.e. is a $\mathbf{C}$-algebra generated by Laurent monomials).

This result is a direct consequence of the Gordon's Lemma.

The following section recalls how we can associate to each finitely generated $\mathbf{C}$-algebra (in particular to R_σ) a coordinate ring, then an affine variety.

2 Some basic results of algebraic geometry.

Let $\mathbf{C}[\xi] = \mathbf{C}[\xi_1, \ldots, \xi_k]$ be the ring of polynomials in k variables over $\mathbf{C}$.

Definition. If $E = (f_1, \ldots f_r) \subset \mathbf{C}[\xi]$, then

$$V(E) = \{x \in \mathbf{C}^k \ : \ f_1(x) = \cdots = f_r(x) = 0\}$$

is the *affine algebraic set* defined by E.

Let $\mathcal{A}$ denote the ideal generated by E, then $V(\mathcal{A}) = V(E)$.

Definition. Let $X \subset \mathbf{C}^k$, then

$$I(X) = \{f \in \mathbf{C}[\xi] \ : \ f|_X = 0\}$$

is called the *vanishing ideal* of X.

Example 6. For a fixed point $x = (x_1, \ldots, x_k) \in \mathbf{C}^k$, consider $E = \{\xi_1 - x_1, \cdots, \xi_k - x_k\}$. Then $V(E) = \{x\}$ and $I(\{x\}) = \mathbf{C}[\xi](\xi_1 - x_1) + \cdots + \mathbf{C}[\xi](\xi_k - x_k)$. It is a maximal ideal $\mathcal{M}_x$ (recall that an ideal $\mathcal{M}$ is maximal if for each ideal $\mathcal{M}'$ such that $\mathcal{M} \subset \mathcal{M}'$ then $\mathcal{M} = \mathcal{M}'$)

Theorem. (Weak version of the Nullstellensatz) : Every maximal ideal in $\mathbf{C}[\xi]$ can be written $\mathcal{M}_x$ for a point x.

Corollary 1. There is a one-to-one correspondence between points in $\mathbf{C}^k$ and maximal ideals $\mathcal{M}$ of $\mathbf{C}[\xi]$.

$$\mathbf{C}^k \longleftrightarrow \{\mathcal{M} \subset \mathbf{C}[\xi], \quad \mathcal{M} \text{ maximal ideal}\}$$

Lemma. Let $\mathcal{A}$ be an ideal of $\mathbf{C}[\xi]$, then $V(\mathcal{A}) = \{x \in \mathbb{C}^k \ : \ \mathcal{A} \subset \mathcal{M}_x\}$.

Definition. Denote the vanishing ideal of $V(\mathcal{A})$ by $\mathcal{A}_V = I(V(\mathcal{A}))$, then $R_V = \mathbb{C}[\xi]/\mathcal{A}_V$ is the *coordinate ring* of the affine algebraic set $V(\mathcal{A})$. It is generated as a $\mathbb{C}$-algebra by the classes $\bar{\xi}_j$ of the coordinate functions ξ_j.

We remark that if $\mathcal{A} = \{0\}$, then $V(\mathcal{A}) = \mathbb{C}^k$ and $R_V = \mathbb{C}[\xi]$. The corollary 1, written for $\mathcal{A} = \{0\}$, is generalized for any ideal in the following way :

Corollary 2. There is a one-to-one correspondence

$$V \longleftrightarrow \{\mathcal{M} \subset R_V,\ \mathcal{M} \text{ maximal ideal}\} =: \mathrm{Specm}(R_V)$$

Defining the Zariski topology on each side (see, for example [4], VI.1), we obtain an homeomorphism

$$V \cong \mathrm{Specm}(R_V)$$

Remark. A finitely generated $\mathbb{C}$-algebra R can be written $\mathbb{C}[\xi_1, \ldots, \xi_k]/\mathcal{A}$, as a coordinate ring, for different k and ideals $\mathcal{A}$. It means that we associate by this way, different affine algebraic sets $V(\mathcal{A}) \in \mathbb{C}^k$. In fact, the corollary 2 shows that these representations $V(\mathcal{A})$ are all homeomorphic to an "abstract affine toric variety" X_σ. We will often write also X_σ for the representations of X_σ.

Remark. It can be useful to consider the spectrum $\mathrm{Spec}(R_V)$ whose elements are prime ideals of R_V. In our application, R_V is a domain, then $\mathrm{Spec}(R_V)$ is an irreducible variety and the prime ideals correspond to irreducible subvarieties of $V(\mathcal{A})$. For such a point of view and discussion, see [5], §1.3.

3 Affine toric varieties.

We are now able to define the affine toric variety associated to a cone σ :

Definition. The affine toric variety corresponding to σ is $X_\sigma := \mathrm{Specm}(R_\sigma)$.

The previous section shows that we can represent the finitely generated $\mathbb{C}$-algebra R_σ as a coordinate ring, according to a choice of generators of S_σ. Let us do this on an example, then give the general situation.

In the case of example 5, let $a_1 = e_1^*, a_2 = e_1^* + e_2^*$ and $a_3 = e_1^* + 2e_2^*$ be a system generators of S_σ. They correspond to monic Laurent monomials $u_1 = z_1, u_2 = z_1 z_2$ and $u_3 = z_1 z_2^2$ by the isomorphism θ. The $\mathbb{C}$-algebra R_σ can be represented as

$$R_\sigma = \mathbb{C}[u_1, u_2, u_3] = \mathbb{C}[\xi_1, \xi_2, \xi_3]/\mathcal{A}_\sigma$$

where the relation $a_1 + a_3 = 2a_2$ provides the relation $u_1 u_3 = u_2^2$ between coordinates. The ideal $\mathcal{A}_\sigma$ is then generated by the binomial relation $\xi_1 \xi_3 = \xi_2^2$ and the affine toric variety corresponding to the cone σ is represented in $\mathbb{C}^3$ as the quadratic cone

$$X_\sigma = \{x = (x_1, x_2, x_3) \in \mathbb{C}^3 \ : \ x_1 x_3 - x_2^2\}$$

It has a singularity at the origin of $\mathbb{C}^3$. The following picture gives the real part of X_σ in $\mathbb{R}^3$.

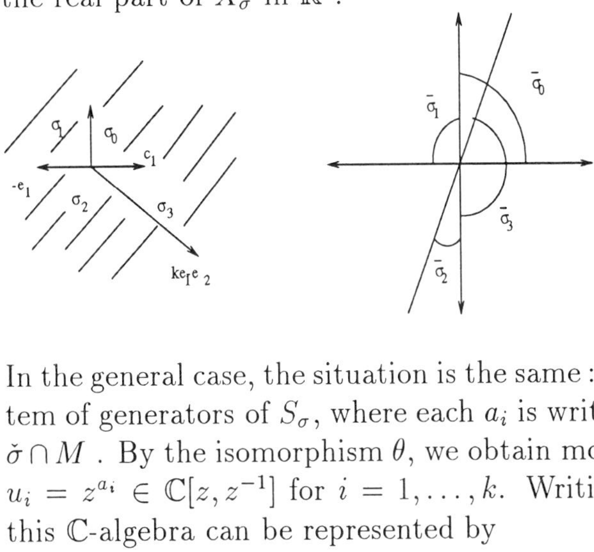

In the general case, the situation is the same : Let $a_1, \ldots, a_k$ be a system of generators of S_σ, where each a_i is written $a_i = (\alpha_i^1, \ldots, \alpha_i^n) \in \check{\sigma} \cap M$. By the isomorphism θ, we obtain monic Laurent monomials $u_i = z^{a_i} \in \mathbb{C}[z, z^{-1}]$ for $i = 1, \ldots, k$. Writing R_σ as $\mathbb{C}[u_1, \ldots, u_k]$, this $\mathbb{C}$-algebra can be represented by

$$R_\sigma = \mathbb{C}[\xi_1, \ldots, \xi_k]/\mathcal{A}_\sigma$$

for some ideal $\mathcal{A}_\sigma$ that we must determinate.

Consider relations between generators of S_σ such that

$$(*) \qquad \sum_{j=1}^{k} \nu_j a_j = \sum_{j=1}^{k} \mu_j a_j \qquad \mu_j, \nu_j \in \mathbf{Z}_{\geq 0} \quad ,$$

we obtain the monomial relations

$$(z^{a_1})^{\nu_1}\cdots(z^{a_k})^{\nu_k}=(z^{a_1})^{\mu_1}\cdots(z^{a_k})^{\mu_k}$$

where $z^{a_i}=(z_1^{\alpha_i^1},\ldots,z_n^{\alpha_i^n})$, i.e. relations

$$u_1^{\nu_1}\cdots u_k^{\nu_k}=u_1^{\mu_1}\cdots u_k^{\mu_k}$$

between the coordinates and finally the *binomial relations*

$$(**)\qquad \xi_1^{\nu_1}\cdots\xi_k^{\nu_k}=\xi_1^{\mu_1}\cdots\xi_k^{\mu_k}$$

that generate $\mathcal{A}_\sigma$.

Theorem. Let σ be a lattice cone in $\mathbb{R}^n$ and $A=(a_1,\ldots,a_k)$ a system of generators of S_σ, the corresponding toric variety X_σ is represented by the affine toric variety $V(\mathcal{A}_\sigma)\subset\mathbb{C}^k$ where $\mathcal{A}_\sigma$ is an ideal of $\mathbb{C}[\xi_1,\ldots,\xi_k]$ generated by finitely many binomials of the form (**) corresponding to relations (*) between elements of A.

Property. If σ is a lattice cone in $\mathbb{R}^n$, then $\dim_{\mathbb{C}} X_\sigma=n$.

The generators $u_1,\ldots,u_k$ of R_σ are the coordinate functions on X_σ in $\mathbb{C}^k$. This means that a point $x=(x_1,\ldots x_k)\in\mathbb{C}^k$ represents a point of X_σ if and only if the relation $x_1^{\nu_1}\cdots x_k^{\nu_k}=x_1^{\mu_1}\cdots x_k^{\mu_k}$ holds for all (ν,μ) appearing in the relation (*).

Example 7. Consider the cone $\sigma=\{0\}$, the dual cone is $\check{\sigma}=(\mathbb{R}^n)^*$. We can take as system of generators of S_σ for example :

$$A_1=(e_1^*,\ldots,e_n^*,-e_1^*,\ldots,-e_n^*)$$

or

$$A_2=(e_1^*,\ldots,e_n^*,-(e_1^*+\cdots+e_n^*)).$$

Let us take the first system of generators. The corresponding monomial $\mathbb{C}$-algebra is

$$\mathbb{C}[z_1,\ldots,z_n,z_1^{-1},\ldots,z_n^{-1}]=\mathbb{C}[\xi_1,\ldots,\xi_{2n}]/\mathcal{A}_\sigma$$

where

$$\mathcal{A}_\sigma=\mathbb{C}[\xi](\xi_1\xi_{n+1}-1)+\cdots+\mathbb{C}[\xi](\xi_n\xi_{2n}-1)$$

hence $X_\sigma = V((\xi_1\xi_{n+1} - 1), \cdots, (\xi_n\xi_{2n} - 1))$.

For $n = 1$, the obtained variety is a complex hyperbola whose asymptotes are the axis $\xi_1 = 0$ and $\xi_2 = 0$. It can be projected bijectively on the axis $\xi_2 = 0$ and the image is $\mathbb{C}^*$:

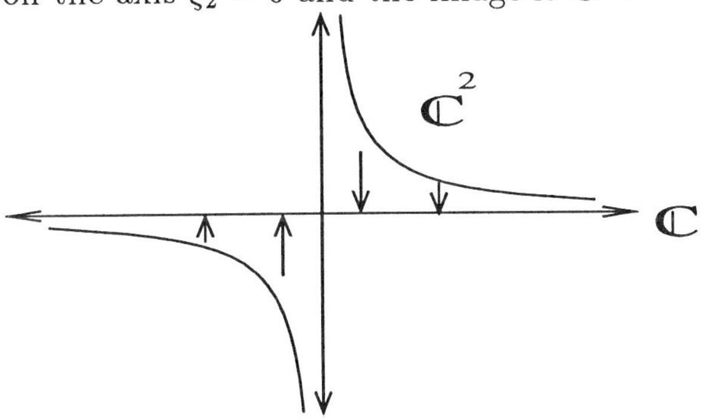

In the general case ($n \geq 1$), and by the same way, X_σ is homeomorphic to

$$\mathbf{T} = \{(z_1, \dots, z_n) \in \mathbb{C}^n \ : \ z_i \neq 0, \quad i = 1, \dots, n\} = (\mathbb{C}^*)^n$$

using the projection $\mathbb{C}^{2n} \mapsto \mathbb{C}^n$ on the first coordinates.

Definition. The set $\mathbf{T} = (\mathbb{C}^*)^n$ is called the complex algebraic n-torus.

Remarks 1. $\mathbf{T}$ includes the real torus as : $\mathbf{T} \cong (S^1)^n \times (\mathbb{R}_{\geq 0})^n$.

2. $\mathbf{T}$ is a closed set in $\mathbb{C}^{2n}$ but, as a subspace of $\mathbb{C}^n$, it is not closed.

3. The second choice of generators A_2 for S_σ provides another realization of X_σ, now in $\mathbb{C}^{n+1}$. We let this as an exercise.

Example 8. Let $\sigma \in \mathbb{R}^2$ be the following cone

S_σ is generated by (e_1^*, e_2^*), $R_\sigma = \mathbb{C}[\xi_1, \xi_2]$, so $\mathcal{A}_\sigma = \{0\}$ and X_σ is

$\mathbb{C}^2$.

Example 9. Let $\tau \in \mathbb{R}^2$ be the following cone

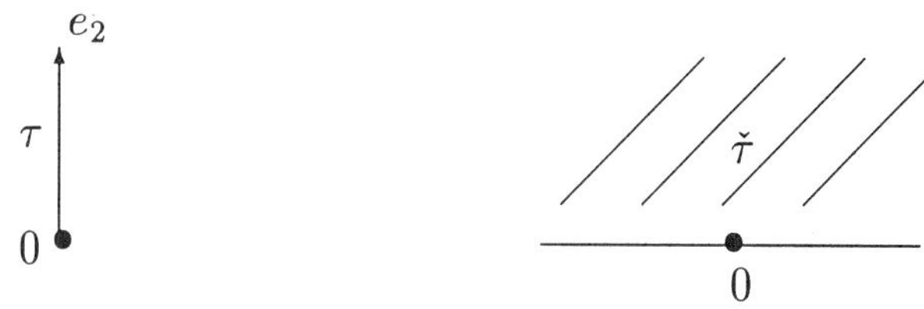

S_τ is generated by $(e_1^*, -e_1^*, e_2^*)$, $R_\tau = \mathbb{C}[u_1, u_2, u_3]$ with $u_2 = u_1^{-1}$. We obtain $R_\tau = \mathbb{C}[\xi_1, \xi_2, \xi_3]/(\xi_1\xi_2 - 1)$ and X_τ is $\mathbb{C}^*_{\xi_1} \times \mathbb{C}_{\xi_2}$.

Example 10. This is the example of arbitrary 2-dimensional affine toric variety.

Let us consider in $\mathbb{R}^2$ the cone generated by e_2 and $pe_1 - qe_2$, for integers $p, q \in \mathbf{Z}_{>0}$ such that $0 < q < p$ and $(p, q) = 1$. In the following picture, $p = 3$, $q = 2$ and N' is pictured by the points $\bullet$.

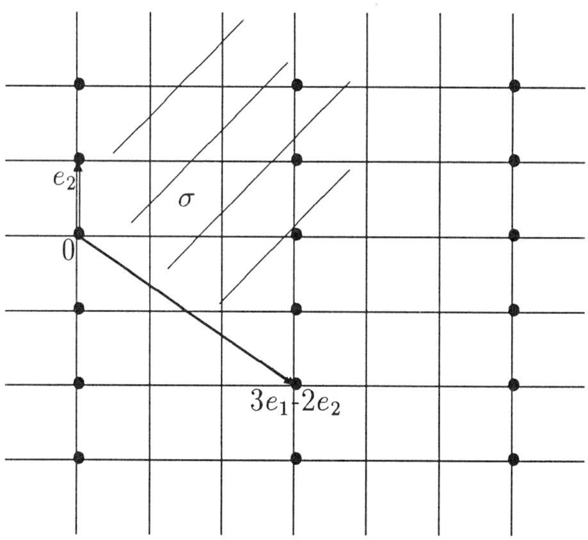

Then $R_\sigma = \mathbb{C}[..., z_1^i z_2^j, ...]$ where the monoids $z_1^i z_2^j$ appear for all i and j such that $j \leq p/q\ i$. Let N' the sublattice of N generated by $pe_1 - qe_2$ and e_2, i.e. by pe_1 and e_2. Let us call σ' the cone σ considered

in N' instead of N. Then σ' is generated by two generators of the lattice N', so $X_{\sigma'}$ is $\mathbb{C}^2$ (cf. example 3).

In such a situation, it is a general result that the inclusion $N' \subset N$ provides a map $X_{\sigma'} \to X_\sigma$. Here the group Γ_p of p-th roots of unity acts on $X_{\sigma'}$ by $\zeta \cdot (u, v) = (\zeta u, \zeta^q v)$ and then $X_\sigma = X_{\sigma'}/\Gamma_p = \mathbb{C}^2/\Gamma_p$. The map $X_{\sigma'} \to X_\sigma$ is the quotient map.

If $n = 2$ (and in general, if σ is simplicial), then singular affine toric varieties are quotient singularities.

4 Toric varieties.

The toric varieties associated to fans will be constructed by gluing affine ones associated to cones. Let us begin by recalling a very simple example, the one of the projective space $\mathbb{P}^2$.

Let us denote by $(t_0 : t_1 : t_2)$ the homogeneous coordinates of the space $\mathbb{P}^2$. It is classically covered by three coordinate charts :

U_0 corresponding to $t_0 \neq 0$ and with affine coordinates $(t_1/t_0, t_2/t_0) = (z_1, z_2)$

U_1 corresponding to $t_1 \neq 0$ and with affine coordinates $(t_0/t_1, t_2/t_1) = (z_1^{-1}, z_1^{-1} z_2)$

U_0 corresponding to $t_0 \neq 0$ and with affine coordinates $(t_0/t_2, t_1/t_2) = (z_2^{-1}, z_1 z_2^{-1})$

Now consider in $\mathbb{R}^2$ the following fan :

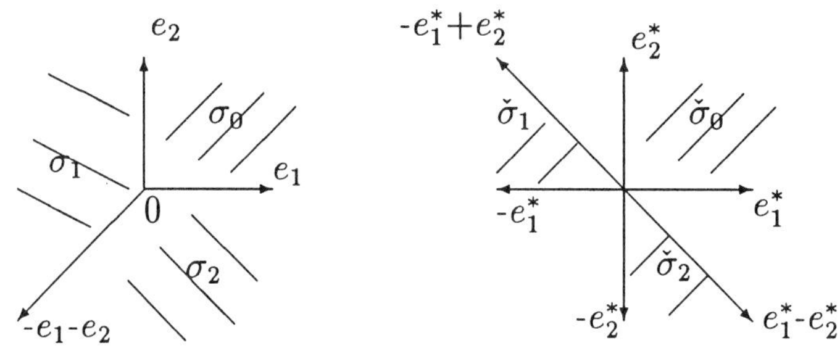

then :

i) S_{σ_0} admits as generators (e_1^*, e_2^*), hence $R_{\sigma_0} = \mathbb{C}[z_1, z_2]$ and $X_{\sigma_0} = \mathbb{C}^2_{(z_1,z_2)}$;

ii) in the same way, S_{σ_1} admits as system of generators $(-e_1^*, -e_1^* + e_2^*)$, hence $R_{\sigma_1} = \mathbb{C}[z_1^{-1}, z_1^{-1}z_2]$ and $X_{\sigma_1} = \mathbb{C}^2_{(z_1^{-1}, z_1^{-1}z_2)}$;

iii) finally, S_{σ_2} admits as system of generators $(-e_2^*, e_1^* - e_2^*)$, hence $R_{\sigma_2} = \mathbb{C}[z_2^{-1}, z_1 z_2^{-1}]$ and $X_{\sigma_2} = \mathbb{C}^2_{(z_2^{-1}, z_1 z_2^{-1})}$.

We see that the three affine toric varieties correspond to the three coordinate charts of $\mathbb{P}^2$. In fact, the structure of the fan provides a gluing between these charts allowing to reconstruct the toric variety $\mathbb{P}^2$ from the U_{σ_i}. Let us explicit the gluing of X_{σ_0} and X_{σ_1} such that $\tau = \sigma_0 \cap \sigma_1$.

For seeing this let us first consider τ as a face of σ_0 :

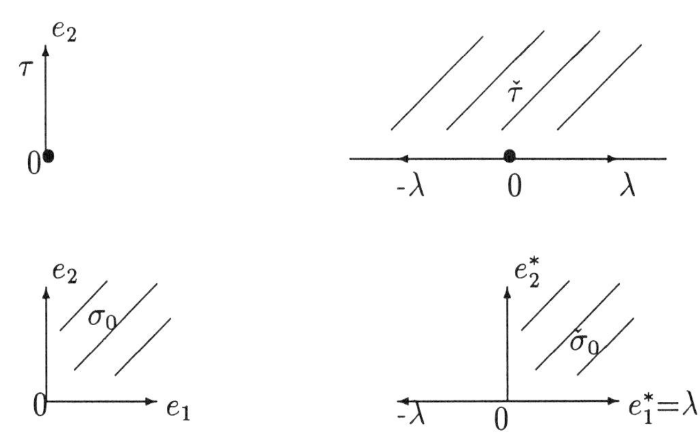

The vector $\lambda = e_1^*$ satisfies :

$$\lambda \in \check{\sigma}_0 \qquad \tau = \sigma_0 \cap \lambda^\perp$$

and we have

$$\check{\tau} = \check{\sigma}_0 + \mathbb{R}_{\geq 0}(-\lambda)$$

In the same way, let us now consider τ as a face of σ_1 :

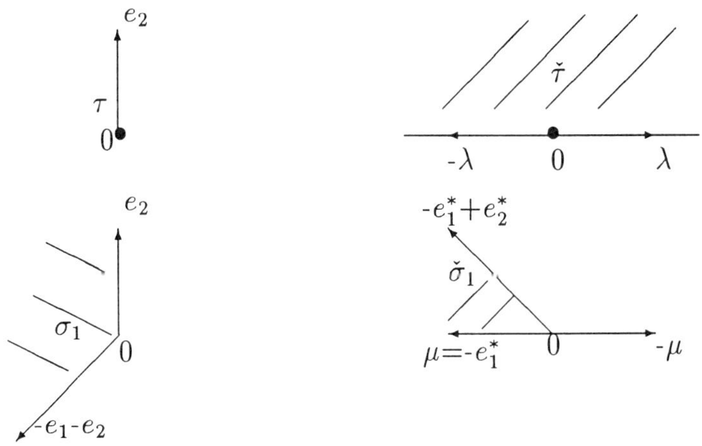

The vector $\mu = -e_1^*$ satisfies :

$$\mu \in \check{\sigma}_1 \qquad \tau = \sigma_1 \cap \mu^\perp$$

and we have

$$\check{\tau} = \check{\sigma}_1 + \mathbb{R}_{\geq 0}(-\mu)$$

Finally consider the origin $\{0\}$ as a face of σ_0 :

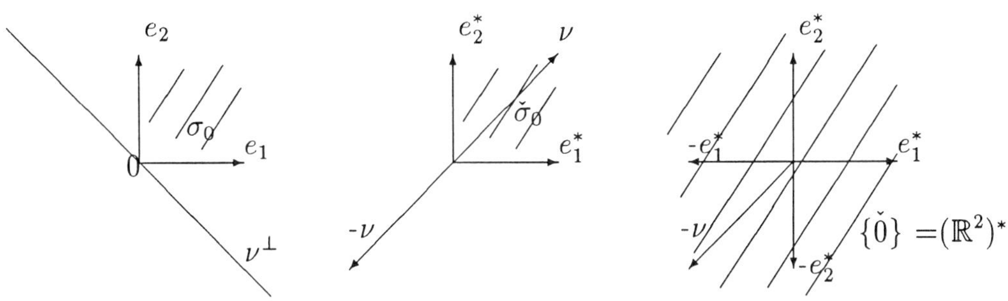

The vector $\nu = e_1^* + e_2^*$ satisfies :

$$\nu \in \check{\sigma}_0 \qquad \{0\} = \sigma_0 \cap \nu^\perp$$

and we have

$$\{\check{0}\} = (\mathbb{R}^2)^* = \check{\sigma}_0 + \mathbb{R}_{\geq 0}(-\nu).$$

The previous three examples are examples of a general result :

Proposition. Let τ be a face of a cone σ and let λ be a vector such that $\lambda \in \check{\sigma}$ and $\tau = \sigma \cap \lambda^\perp$, then $\check{\tau} = \check{\sigma} + \mathbb{R}_{\geq 0}(-\lambda)$.

The proof of this fact is a funny demonstration using properties of duality and orthogonality.

Corollary. Let τ be a face of a cone σ and let λ be a vector such that $\lambda \in \check{\sigma}$ and $\tau = \sigma \cap \lambda^\perp$, then $S_\tau = S_\sigma + \mathbb{Z}_{\geq 0}(-\lambda)$.

The monoid S_τ is thus obtained from S_σ by adding only one generator $-\lambda$. As λ is an element of a system of generators $(a_1, \dots, a_k)$ for S_σ, we may suppose that $\lambda = a_k$. So for obtaining the relations between the generators of S_τ, we have to add only the supplementary relation $a_k + a_{k+1} = 0$ to the previous ones between the generators of S_σ.

This relation corresponds to the multiplicative relation $u_k u_{k+1} = 1$ in R_τ and it is the only supplementary relation we need defining R_τ from R_σ. As the generators u_i are precisely the coordinate functions on the toric varieties X_σ and X_τ, this means that the projection $\mathbb{C}^{k+1} \to \mathbb{C}^k$:

$$(x_1, \dots, x_k, x_{k+1}) \mapsto (x_1, \dots, x_k)$$

identifies X_τ with the open subset of X_σ defined by $x_k \neq 0$. This can be written :

Proposition. There is a natural identification

$$X_\tau \cong X_\sigma \setminus (u_k = 0)$$

This allows us to glue together X_σ and $X_{\sigma'}$ such that $\sigma \cap \sigma' = \tau$. Writing $(v_1, \dots, v_l)$ the coordinates on $X_{\sigma'}$, there is an homeomorphism

$$X_\tau \cong X_{\sigma'} \setminus (v_l = 0)$$

and we obtain a gluing map

$$\psi_{\sigma,\sigma'} : X_\sigma \setminus (u_k = 0) \xrightarrow{\cong} X_\tau \xrightarrow{\cong} X_{\sigma'} \setminus (v_l = 0).$$

Theorem. (Definition of Toric Varieties). Let Δ be a fan in $\mathbb{R}^n$. Consider the disjoint union $\cup_{\sigma\in\Delta} X_\sigma$ where two points $x \in X_\sigma$ and $x' \in X_{\sigma'}$ are identified if $\psi_{\sigma,\sigma'}(x) = x'$. The resulting space X_Δ is called a *toric variety*. It is a topological space endowed with an open covering by the affine toric varieties X_σ for $\sigma \in \Delta$. It is an algebraic variety whose charts are defined by binomial relations.

Example 11. Let us return to the example of the projective space $\mathbb{P}^2$, we have, using the previous notations :

$S_\tau = S_{\sigma_0} + \mathbb{Z}_{\geq 0}(-e_1^*)$ and $X_\tau = X_{\sigma_0} \setminus (z_1 = 0) = \mathbb{C}_{z_2} \times \mathbb{C}^*_{z_1}$. Recall that $X_{\sigma_0} = \mathbb{C}^2_{(z_2,z_1)}$.

In the same way, $S_\tau = S_{\sigma_1} + \mathbb{Z}_{\geq 0}(e_1^*)$ and $X_\tau = X_{\sigma_1} \setminus (z_1^{-1} = 0) = \mathbb{C}_{z_1^{-1}z_2} \times \mathbb{C}^*_{z_1^{-1}}$ with $X_{\sigma_1} = \mathbb{C}^2_{(z_1^{-1}z_2, z_1^{-1})}$.

We can glue together X_{σ_0} and X_{σ_1} along X_τ using the change of coordinates $(z_2, z_1) \mapsto (z_1^{-1}z_2, z_1^{-1})$. We obtain $\mathbb{P}^2 \setminus \{(0:0:1)\}$.

Gluing this space by the same procedure with $X_{\sigma_2} = \mathbb{C}^2_{(z_2^{-1}, z_1 z_2^{-1})}$, we obtain the total space $\mathbb{P}^2$.

In fact, we have shown that, for cones $\tau < \sigma$, we have inclusions :

$$\begin{array}{ccc} \tau & \overset{\cdot}{\hookrightarrow} & \tau \\ \check{\tau} & \hookleftarrow & \check{\sigma} \\ R_\tau & \hookleftarrow & R_\sigma \\ X_\tau & \hookrightarrow & X_\sigma \end{array}$$

Before giving more examples, let us show a fundamental result :

Proposition. Every toric variety contains the torus $\mathbf{T}$ as a Zariski open dense subset.

Proof : The torus $\mathbf{T}$ corresponds to the zero cone, which is a face of every $\sigma \in \Delta$, i.e. $\mathbf{T} = X_{\{0\}}$. Let us explicit the embedding of the torus into every affine toric variety X_σ. Let $(a_1, \ldots, a_k)$ be a system of generators for the monoid S_σ and let $V(\mathcal{A}) \subset \mathbb{C}^n$ be a representation of X_σ. With the previous coordinates of $\mathbb{R}^n$, each a_i is written $a_i = (\alpha_i^1, \dots \alpha_i^n)$ with $\alpha_i^j \in \mathbb{Z}$ and $t \in \mathbf{T}$ is written $t = (t_1, \dots, t_n)$ with $t_j \in \mathbb{C}^*$. The embedding $\mathbf{T} \hookrightarrow X_\sigma$ is given by

$$\begin{aligned} t &= (t_1, \dots, t_n) \mapsto (t^{a_1}, \dots, t^{a_k}) \in V(\mathcal{A}) \cap (\mathbb{C}^*)^k \\ &\quad \text{where } t^{a_i} = t_1^{\alpha_i^1} \cdots t_n^{\alpha_i^n} \in \mathbb{C}^* \quad . \end{aligned}$$

By the previous identifications, all the tori corresponding to affine toric varieties X_σ in X_Δ glue together in an open dense subset in X_Δ.

In the case of example 5, the embedding is given by

$$(t_1, t_2) \mapsto (t_1, t_1 t_2, t_1 t_2^2) \in V(\mathcal{A}) \cap (\mathbb{C}^*)^3$$

Here are some of the classical examples of toric varieties :

Example 12 Consider the following fan :

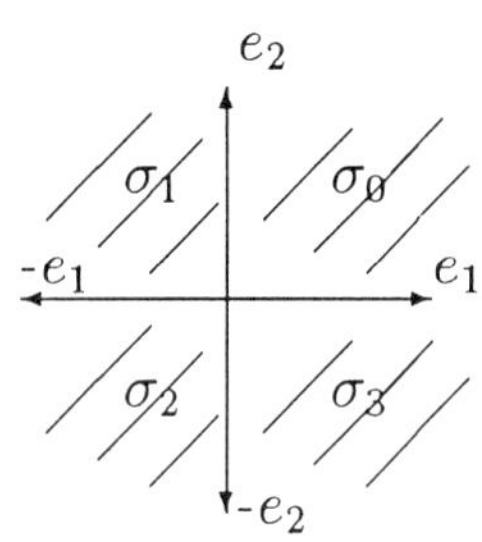

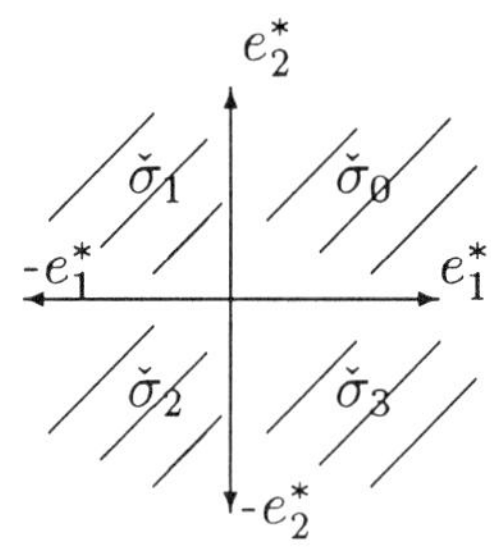

which gives the following monoids :

$$\begin{array}{ccc} S_{\sigma_1} \text{ gen. by } (-e_1^*, e_2^*) & \leftrightarrow & S_{\sigma_0} \text{ gen. by } (e_1^*, e_2^*) \\ \updownarrow & & \updownarrow \\ S_{\sigma_2} \text{ gen. by } (-e_1^*, -e_2^*) & \leftrightarrow & S_{\sigma_3} \text{ gen. by } (e_1^*, -e_2^*) \end{array}$$

and the following $\mathbb{C}$-algebra :

$$\begin{array}{ccc} R_{\sigma_1} = \mathbb{C}[z_1^{-1}, z_2] & \leftrightarrow & \mathbb{C}[z_1, z_2] = R_{\sigma_0} \\ \updownarrow & & \updownarrow \\ R_{\sigma_2} = \mathbb{C}[z_1^{-1}, z_2^{-1}] & \leftrightarrow & \mathbb{C}[z_1, z_2^{-1}] = R_{\sigma_3} \end{array}$$

The gluing of X_{σ_1} and X_{σ_0} gives $\mathbb{P}^1 \times \mathbb{C}$ with coordinates $((t_0 : t_1), z_2)$ where $(z_1 = t_0/t_1)$,

The gluing of X_{σ_2} and X_{σ_3} gives $\mathbb{P}^1 \times \mathbb{C}$ with coordinates $((t_0 : t_1), z_2^{-1})$,

The gluing of these two gives $X_\Delta = \mathbb{P}^1 \times \mathbb{P}^1$ with coordinates $((t_0 : t_1), (s_0 : s_1))$ where $(z_2 = s_0/s_1)$.

Example 13. Consider the following fan :

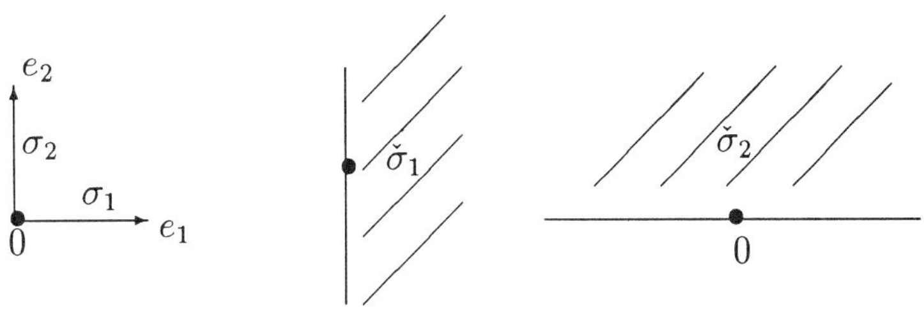

then S_{σ_1} is generated by $(e_1^*, e_2^*, -e_2^*)$. The monoid S_{σ_2} is generated by $(e_1^*, -e_1^*, e_2^*)$ and $S_{\{0\}}$ is generated by $(e_1^*, -e_1^*, e_2^*, -e_2^*)$. The corresponding $\mathbb{C}$-algebras are respectively $R_{\sigma_1} = \mathbb{C}[z_1, z_2, z_2^{-1}]$, $R_{\sigma_2} = \mathbb{C}[z_1, z_1^{-1}, z_2]$ and $R_{\{0\}} = \mathbb{C}[z_1, z_1^{-1}, z_2, z_2^{-1}]$. The corresponding affine toric varieties are $X_{\sigma_1} = \mathbb{C}_{z_1} \times \mathbb{C}^*_{z_2}$, $X_{\sigma_2} = \mathbb{C}^*_{z_1} \times \mathbb{C}_{z_2}$ and $X_{\{0\}} = \mathbb{C}^*_{z_1} \times \mathbb{C}^*_{z_2}$. The gluing of the affine toric X_{σ_1} and X_{σ_2} along $X_{\{0\}}$ gives $X_\Delta = \mathbb{C}^2 - \{0\}$.

Example 14. Consider the following fan :

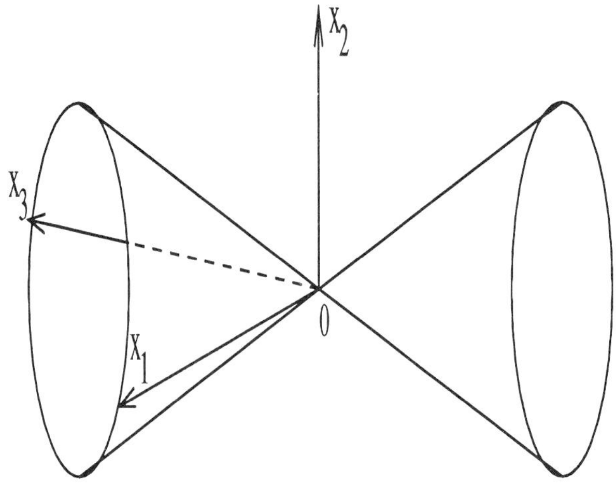

Then the monoids S_{σ_i} are generated :

$$\begin{array}{ccc} S_{\sigma_1} \text{ by } (-e_1^*, e_2^*) & \leftrightarrow & S_{\sigma_0} \text{ by } (e_1^*, e_2^*) \\ \updownarrow & & \updownarrow \\ S_{\sigma_2} \text{ by } (-e_1^* - ke_2^*, -e_2^*) & \leftrightarrow & S_{\sigma_3} \text{ by } (e_1^* + ke_2^*, -e_2^*) \end{array}$$

and the corresponding affine varieties are

$$\begin{array}{ccc} X_{\sigma_1} = \mathbb{C}^2_{(z_1^{-1}, z_2)} & \leftrightarrow & X_{\sigma_0} = \mathbb{C}^2_{(z_1, z_2)} \\ \updownarrow & & \updownarrow \\ X_{\sigma_2} = \mathbb{C}^2_{(z_1^{-1} z_2^{-k}, z_2^{-1})} & \leftrightarrow & X_{\sigma_3} = \mathbb{C}^2_{(z_1 z_2^k, z_2^{-1})} \end{array}$$

The gluing of X_{σ_1} and X_{σ_0} gives $\mathbb{P}^1 \times \mathbb{C}$ with coordinates $((t_0 : t_1), z_2)$ where $z_1 = t_0/t_1$, the gluing of X_{σ_2} and X_{σ_3} gives $\mathbb{P}^1 \times \mathbb{C}$ with coordinates $((s_0 : s_1), z_2^{-1})$ where $z_1 z_2^k = s_0/s_1$.

These two gluing, glued together, provide a $\mathbb{P}^1$-bundle over $\mathbb{P}^1$ (gluing the second coordinates), which is a rational ruled surface denoted $\mathbf{F}_k$ and called Hirzebruch surface. It is the hypersurface in $\mathbb{P}^1 \times \mathbb{P}^2$ defined by

$$\{(\lambda_0 : \lambda_1), (\mu_0 : \mu_1 : \mu_2) \ : \ \lambda_0^k \mu_0 = \lambda_1^k \mu_1\}$$

Example 15. Consider the following fan :

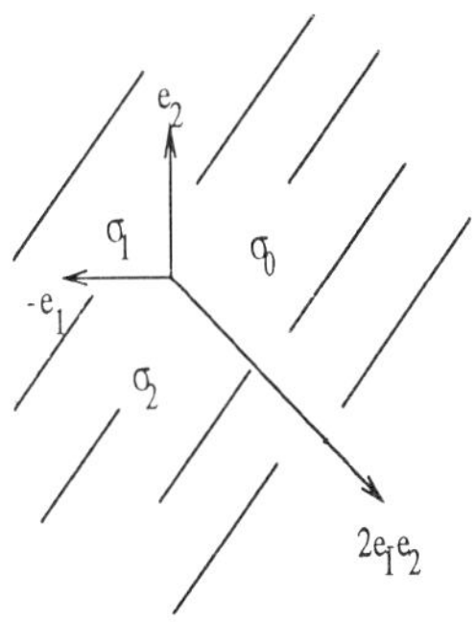

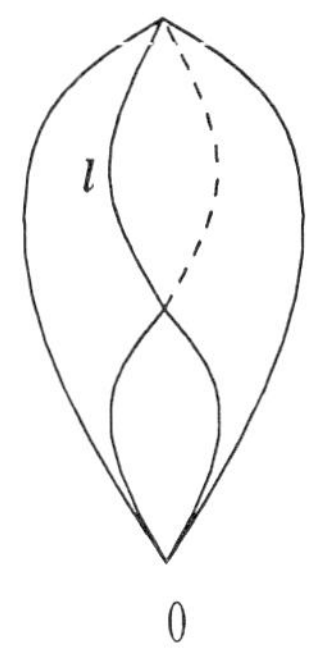

Then X_{σ_0} is the affine quadratic cone (cf example 5), X_{σ_1} and X_{σ_2} are affine planes (example 8). The affine quadratic cone is completed by a "circle at infinity" that represents a complex projective line. The real picture of X_Δ is a pinched torus.

Example 16. Generalizing the example of $\mathbb{P}^2$, we can consider the fan Δ whose cones are generated by all proper subsets of $(v_0, \dots, v_n) = (e_1, \dots, e_n, -(e_1 + \dots + e_n))$, i.e. σ_0 is generated by $(e_1, \dots, e_n)$ and, for $i = 1, \dots, n$, the cone σ_i is generated by $(e_1, \dots, e_{i-1}, e_{i+1}, \dots, e_n, -(e_1 + \dots + e_n)$. For $i = 0, \dots, n$, the affine toric varieties X_{σ_i} are copies of $\mathbb{C}^n$ giving the classical charts of $\mathbb{P}^n$.

Example 17. Let $d_0, \dots, d_n$ be positive integers. Consider the same fan than the previous one (example 16) but consider the lattice N' generated by the vectors $(1/d_i) \cdot v_i$, for $i = 0, \dots, n$. Then the resulting toric variety is

$$\mathbb{P}(d_0, \dots, d_n) = \mathbb{C}^{n+1} - \{0\}/\mathbb{C}^*$$

where $\mathbb{C}^*$ acts by $\zeta \cdot (x_0, \dots, x_n) = (\zeta^{d_0} x_0, \dots, \zeta^{d_n} x_n)$. It is called twisted or weighted projective space.

Definition. A cone σ defined by the set of vectors $x_1, \dots x_r$ is a *simplex* if all the vectors x_i are linearly independent. A fan Δ is *simplicial* if all cones of Δ are simplices.

Definition. A vector $x \in \mathbb{Z}^n$ is *primitive* if its coordinates are co-prime. A cone is *regular* if the vectors $(x_1, \dots, x_r)$ spanning the cone are primitive and there exists primitive vectors $x_{r+1}, \dots, x_n$ such that

$\det(x_1, \ldots, x_n) = \pm 1$. In another words, the vectors $(x_1, \ldots, x_r)$ can be completed in a basis of the lattice N. A fan is *regular* if all its cones are regular ones.

Definitions 1. A fan Δ is *complete* if its cones cover $\mathbb{R}^n$, i.e. $|\Delta| = \mathbb{R}^n$.

2. A fan is *polytopal* if there exists a polytope P such that $0 \in P$ and Δ is spanned by the faces of P (let us recall that a polytope is the convex hull of a finite number of points).

Remarks 1. Every complete fan in $\mathbb{R}^2$ is polytopal,

2. Not every complete fan is isomorphic to a polytopal one. For example take the cube in $\mathbb{R}^3$ with all coordinates ± 1. The faces of the cube provide a polytopal fan. Now replace the point $(1, 1, 1)$ by $(1, 2, 3)$ and consider the corresponding fan. It is clearly not isomorphic to a polytopal one : there exists 4 points not lying in the same affine plane.

Theorem. 1. The fan Δ is complete if and only if X_Δ is compact.

2. The fan Δ is regular if and only if X_Δ is smooth.

3. The fan Δ is polytopal if and only if X_Δ is projective.

Let us give some precisions about these results :

An affine toric variety X_σ is smooth if and only if $X_\sigma = \mathbb{C}^k \times (\mathbb{C}^*)^{n-k}$ where $k = \dim \sigma$.

If the fan Δ is simplicial, then X_Δ is an orbifold (or V-manifold) : it has only quotient singularities.

Also we remark that if Δ is complete, then X_Δ is a compactification of $\mathbf{T} = (\mathbb{C}^*)^n$.

Resolution of singularities ([5], §2.6)

We give only examples. The general way of resolving singularities follows these examples (see for example [5], §2.6)

Example 18. Consider the following fan

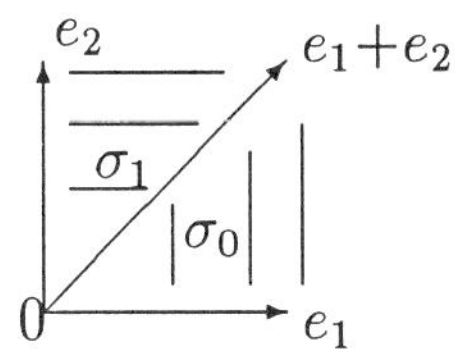

The corresponding toric variety is a blow-up of a point in $\mathbb{C}^2$.

Example 19. Consider the following fan (cone) Δ and its subdivision Δ' :

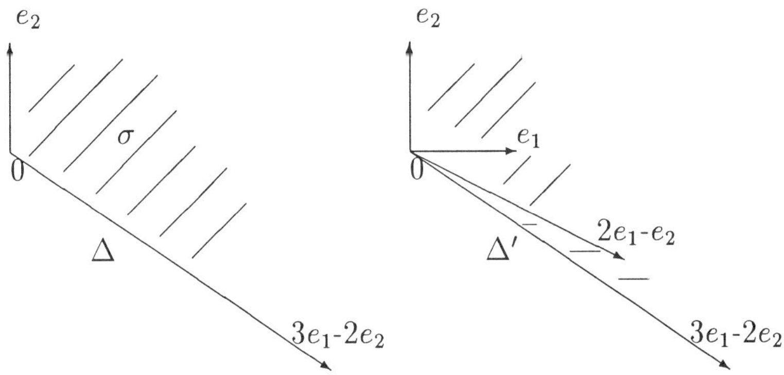

The fan Δ' is a regular fan, hence $X_{\Delta'}$ is a smooth toric variety. The identity map of N provides a map $X_{\Delta'} \to X_\Delta$ which is birational proper. It is an isomorphism on the open torus $\mathbf{T}$ contained in each. This is the first example (and standard one) of resolution of singularities.

The procedure is the following : beginning with the cone σ generated by the two vectors $v = e_2$ and $v' = 3e_1 - 2e_2$, we add primitive vectors (here $v_1 = e_1$ and $v_2 = 2e_1 - e_2$) such that, with $v_0 = v$ and $v_3 = v'$, we have

$$\lambda_i v_i = v_{i-1} + v_{i+1} \qquad i = 1, 2$$

For $i = 1, 2$, the vectors v_i correspond to exceptional divisors $E_i \cong \mathbb{P}^1$ in $X_{\Delta'}$ and their self-intersection is $(E_i, E_i) = -\lambda_i$. In this particular case, we obtain two exceptional divisors with self-intersection -2.

5 The torus action and the orbits.

The torus $\mathbf{T} = (\mathbb{C}^*)^n$ is a group operating on itself by multiplication. The action of the torus on each affine toric variety X_σ is described as follows :

Let $(a_1, \ldots, a_k)$ be a system of generators for the monoid S_σ. With the previous coordinates of $\mathbb{R}^n$, each a_i is written $a_i = (\alpha_i^1, \ldots \alpha_i^n)$ with $\alpha_i^j \in \mathbb{Z}$ and $t \in \mathbf{T}$ is written $t = (t_1, \ldots, t_n)$ with $t_j \in \mathbb{C}^*$. A point $x \in X_\sigma$ is written $x = (x_1, \ldots, x_k) \in \mathbb{C}^k$. The action of $\mathbf{T}$ on X_σ is given by :

$$\begin{array}{rcl} \mathbf{T} \times X_\sigma & \to & X_\sigma \\ (t, x) & \mapsto & t \cdot x = (t^{a_1} x_1, \ldots, t^{a_k} x_k) \end{array}$$

where $t^{a_i} = t_1^{\alpha_i^1} \cdots t_n^{\alpha_i^n} \in \mathbb{C}^*$.

Now let Δ be a fan in $\mathbb{R}^n$ and let τ be a face of a cone $\sigma \in \Delta$. The identification $X_\tau \cong X_\sigma \setminus (u_k = 0)$ is compatible with the torus action, which implies that the gluing maps $\psi_{\sigma,\sigma'}$ respect also this torus action. We obtain the :

Theorem. Let Δ be a fan in $\mathbb{R}^n$, the torus action on the affine toric varieties X_σ, for $\sigma \in \Delta$, provide a torus action on the toric variety X_Δ.

It is clear that the embedded tori in each X_σ correspond each other by the gluing maps. We obtain an open embedding of the torus $\mathbf{T} = (\mathbb{C}^*)^n$ in the toric variety X_Δ.

Let $\Delta = \{0\}$, we seen that $X_\Delta = (\mathbb{C}^*)^n$ is the torus. There is only one orbit which is the entire space and is the orbit of the point whose coordinates u_i are $(1, \ldots, 1)$ in $\mathbb{C}^n$. In fact, for every toric variety, the apex $\sigma = \{0\}$ of Δ provides an open dense orbit which is the embedded torus $\mathbf{T} = (\mathbb{C}^*)^n$. Let us describe the other orbits.

Theorem. (see [5], §2.1 and 3.1) Let Δ be a fan in $\mathbb{R}^n$, to each $\sigma \in \Delta$, we can associate a distinguished point $x_\sigma \in X_\sigma \subset X_\Delta$ and the orbit $O_\sigma \subset X_\sigma$ of x_σ satisfying :

1) $X_\sigma = \coprod_{\tau<\sigma} O_\tau$,

2) if V_τ denotes the closure of the orbit O_τ, then $V_\tau = \coprod_{\tau<\sigma} O_\sigma$,

3) $O_\tau = V_\tau \setminus \bigcup_{\substack{\tau<\sigma \\ \tau\neq\sigma}} V_\sigma$.

Let τ be a face of a cone σ, then $O_\sigma \subset \overline{O_\tau}$. The image of $V_\tau = \overline{O_\tau}$ in a representation of X_σ can be determined in the following way :

Consider a system of generators $(a_1, \dots a_k)$ of the monoid S_σ, denote I the set of indices $1 \leq i \leq k$ such that $a_i \notin \tau^\perp$. In other words, if $(x_1, \dots, x_s)$ denote the vectors that span τ, we have

$$i \in I \iff \forall j, \quad 1 \leq j \leq s \quad \langle a_i, x_j\rangle \neq 0$$

In X_σ with coordinates $u_i = z^{a_i}$, then V_τ is defined by $u_i = 0$ if $i \in I$.

Let us give two examples :

Example 20. In the case of example 5, X_σ has coordinates $(u_1, u_2, u_3) = (z_1, z_1z_2, z_1z_2^2)$ and S_σ is generated by $a_1 = e_1^*$, $a_2 = e_1^* + e_2^*$ and $a_3 = e_1^* + 2e_2^*$. Let us consider the edge τ_1 generated by e_2, then

$$i \in I \iff \langle a_i, e_2\rangle \neq 0$$

hence $I = \{2, 3\}$. In X_σ the set V_{τ_1} is defined by $u_2 = 0$, $u_3 = 0$. In $\mathbb{C}^3 = \mathbb{C}^3_{(\xi_1,\xi_2,\xi_3)}$ it is $V_{\tau_1} = \mathbb{C}_{\xi_1} \times \{0\} \times \{0\}$.

Consider the edge τ_2 generated by $2e_1 - e_2$, then

$$i \in I \iff \langle a_i, 2e_1 - e_2\rangle \neq 0$$

hence $I = \{1, 2\}$. In X_σ, the set V_{τ_2} is defined by $u_1 = 0$, $u_2 = 0$. In $\mathbb{C}^3_{(\xi_1,\xi_2,\xi_3)}$, we have $V_{\tau_2} = \{0\} \times \{0\} \times \mathbb{C}_{\xi_3}$.

The cone σ is a face of itself. For this face, $I = \{1, 2, 3\}$ and, in X_σ, the set V_σ is defined by $u_1 = 0$, $u_2 = 0$, $u_3 = 0$. Hence $V_\sigma = O_\sigma$ is the origin $(0, 0, 0) \in \mathbb{C}^3$.

We can conclude by the list of orbits in this example :

$$\begin{array}{lcl} O_\sigma & = & \{(0,0,0)\} \\ O_{\tau_1} & = & \mathbb{C}^*_{\xi_1} \times \{0\} \times \{0\}, \text{ orbit of the distinguished point } x_{\tau_1} = (1,0,0) \\ O_{\tau_2} & = & \{0\} \times \{0\} \times \mathbb{C}^*_{\xi_3}, \text{ orbit of the distinguished point } x_{\tau_2} = (0,0,1) \\ O_{\{0\}} \cdot & = & (\mathbb{C}^*)^2, \text{orbit of the distinguished point} x_{\{0\}} = (1,1,1) \end{array}$$

Example 21. Orbits in $\mathbb{P}^2$ (see the pictures in §4).

With the notations and the pictures of §4, let us consider the image of $V_\tau = \overline{O_\tau}$ in X_{σ_0} and U_{σ_1}. The monoid S_{σ_0} is generated by $a_1 = e_1^*$ and $a_2 = e_2^*$. In X_{σ_0} with coordinates $(u_1, u_2) = (z_1, z_2)$, we have :

$$i \in I \iff \langle a_i, e_2 \rangle \neq 0$$

hence $I = \{2\}$ and, in $X_{\sigma_0} = \mathbb{C}^2_{(u_1,u_2)}$, V_τ is defined by $u_2 = 0$. Hence V_τ is $\mathbb{C}_{\xi_1} \times \{0\}$ and $O_\tau = \mathbb{C}^*_{z_1} \times \{0\}$ is the orbit of $\{x_\tau\} = (1,0)$. This point is a representation of the point $(1:1:0)$ of $\mathbb{P}^2$.

The monoid S_{σ_1} is generated by $a_1 = -e_1^*$ and $a_2 = -e_1^* + e_2^*$. In X_{σ_1} with coordinates $(u_1, u_2) = (z_1^{-1}, z_1^{-1} z_2)$, we have :

$$i \in I \iff \langle a_i, e_2 \rangle \neq 0$$

hence $I = \{2\}$ and, in $X_{\sigma_1} = \mathbb{C}^2_{(u_1,u_2)}$, V_τ is defined by $u_2 = z_1^{-1} z_2 = 0$. Hence V_τ is $\mathbb{C}_{(z_1^{-1})} \times \{0\}$. The orbit $O_\tau = \mathbb{C}^*_{(z_1^{-1})} \times \{0\}$ is the same than before, i.e. the orbit of $\{x_\tau\}$.

The projective space is the union of 7 orbits of the torus action :

- $O_{\{0\}} = (\mathbb{C}^*)^2$,

- 3 orbits homeomorphic to $\mathbb{C}^*$ corresponding to the three edges and whose images in each X_{σ_i} are described in the same way than O_τ. They are the orbits of the points $(1:1:0)$, $(1:0:1)$ and $(0:1:1)$ of $\mathbb{P}^2$.

- 3 fixed points $\{x_{\sigma_i}\}$, $i = 1,2,3$ corresponding to the 2-dimensional cones σ_i. They are fixed points of the torus action and are the points $(1:0:0)$, $(0:1:0)$ and $(0:0:1)$.

These examples provide examples of general situations :

Properties. 1. If $\dim_{\mathbf{R}} \sigma = n$, then O_σ is a fixed point $\{x_\sigma\}$. Consider a representation of X_σ in $\mathbb{C}^k$, then $O_\sigma = \{x_\sigma\}$ corresponds to the origin of $\mathbb{C}^k$.

2. If $\dim_{\mathbf{R}} \sigma = k$, then $O_\sigma \cong (\mathbb{C}^*)^{n-k}$.

3. Let τ_i be an edge (1-dimensional cone) in Δ, then $O_{\tau_i} \cong (\mathbb{C}^*)^{n-1}$. If $\dim_{\mathbf{R}} \Delta = n$, then V_{τ_i} is a codimension one variety in X_Δ. We will see that V_{τ_i} is a Weil divisor.

4. The distinguished point $\{x_\sigma\}$ corresponding to each cone σ can be defined in a direct way (cf [5]).

6 Divisors.

In this section, we will denote by X a complex algebraic variety.

A *Weil divisor* is an element of the free abelian group $W(X)$ generated by the irreducible closed subvarieties of (complex) codimension 1 in X. Such a divisor can be written :

$$\sum n_i A_i - \sum m_j B_j \qquad \text{with } n_i, m_j > 0$$

where the A_i and B_j are subvarieties of codimension 1 in X.

For example, in the space $\mathbf{C}^2$ with coordinates (z_1, z_2), let us consider the axis $z_1 = 0$ denoted by A, and the axis $z_2 = 0$ denoted by B. An example of Weil divisor is given by $2A - B$.

Let us denote by $\mathcal{R}(U)$ the set of rational functions in the open set U in X. A *Cartier divisor* is given by a covering $X = \bigcup U_\alpha$ of X and by nonzero rational functions $f_\alpha \in \mathcal{R}(U_\alpha)$ such that for any α and β, we have $f_\alpha / f_\beta \in \mathcal{O}^*(U_\alpha \cap U_\beta)$ (nowhere zero holomorphic function). We denote by $D = (U_\alpha, f_\alpha)$ such a divisor. The set of Cartier divisors is a group denoted by $C(X)$.

Let us consider the example of $X = \mathbf{C}^2$ covered by only one open set $U = \mathbf{C}^2$ and consider, in U, the rational function $f(z_1, z_2) = z_1^2/z_2$, we obtain a Cartier divisor $D = (U, f)$.

Recall that a an algebraic complex variety is *normal* if the local ring at every point is an integrally closed ring. In particular, toric varieties are normal, each ring R_σ is integrally closed.

Proposition. For a normal variety X, there is an inclusion

$$C(X) \hookrightarrow W(X)$$

Let us explicit this inclusion in the previous example : If $A = \{f = 0\}$ is the set of zeroes of f counted with multiplicities and $B = \{1/f = 0\}$ is the set of poles of f counted with multiplicities, then the previous Weil divisor $2A - B$ corresponds to the previous Cartier divisor $D = (U, f)$.

In general this inclusion is defined by

$$[D] = \sum_{\operatorname{codim}(V,X)=1} \operatorname{ord}_V(D) \cdot V$$

where $\operatorname{ord}_V(D)$ is the vanishing order of an equation for D in the local ring along the subvariety V. If X is normal, then local rings are discrete valuation rings and the order is the naive one.

In fact, the previous example is an example of *principal divisor* : The subgroup of principal divisors, denoted by $P(X)$, is the subgroup of Cartier divisors corresponding to the nonzero rational functions. Let us consider the quotients :

$$\mathcal{C}(X) = C(X)/P(X) \qquad \text{and} \qquad \mathcal{W}(X) = W(X)/P(X)$$

There is an inclusion $\mathcal{C}(X) \hookrightarrow \mathcal{W}(X)$, which is not an equality as shown by the example of the toric variety of example 10 (with q=1) : let X be the quotient variety of $\mathbf{C}^2$ by the subgroup G of p-th roots of unity. Then, we have :

$$\mathcal{C}(X) = \{0\} \hookrightarrow \mathcal{W}(X) = \mathbf{Z}_p \ .$$

Let $X = X_\Delta$ be a toric variety. The Weil and Cartier divisor classes, invariant by the action of the torus $\mathbf{T}$ will be denoted respectively $C^{\mathbf{T}}(X)$ and $W^{\mathbf{T}}(X)$. In the same way, the subgroup of the invariant principal divisors will be denoted $P^{\mathbf{T}}(X)$. We define $\mathcal{C}^{\mathbf{T}}(X) = C^{\mathbf{T}}(X)/P^{\mathbf{T}}(X)$ et $\mathcal{W}^{\mathbf{T}}(X) = W^{\mathbf{T}}(X)/P^{\mathbf{T}}(X)$. There is still an inclusion

$$\mathcal{C}^{\mathbf{T}}(X) \hookrightarrow \mathcal{W}^{\mathbf{T}}(X)$$

Let Δ be a fan containing q edges and let X_Δ be the associated toric variety. Let τ_i be a edge of Δ and denote by $D_i = V_{\tau_i}$ the closure of the orbit O_{τ_i} associated to τ_i, then D_i is an invariant Weil divisor and all such divisors are on the form

$$\sum_{i=1}^{q} \lambda_i D_i \qquad \lambda_i \in \mathbb{Z} .$$

We obtain :

Lemma. The group of invariant Weil divisors is homeomorphic to :

$$W^{\mathbf{T}}(X) \cong \bigoplus_{i=1}^{q} \mathbb{Z}[D_i]$$

If $u \in M$, then there is a surjective homomorphism

$$\begin{array}{rccl} \operatorname{div}: & M & \to & C^{\mathbf{T}}(X) \\ & u & \mapsto & \operatorname{div}(u) = \sum_{i=1}^{q} \langle u, v_i \rangle D_i \end{array}$$

where v_i is the first lattice point on the edge τ_i. This implies :

Lemma. Let $u \in M$ and v_i the first lattice point of the edge τ_i, then

$$\operatorname{ord}_{V_{\tau_i}}(\operatorname{div}(u)) = \langle u, v_i \rangle$$

Example 22. In the case of example 5, there are two invariant Weil divisors corresponding to the two edges of the cone σ : D_1 corresponding to the edge τ_1 of e_2 and D_2 corresponding to the edge

τ_2 spanned by $2e_1 - e_2$. If $u \in M$ has coordinates (a, b) in $(\mathbb{C}^*)^2$, then $\mathrm{div}(u) = bD_1 + (2a - b)D_2$ and all invariant Cartier divisors are on this form. For example, $2D_1$ and $2D_2$ are such Cartier divisors but D_1 and D_2 are not.

The two divisors $2D_1$ and $2D_2$ are principal divisors, so we obtain : $\mathcal{C}^{\mathbf{T}}(X) = 0$ and $\mathcal{W}^{\mathbf{T}}(X) = \mathbb{Z}_2$.

Example 23. Let σ be the cone spanned by $x_1 = 2e_1 - e_2$ and $x_2 = -e_1 + 2e_2$. Each of these two vectors span a edge τ_i and the two corresponding Weil divisors are denoted D_1 and D_2. Then $\lambda_1 D_1 + \lambda_2 D_2$ is a Cartier divisor if and only if $\lambda_1 = \lambda_2 \bmod 3$ (Exercise).

7 Divisors, homology and cohomology.

In this section we will consider the general case of a complex algebraic variety.

Let n denote the (complex) dimension of X. A Weil divisor is a cycle in X. The application which associates, to each Weil divisor, its homology class defines in an evident way an homomorphism $\kappa : W(X) \longrightarrow H_{2n-2}(X)$. The image of a principal divisor is zero, so we obtain an homomorphism, still denoted

$$\kappa : \mathcal{W}(X) \longrightarrow H_{2n-2}(X) .$$

In other hand, for a normal variety, there is an isomorphism (cf. [6], II, Prop. 6.15)

$$\alpha : \mathcal{C}(X) \xrightarrow{\cong} \mathrm{Pic}(X)$$

between the group of classes of Cartier divisors and the Picard group of X, denoted $\mathrm{Pic}(X)$. This one is the group of isomorphy classes of line bundles (or isomorphy classes of invertible sheaves) on X. The isomorphism α is given by the map which associates, to the divisor $D = (U_\alpha, f_\alpha)$, the line bundle $\mathcal{O}(D)$ whose transition functions $U_\alpha \to U_\beta$ are given by f_α / f_β. Reciprocally, given an invertible sheaf, we associate the class of the divisor of a global rational and non trivial section.

By composition of α with the morphism $\mathrm{Pic}(X) \to H^2(X)$ which associates to each line bundle ξ on X, its Chern class $c^1(\xi)$, we obtain a morphism denoted

$$c^1 : \mathcal{C}(X) \longrightarrow H^2(X) .$$

8 Poincaré homomorphism.

The toric varieties are examples of pseudovarieties of (real) even dimension. By definition, a pseudovariety X of (real) dimension $2n$ is a connected topological space such that there is a closed subspace Σ such that :

(a) $X - \Sigma$ is an oriented smooth variety, of dimension $2n$, dense in X,

(b) $\dim \Sigma \leq 2n - 2$.

A $2n$-pseudovariety admits a fundamental class in integer homology $[X] \in H_{2n}(X)$. The Poincaré morphism

$$H^i(X) \longrightarrow H_{2n-i}(X)$$

is the cap-product by the fundamental class. If X is smooth, it is an isomorphism.

An example of pseudovariety for which the Poincaré homomorphism is not an isomorphism is given by the toric variety of example 10 (with q=1). We have $H^2(X) = 0$ and $H_2(X) = \mathbf{Z}_p$.

Theorem. Let X be a normal compact pseudovariety, there is a commutative diagram :

$$\begin{array}{ccc} \mathcal{C}(X) & \hookrightarrow & \mathcal{W}(X) \\ \downarrow c^1 & & \downarrow \kappa \\ H^2(X) & \stackrel{\cap[X]}{\longrightarrow} & H_{2n-2}(X) \end{array}$$

where the horizontal down arrow is the Poincaré morphism of the pseudovariety X.

If X is a compact toric variety, we can prove the following result :

Theorem. [2] Let $X = X_\Delta$ be a compact toric variety, there is a commutative diagram :

$$\begin{array}{ccc} \mathcal{C}^{\mathbf{T}}(X) & \hookrightarrow & \mathcal{W}^{\mathbf{T}}(X) \\ \downarrow \cong & & \downarrow \cong \\ H^2(X) & \stackrel{\cap[X]}{\longrightarrow} & H_{2n-2}(X) \end{array}$$

where the vertical isomorphisms correspond to the morphisms c_1 and κ of the previous theorem.

We obtain, in an evident way, an interpretation of the Poincaré morphism in terms of divisors, for the compact toric varieties. In particular, the Poincaré morphism $H^2(X) \longrightarrow H_{2n-2}(X)$ is injective.

This theorem can be generalized to any toric variety (see [2]) and the Poincaré homomorphism can be described in terms of Cartier and Weil divisors : Let X be a degenerated toric variety, i.e. containing a subtorus $\mathbf{T}''$ of $\mathbf{T}$ in factor : $X = Y \times \mathbf{T}''$ where Y is a toric variety relatively to the torus $\mathbf{T}'$ such that $\mathbf{T} = \mathbf{T}' \times \mathbf{T}''$. We have the following result :

Theorem. [2] Let $X = X_\Delta$ be a n-dimensional toric variety containing a toric factor $\mathbf{T}''$ of dimension $n-d$, then we have the following isomorphisms :

i) $H^1(X) \cong H^{\text{cld}}_{2n-1}(X) \cong H^1(\mathbf{T}'') \cong H^{\text{cld}}_{2n-2d-1}(\mathbf{T}'') \cong \mathbb{Z}^{n-d}$;

the homomorphisms c^1 and κ are injective and there are isomorphisms

ii) $$H^2(X) \cong \mathcal{C}^{\mathbf{T}}(X) \oplus H^2(\mathbf{T}'') \cong \mathcal{C}^{\mathbf{T}}(X) \oplus \mathbb{Z}^b\,;$$

iii) $$H^{\text{cld}}_{2n-2}(X) \cong \mathcal{W}^{\mathbf{T}}(X) \oplus H^{\text{cld}}_{2n-2d-2}(\mathbf{T}'') \cong \mathcal{W}^{\mathbf{T}}(X) \oplus \mathbb{Z}^b$$

with $b := \binom{n-d}{2}$, such that the following diagram commutes :

$$\begin{array}{ccc} \mathcal{C}^{\mathbf{T}}(X) \oplus H^2(\mathbf{T}'') & \longrightarrow & \mathcal{W}^{\mathbf{T}}(X) \oplus H^{\text{cld}}_{2n-2d-2}(\mathbf{T}'') \\ {\scriptstyle c^1 \oplus \text{pr}^*} \Big\downarrow {\scriptstyle \cong} & & {\scriptstyle \kappa \oplus \text{pr}^*} \Big\downarrow {\scriptstyle \cong} \\ H^2(X) & \xrightarrow{\cap [X]} & H^{\text{cld}}_{2n-2}(X) \quad . \end{array}$$

This diagram can be completed by the intersection homology of X_Δ which admits also an interpretation in terms of divisors (see [3]). Let us give the (simpler) compact case :

Let $\bar{p}$ be a perversity, we define $V_{\bar{p}}$ as the open invariant subset of X_Δ, union of orbits B such that $\text{codim}_{\mathbb{C}} B \leq \max\{i\,;\ \bar{p}(2i) \leq 1\}$. Let

$$\mathcal{C}^{\mathbf{T}}_{\bar{p}}(X) := \{[D] \in \mathcal{W}^{\mathbf{T}}(X)\,;\ D|_{V_{\bar{p}}} \in \mathcal{C}^{\mathbf{T}}(V_{\bar{p}})\}$$

be the group of invariant Weil divisors on X whose restriction to $V_{\bar{p}}$ is a Cartier divisor. This group is clearly isomorphic to $\mathcal{C}^{\mathbf{T}}(V_{\bar{p}})$.

Theorem. Let X be a n-dimensional compact toric variety, then :

i) We have :

$$IH^{\bar{p},\text{cld}}_{2n-1}(X) = 0\,.$$

ii) there is a natural isomorphism $\mathcal{C}^{\mathbf{T}}_{\bar{p}}(X) \xrightarrow{\cong} IH^{\bar{p},\text{cld}}_{2n-2}(X)$ such that the following diagram commutes :

$$\begin{array}{ccccc} \mathcal{C}^{\mathbf{T}}(X) & \xrightarrow{\subset} & \mathcal{C}^{\mathbf{T}}_{\bar{p}}(X) & \xrightarrow{\subset} & \mathcal{W}^{\mathbf{T}}(X) \\ {\scriptstyle c^1} \Big\downarrow {\scriptstyle \cong} & & \Big\downarrow {\scriptstyle \cong} & & {\scriptstyle \kappa} \Big\downarrow {\scriptstyle \cong} \\ H^2(X) & \xrightarrow{\alpha} & IH^{\bar{p},\text{cld}}_{2n-2}(X) & \xrightarrow{\omega} & H^{\text{cld}}_{2n-2}(X) \quad . \end{array}$$

where the composition of the two lower horizontal arrows is the Poincaré homomorphism.

9 Characteristic classes.

Let X_Δ be a smooth toric variety. The Poincaré homomorphism is an isomorphism between $H^k(X_\Delta)$ and $H_{2n-k}(X_\Delta)$ for every k. The Chern characteristic classes of X_Δ are usually defined in cohomology but their image in homology can be easily described in terms of the orbits. In fact, the total homology Chern class of X_Δ is :

$$\begin{aligned} c(X_\Delta) &= \Pi_{i=1}^{q}(1 + D_i) \\ &= \Sigma_{\sigma\in\Delta}[V_\sigma] \end{aligned}$$

where $D_i = V_{\tau_i}$ are the divisors corresponding to the edges of Δ. The intersection product is given by

$$D_i \cdot V_\sigma = \begin{cases} V_\gamma & \text{if } \sigma \text{ and } \tau_i \text{ span a cone } \gamma \text{ in } \Delta \\ 0 & \text{in the other case.} \end{cases}$$

In [1] it is shown that this result is also true for singular toric varieties. More precisely, it is well known that there is no cohomology Chern classes for a singular algebraic variety. In homology we can define the Schwartz-MacPherson classes which generalize homology Chern classes and we obtain the following result :

Theorem. [1] Let X_Δ be any toric variety, the total Schwartz-MacPherson class of X_Δ is given by :

$$c(X_\Delta) = \sum_{\sigma\in\Delta}[V_\sigma]$$

References

[1] **G. Barthel, J.-P. Brasselet, K.-H. Fieseler** *Classes de Chern des variétés toriques singulières.* C.R.Acad.Sci. Paris 315 (1992), 187–192.

[2] **G. Barthel, J.-P. Brasselet, K.-H. Fieseler et L. Kaup** *Diviseurs invariants et homomorphisme de Poincaré des variétés toriques complexes.* To appear in Tôhoku Math. Journal.

[3] **G. Barthel, J.-P. Brasselet, K.-H. Fieseler et L. Kaup** *Divisoren und Schnitthomologie torisher Varietäten.* To appear in Banach Institute Publications.

[4] **G. Ewald** *Combinatorial convexity and Algebraic Geometry.* Graduate Texts in Mathematics. To appear.

[5] **W. Fulton** *Introduction to Toric Varieties.* Annals of Math. Studies, Princeton Univ. Press 1993.

[6] **R. Hartshorne** *Algebraic Geometry.* Graduate Texts in Math. **52**, Springer Verlag, New York etc., 1977.

[7] **T. Oda** *Convex Bodies and Algebraic Geometry.* Ergebn. Math.Grenzgeb. (3. Folge), Bd. **15**, Springer-Verlag, Berlin etc., 1988.

Integration of Algebraic Functions and the Riemann-Kempf Singularity Theorem

Xavier Gómez–Mont

Centro de Investigación en Matemáticas, A. C.
Apartado Postal 402
36000 Guanajuato, Gto.
México

e-mail: gmont@fractal.cimat.mx

Introduction

These notes begin with the problem of integrating algebraic functions like

$$\int \frac{1}{\sqrt{1-x^6}}dx.$$

By extending the domain of definition from the real to the complex numbers, the problem becomes the integration of a multivalued algebraic function defined on the Riemann sphere, which we then transform to the integration of a rational differential 1-form on a compact Riemann surface C. The integration of such a differential form is a multivalued function due to the presence of poles with non-zero residues that incorporate a logarithmic term to the function and the non-trivial topology of C, that is measured by the genus $g \geq 0$. In order to concentrate on the serious part of the multivaluedness of these functions, one restricts to integrate Abelian differentials, which means that no poles are allowed. These differentials form a vector space of dimension g, and by choosing a basis one wants to understand the integrals of them together

$$\int_p^* (\omega_1, \ldots, \omega_g) : C \to \mathbb{C}^g$$

Of course the above function is multivalued, but it has a 'mild' multivaluedness, since the difference between two branches form an additive subgroup Λ of $\mathbb{C}^g$, called the periods of C, which is discrete with a compact fundamental region. The quotient group $\mathbb{C}^g/\Lambda$ is called the **Jacobian variety** $J(C)$ and the map

$$\int_p^* (\omega_1, \ldots, \omega_g) : C \to J(C) = \mathbb{C}^g/\Lambda$$

is called the **universal Abelian integral**. This is an algebraic map. The Abelian differential forms on $J(C)$ may be written as $\sum_{j=1}^g a_j dz_j$ where a_j are constants and its integral correspond to the multivalued functions on $J(C)$ induced by the linear function $\sum_{j=1}^g a_j z_j$ on $\mathbb{C}^g$. These 'constant' differential forms pull back via the universal Abelian integral to the Abelian differentials on C. Hence the integrals of Abelian differentials on C may be decomposed as the composition of an algebraic map and a 'multivalued linear function' on $J(C)$. The fibres of the integral of an Abelian differential

$$\int \sum_{j=1}^{g} a_j dz_j$$

on $J(C)$ corresponds to parallel hyperplanes in $\mathbb{C}^g$ projected to $J(C)$, and may be understood as the interplay of the hyperplane with the periods. We then want to intersect this codimension 1 foliation in $J(C)$ with the image of the universal Abelian integral, which as we mentioned, is an algebraic map. This reduces the problem of understanding the integrals of Abelian differentials to understanding an algebraic map and linear foliations of codimension 1 in $J(C)$ and how they intersect.

We then explain the connection found by Abel between the integration of Abelian differentials and function theory of C. Since the Jacobian variety $J(C)$ is a group, we can construct by addition on the image the nth-Abelian map

$$\int_{np} : C^{(n)} \to J(C) = \mathbb{C}^g/\Lambda$$

$$\int_{np}(c_1 + \ldots + c_n) := \int_p^{c_1}(\omega_1, \ldots, \omega_g) + \ldots + \int_p^{c_n}(\omega_1, \ldots, \omega_g)$$

defined on the symmetric product $C^{(n)} := C^n/Sym(n)$ of C. Abel's Theorem relates the fibers of these maps to rational functions on C. Where the nth Abelian map is injective, it means that there is no rational function associated, but where it is not injective, then the fibre may be identified with a projective space of a finite dimensional family of rational functions on C of degree n. If we denote the image variety by

$$\mathcal{W}_n := \int_{np}(C^{(n)}) \subset J(C)$$

then there is a relationship between the singularities of $\mathcal{W}_n$ and the fibers of $\int_{np}$, which is the content of the Riemann-Kempf singularity theorem. From the degree of the singularity of $\mathcal{W}_n$ at v, we can read off the dimension of the fibre $\int_{np}^{-1}(v)$ and hence of the associated family of rational functions. Furthermore, one can give a geometric description of the tangent cone to $\mathcal{W}_n$ at v in terms of the linear geometry of the canonical embedding of C in $\mathbb{P}^{g-1}$.

This theorem was first noted by Bernhard Riemann (1826-1866) who analyzed $\mathcal{W}_{g-1}$. In this case, $\mathcal{W}_{g-1}$ is a hypersurface, and hence defined by one equation in $J(C)$. This equation may be written as a holomorphic function on $\mathbb{C}^g$ and it is Riemann's famous Theta function.

The jump in the level of discussion, from the explicit integration of algebraic functions to the analysis of the singularities of Riemann's Theta function, has always impressed me. The ability of Riemann of expressing his result with the modest machinery at hand, is also remarkable. The modern algebraic proof by George Kempf [4], using Grothendieck's variational machinery, is also a significant contribution to understanding the Riemann-Kempf theorem. I recommend [4] for deeper reading.

There are 3 sections. In section 1 we show how to pass from an algebraic function to the Riemann surface that it defines. In section 2 we explain Abel's theorem, and finally in section 3 we state the Riemann-Kempf singularity theorem.

These notes were written for the Algebraic Geometry Summer School in Bilkent University, Ankara, Turkey, from the 7 till the 18 of August of 1995. I would like to thank my friend professor Sinan Sertöz for the excellent environment during the school, as well as Professor İhsan Doğramacı for being an extraordinary benefactor of the Arts and Sciences.

1 Integration of Algebraic Functions

1.1 Integration of Elementary Algebraic Functions

The integral of a rational function of 1 variable can always be carried out:

Example 1.1 *If $X(x)$ is a rational function of $x \in \mathbb{R}$ then*

$$\int X dx = Rat(x) + \beta \sum Log(Rat(x)) + \beta \sum ArcTan(Rat(x))$$

for some rational functions $Rat(x)$ and real numbers β.

Proof: Let the rational function X be

$$X(x) = \frac{a_m x^m + \ldots + a_0}{b_n x^n + \ldots + b_0} = Pol(x) + \frac{c_{n-1}x^{n-1} + \ldots + c_0}{b_n x^n + \ldots + b_0}$$

where we have used the Euclidean Algorithm of division. If

$$b_n x^n + \ldots + b_0 = \Pi_j b_n (x - r_j)^{n_j} \Pi_k [(x - \alpha_k)(x - \bar{\alpha}_k)]^{m_k},\ r_j \in \mathbb{R},\ \alpha_k \in \mathbb{C} - \mathbb{R}$$

is the factorization of the denominator of X, then we can further expand X into partial fractions as

$$X(x) = Pol(x) + \sum_{j, 0<j_1 \leq n_j} \frac{d_{j,j_1}}{(x - r_j)^{j_1}} + \sum_{k, 0<k_1 \leq m_k} [\frac{e_{k,k_1}}{(x - \alpha_k)^{k_1}} + \frac{\bar{e}_{k,k_1}}{(x - \bar{\alpha}_k)^{k_1}}]$$

with $d_{j,j_1} \in \mathbb{R}$ and $e_{k,k_1} \in \mathbb{C}$. So

$$\begin{aligned} \int X(x) dx &= Rat(x) + \sum_j \int \frac{d_{j,1}}{(x - r_j)} dx + \sum_k \int [\frac{e_{k,1}}{(x - \alpha_k)} + \frac{\bar{e}_{k,1}}{(x - \bar{\alpha}_k)}] dx \\ &= Rat(x) + \sum_j d_{j,1} Log(x - r_j) + \sum_k Re(e_{k,1}) Log(|x - \alpha_k|^2) \\ &\quad + 2 \sum_k Im(e_{k,1}) Arg(\alpha_k - x) \end{aligned}$$

And finally

$$Arg(\alpha_k - x) = ArcTan(\frac{Im(\alpha_k)}{Re(\alpha_k) - x})$$

□

More generally we have:

Example 1.2 *If $X(x)$ is a rational function of x and $\sqrt{1 - x^2}$ then*

$$\begin{aligned} \int X(x) dx &= Rat(x, \sqrt{1 - x^2}) + \sum \beta Log(Rat(x, \sqrt{1 - x^2})) \\ &\quad + \sum \beta ArcTan(Rat(x, \sqrt{1 - x^2})) \end{aligned}$$

Figure 1: Rational parametrization of $x^2 + y^2 - 1 = 0$

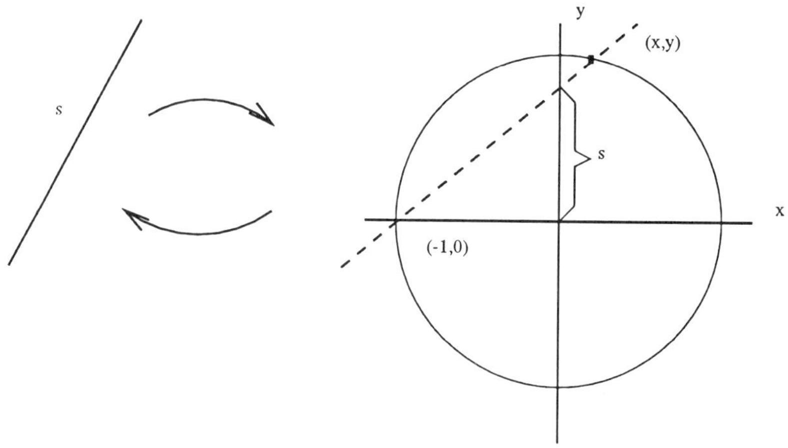

Proof: $y = \sqrt{1-x^2}$ if and only if $x^2 + y^2 = 1$. Let C be the unit circle. There is a rational parametrization of C by rational functions:

$$s \in \mathbb{R} \to (x,y) = (\frac{-s^2+1}{s^2+1}, \frac{2s}{s^2+1}) \in C \qquad (x,y) \in C \to s = \frac{y}{x+1}$$

Doing a change of variables in the integral and applying Example 1.1 we obtain:

$$\int X(x, \sqrt{1-x^2})dx = \int X(x,y)dx = \int X(\frac{-s^2+1}{s^2+1}, \frac{2s}{s^2+1}))(\frac{-4s}{(s^2+1)^2})ds$$

$$= Rat(s) + \sum \beta Log(Rat(s)) + \sum \beta ArcTan(Rat(s))$$

$$= Rat(\frac{y}{x+1}) + \sum \beta Log(Rat(\frac{y}{x+1})) + \sum \beta ArcTan(Rat(\frac{y}{x+1}))$$

$$= Rat(\frac{\sqrt{1-x^2}}{x+1}) + \sum \beta Log(Rat(\frac{\sqrt{1-x^2}}{x+1})) + \sum \beta ArcTan(Rat(\frac{\sqrt{1-x^2}}{x+1}))$$

□

Example 1.3 *If $X(x)$ is a rational function of x and $\sqrt{x^2+bx+c}$ then*

$$\int X(x)dx = Rat(x,\sqrt{x^2+bx+c}) + \sum \beta Log(Rat(x,\sqrt{x^2+bx+c}))$$

$$+\sum \beta ArcTan(Rat(x,\sqrt{x^2+bx+c}))$$

Proof: Parametrize $y^2 = x^2+bx+c$ with a rational function of $s \in \mathbb{R}$ and repeat the argument in Exercise 1.2. $\square$

Trying to integrate more complicated algebraic functions, we almost never get such simple answers, because $\int X(x)dx$ is a much too complicated function to admit a representation in terms of elementary expressions. Classically the elliptic integrals

$$\int \frac{1}{\sqrt{1-x^4}}dx$$

were found not to be integrable in terms of elementary functions. To see this, it was convenient to extend x to a complex variable.

1.2 Basic Facts of Functions of a Complex Variable and Their Integrals

$\mathbb{R}^2$ and $\mathbb{C}$ are naturally and canonically isomorphic, via $(x,y) \to x+iy$.

The $\mathbb{C}$-linear map $\mathbb{C} \to \mathbb{C}$ obtained by multiplying with the complex number $a+ib$, corresponds to the $\mathbb{R}$-linear map $\mathbb{R}^2 \to \mathbb{R}^2$ with matrix representation $\begin{pmatrix} a & -b \\ b & a \end{pmatrix}$.

The $\mathbb{R}$-linear map $\mathbb{R}^2 \to \mathbb{R}^2$ with matrix representation $\begin{pmatrix} a & c \\ b & d \end{pmatrix}$ corresponds to a $\mathbb{C}$-linear map $\mathbb{C} \to \mathbb{C}$ if and only if $a = d$ and $c = -b$.

A C^1-function $f = (f^1, f^2) : U \to \mathbb{R}^2$, U open in R^2 is **holomorphic** if and only if its linear approximations at every point of U correspond to $\mathbb{C}$ linear map, i.e. if and only if it satisfies the Cauchy-Riemann differential equations

$$\frac{\partial f^1}{\partial x} = \frac{\partial f^2}{\partial y} \qquad \frac{\partial f^1}{\partial y} = -\frac{\partial f^2}{\partial x}$$

Let U be an open set in $\mathbb{R}^2$ and $A, B : U \to \mathbb{R}$ be C^1-functions on U. A **real 1-form** is an expression of the form $\omega = Adx + Bdy$. If $\gamma = (\gamma^1, \gamma^2) : [a, b] \to U$ is a piecewise smooth curve in U, the **integral of the 1-form ω over the curve γ** is by definition:

$$\int_\gamma \omega = \int_a^b [A(\gamma(t))\frac{d\gamma^1}{dt}(t) + B(\gamma(t))\frac{d\gamma^2}{dt}(t)]dt \qquad (1.1)$$

If $A, B : U \to \mathbb{C}$ are C^1-functions on U, then $\omega = Adx + Bdy$ is a **(complex valued) 1-form**, and the integral of ω over the curve γ is defined again by (1.1).

We can write complex valued differential forms in a different way using a new basis:

$$\mathbb{C}dx \oplus \mathbb{C}dy = \mathbb{C}dz \oplus \mathbb{C}d\bar{z}$$

where we have the linear relations between the basis:

$$dz = dx + idy \qquad d\bar{z} = dx - idy$$

$$dx = \frac{dz + d\bar{z}}{2} \qquad dy = \frac{dz - d\bar{z}}{2i}$$

By definition, we then have

$$\int_\gamma Adz + Bd\bar{z} = \int_\gamma A(dx + idy) + B(dx - idy)$$

The **differential** $d\omega$ **of a** C^1 **1-form** ω is the 2-form

$$d\omega = [\frac{\partial B}{\partial x} - \frac{\partial A}{\partial y}]dx \wedge dy$$

Recall that 2 forms may be integrated over open subsets of $\mathbb{R}^2$.

Stokes' Theorem *If ω is a C^1-form on the open set U in $\mathbb{R}^2$ and if V is an open set with compact closure in U and having a boundary ∂V formed by a finite number of smooth curves, then*

$$\int_{\partial V} \omega = \iint_V d\omega$$

Lemma 1.4 *Let $f = f^1 + if^2$ be a complex valued C^1-function on $U \subset \mathbb{R}^2$, then $d(fdz) = 0$ if and only if f is holomorphic in U.*

Proof:

$$d(fdz) = d((f^1 + if^2)(dx + idy)) = d((f^1 + if^2)dx + (-f^2 + if^1)dy)$$

$$= [\frac{\partial}{\partial x}(-f^2 + if^1) - \frac{\partial}{\partial y}(f^1 + if^2)]dx \wedge dy =$$
$$[(-\frac{\partial f^2}{\partial x} - \frac{\partial f^1}{\partial y}) + i(\frac{\partial f^1}{\partial x} - \frac{\partial f^2}{\partial y})]dx \wedge dy$$

Hence the result follows from the Cauchy-Riemann equations. □

A **holomorphic 1-form** fdz on $U \subset \mathbb{C}$ is a differential form with f a holomorphic function on U. The Cauchy-Riemann differential equations (embedded in Lemma 1.4) and Stokes' Theorem blend to give:

Cauchy's Lemma *If ω is a holomorphic 1-form on U and V is an open set with compact closure in U and boundary a finite number of C^1-curves we have*

$$\int_{\partial V} \omega = 0$$

Corollary 1.5 *If fdz is a holomorphic 1-form on a disc $D \subset \mathbb{C}$ and $\gamma : [a,b] \to D$ is a C^1 curve in D then $\int_\gamma fdz$ depends only on the extreme points $\gamma(a), \gamma(b)$ of γ.*

We say that 2 curves $\gamma_0, \gamma_1 : [a,b] \to U$ are **homotopic** if:

1) They have the same extreme points $\gamma_0(a) = \gamma_1(a)$ and $\gamma_0(b) = \gamma_1(b)$;

2) There is a continuous deformation of γ_0 to γ_1 keeping the extreme points fixed: i.e. there exists a continuous map $\Gamma : [a,b] \times [0,1] \to U$ with $\Gamma(\ ,0) = \gamma_0$, $\Gamma(\ ,1) = \gamma_1$ and for every $s \in [0,1]$ we have $\Gamma(a,s) = \gamma_0(a)$ and $\Gamma(b,s) = \gamma_0(b)$.

Corollary 1.6 *If fdz is a holomorphic 1-form on an open set $U \subset \mathbb{C}$ and $\gamma_0, \gamma_1 : [a,b] \to D$ are C^1 curves in U which are homotopic then*

$$\int_{\gamma_0} fdz = \int_{\gamma_1} fdz$$

We may define the **integral of a holomorphic 1-form** $f(z)dz$ as the function

$$z \to \int_p^z f(w)dw$$

where we carry out the integration over a path from p to a point z_0 and then make small extensions of this path to nearby point z. We have

$$\frac{d}{dz}\int_p^z f(w)dw = f(z)$$

If we choose another path of integration from p to z_0 the function we obtain differs from the preceding one by adding a constant.

Cauchy's integral formula : *Let $f : U \to \mathbb{C}$ be a holomorphic function, $\{|z - z_0| \leq r\} \subset U$, then for $|z - z_0| < r$ we have*

$$f(z) = \frac{1}{2\pi i}\int_{|z-z_0|=r} \frac{f(w)}{w - z} dw$$

Corollary 1.7 *Under the above hypothesis we have*

*1)***Power series expansion:** *For $|z - z_0| < r$ we have*

$$f(z) = \sum_{n=0}^{\infty} a_n(z - z_0)^n \qquad a_n = \frac{1}{n!}f^n(z_0) := \frac{1}{n!}\frac{d^n f}{dz^n}(z_0)$$

2) **Mean Value Property:**

$$f(z_0) = \frac{1}{2\pi}\int_0^{2\pi} f(z_0 + re^{i\theta})d\theta$$

Maximum Modulus Principle *If $f : U \to \mathbb{C}$ is holomorphic and W is a compact subset of V, then*

$$max\{|f(z)| \mid z \in W\} = max\{|f(z)| \mid z \in \partial W\}$$

Corollary 1.8 (Laurent series expansion) *If $f : D^* \to \mathbb{C}$ is a holomorphic function defined in the punctured disc $D^* = \{0 < |z - z_0| < r\}$ then for $z \in D^*$ we have*

$$f(z) = \sum_{n=-\infty}^{\infty} a_n (z - z_0)^n$$

Under the above assumptions, we say that z_0 is a **removable singularity of** f if $a_n = 0$ for all $n < 0$. In this case by defining $f(z_0) = a_0$ we have a holomorphic function on $D = \{|z - z_0| < r\}$. z_0 is a **pole** of f if $a_n = 0$ for $n < -N < 0$ and $a_{-N} \neq 0$. We say in this case that $N = ord(f, z_0)$ is the **order of the pole** of f at z_0. If $a_n \neq 0$ for an infinite number of negative n, then we say that f has an **essential singularity** at z_0 (i.e. an infinite number of non-zero terms in the negative Laurent series expansion). A function that is holomorphic except for a discrete set of poles is a **meromorphic function**.

The term a_{-1} is the **residue of** f **at** z_0, and it is denoted by $Res(f, z_0)$.

Example 1.9

The exponential function $e : \mathbb{C} \to \mathbb{C}^* := \mathbb{C} - \{0\}$ is defined by

$$e^z := \sum_{n=0}^{\infty} \frac{1}{n!} z^n = e^{Re(z)}[Cos(Im(z)) + iSin(Im(z))]$$

Example 1.10

The Logarithmic 'function' is the multivalued function defined by the formula

$$Log(z) = \int_1^z \frac{1}{w} dw$$

$Log : \mathbb{C}^* \to \mathbb{C}$. It is easy to see that the logarithmic function is the

Figure 2: The exponential and Logarithmic Functions

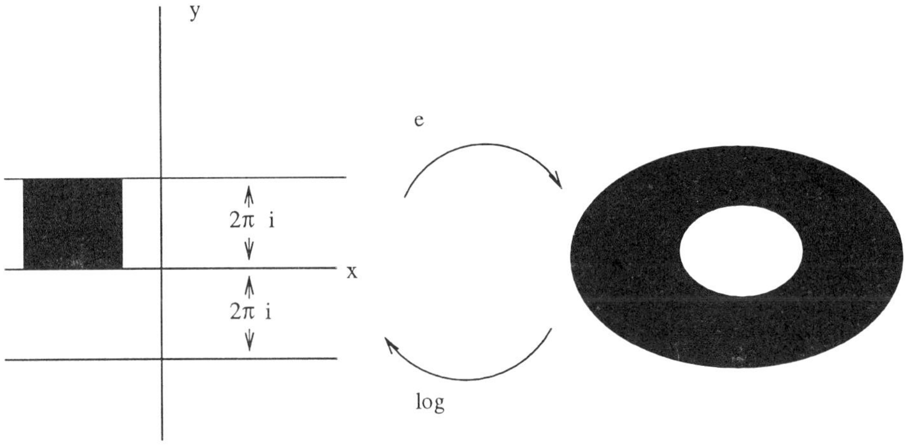

inverse of the exponential function (i.e. $\frac{d}{dz}e^{Log(z)} = 1$), and that the difference between 2 branches is an integer multiple of $2\pi i$.

We also have the related multivalued holomorphic function $Log(z - z_0) : \mathbb{C} - \{z_0\} \to \mathbb{C}$ satisfying

$$Log(z - z_0) = \int_1^z \frac{1}{w - z_0} dw$$

Its inverse is the function $e^w + z_0$.

Corollary 1.11 *Let $f : D^* \to \mathbb{C}$ be a holomorphic function defined in the punctured disc $D^* = \{0 < |z - z_0| < r\}$ with Laurent expansion $f(z) = \sum_{-\infty}^{\infty} a_n (z - z_0)^n$ then $\int f(z)dz$ is a multivalued holomorphic function admitting a representation*

$$\int f(z)dz = \sum_{n \in \mathbb{Z} - \{-1\}} \frac{1}{n+1} a_n (z - z_0)^{n+1} + a_{-1} Log(z - z_0) + Constant$$

Two different branches of $\int f(z)dz$ differ by $2\pi i a_{-1}$(integer), where a_{-1} is the residue of f at z_0. If this residue is 0, then $\int f(z)dz$ is a holomorphic function on D^*.

Figure 3: A multiple connected domain

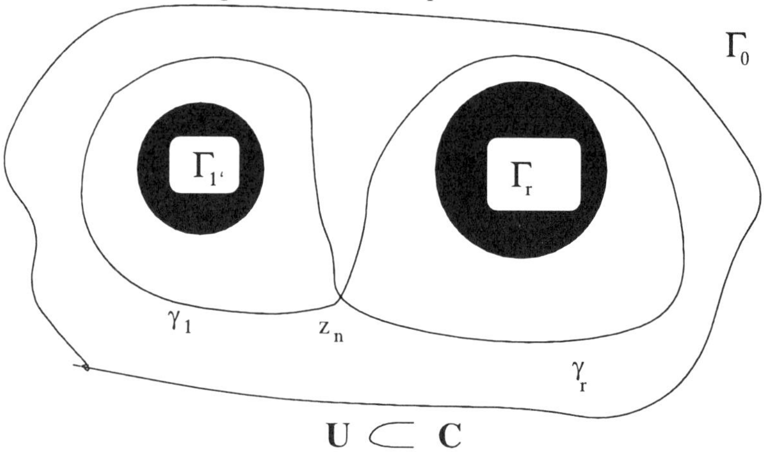

Proof: One may permute the summation of the Laurent series with the integral, due to uniform convergence of the series. Then one integrates the monomials one by one, giving another monomial except for $n = -1$ where one gets a logarithm. □

Let $U \subset \mathbb{C}$ be a multiple connected domain in the complex plane, that is, such that its boundary ∂U is the union of a finite number of simple closed curves $\Gamma_0, \ldots, \Gamma_r$, where Γ_0 corresponds to the exterior boundary. A closed loop is a continuous map from the circle S^1 to U that usually takes a marked point $* \in S^1$ to some marked point $z_0 \in U$. Two closed loops based at z_0 are **homotopic** if we may deform one to the other by means of a continuous 1-parameter family of closed loops based at z_0. The **fundamental group** $\pi_1(U, z_0)$ of U is the group of homotopy classes of closed loops in U based at z_0 and it is a free group generated by loops $\gamma_1, \ldots, \gamma_r$ which go once around only one component Γ_k. The function $\int f(z)dz$ is again a multivalued function. The difference between any two branches is a number $\sum_{k=1}^{r}(integer)\beta_k$ where $\beta_k = \int_{\gamma_k} f(z)dz$ are the **periods** of the integral around the holes.

So we see that $\int f(z)dz$ is multivalued because of 2 different reasons: either because f has a pole with a non-zero residue, or because of the non-trivial topology of U.

We will end this subsection with

Residue Theorem *Let $f : U - \{z_1, \dots, z_r\} \to \mathbb{C}$ be a holomorphic function and $W \subset U$ be an open set containing $\{z_1, \dots, z_r\}$, with compact closure in U and whose boundary consists of a finite number of piecewise smooth curves, then*

$$\int_{\partial W} f(z)dz = 2\pi i \sum_{k=1}^{r} Res(f, z_k)$$

Proof: Apply Cauchy's Lemma to $W - \cup_{k=1}^{r}\{|z - z_k| < \epsilon\}$ and Corollary 1.11 to each punctured disc $\{0 < |z - z_0| < \epsilon\}$. □

1.3 Geometric Facts of Functions of a Complex Variable

Non-zero $\mathbb{C}$ linear maps on $\mathbb{C}$ may be determined as those $\mathbb{R}$-linear maps that are orientation preserving and conformal, in the sense that they preserve angles. Since holomorphic functions may be approximated by $\mathbb{C}$-linear maps, holomorphic maps preserve angles at those points where the derivative is non-vanishing.

Proposition 1.12 *Let $f : U \to \mathbb{C}$ be a holomorphic map and $z_0 \in U$:*

1) If $f'(z_0) \neq 0$, then there exists a neighborhood U_0 of z_0 and U_1 of $f(z_0)$ such that the restriction of f induces a homeomorphism $f : U_0 \to U_1$ with a holomorphic inverse (i.e. f is a local biholomorphism).

2) If $f'(z_0) = \dots = f^{k-1}(z_0) = 0$ and $f^k(z_0) \neq 0$, then there exist neighborhoods U_0 of z_0, U_1 of $f(z_0)$ and a change of variable w in U_0 such that the restriction of f to these neighborhoods and variables has the form $f(w) = a_0 + w^k$.

Proof: 1) Since $det \begin{pmatrix} a & -b \\ b & a \end{pmatrix} = a^2 + b^2$, we may apply the Implicit

Figure 4: Local behaviour of holomorphic maps

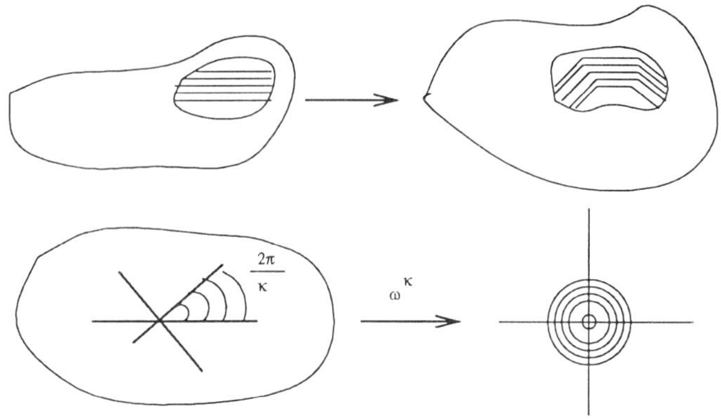

Function Theorem in $\mathbb{R}^2$ to obtain the result.

2) Expanding f in power series

$$f(z) = a_0 + \sum_{n \geq k} a_n (z - z_0)^n = a_0 + (z - z_0)^k [\sum_{n \geq k} a_n (z - z_0)^{n-k}]$$

Let $h(z) = [\sum_{n \geq k} a_n (z - z_0)^{n-k}]^{1/k} = e^{Log[\sum_{n \geq k} a_n (z-z_0)^{n-k}]^{1/k}}$ and set the new variable $w = (z - z_0)h(z)$. Then we have $f(w) = a_0 + (z - z_0)^k h(z)^k = a_0 + w^k$ □

Geometrically this means that locally holomorphic maps are very simple:

A point z_0 of type 2) will be called a **branch point with ramification index** $k := \nu(f, z_0)$.

Example 1.13

The Riemann sphere $\bar{\mathbb{C}} = \mathbb{C} \cup \{\infty\}$ is obtained by gluing a point to the complex plane, as in stereographic projection:

Mathematically, it is simpler to consider 2 copies of the complex plane $(\mathbb{C}, z)$ and $(\mathbb{C}, w)$ each with its own variable and glue $(\mathbb{C}^*, z) \to (\mathbb{C}^*, w)$ with the maps $z \to w = \frac{1}{z}$ and $w \to z = \frac{1}{w}$.

Figure 5: The Riemann Sphere

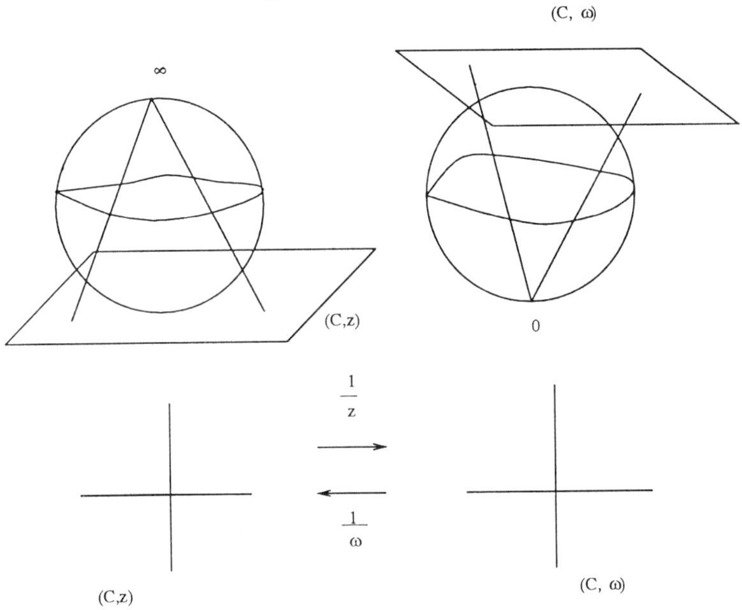

Let U be an open set in $\mathbb{C}$ and $f : U \to \bar{\mathbb{C}}$. f is said to be a **holomorphic map to the Riemann sphere** if:

1) $f : U - f^{-1}(\infty) \to (\mathbb{C}, z)$ is a holomorphic function;

2) $\frac{1}{f} : U - f^{-1}(0) \to (\mathbb{C}, w)$ is a holomorphic function.

Recall that a **meromorphic function** f on the open set U of the complex plane is a holomorphic function $f : U - \Lambda \to \mathbb{C}$, where Λ is a discrete set of points of U, such that f has a pole or a removable singularity at each point of Λ.

Theorem 1.14 *There is a one to one correspondence between meromorphic functions on $U \subset \mathbb{C}$ and holomorphic maps $U \to \bar{\mathbb{C}}$.*

Proof: 1) Assume f is meromorphic on U and let Λ be the poles of f. If $z_0 \in \Lambda$ then

$$f(z) = \sum_{n=-k}^{\infty} a_n (z-z_0)^n = (z-z_0)^{-k} \sum_{n=-k}^{\infty} a_n (z-z_0)^{n+k} = (z-z_0)^{-k} h(z)$$

In the other coordinate of $\bar{\mathbb{C}}$, f is represented by

$$\frac{1}{f(z)} = (z - z_0)^k \frac{1}{h(z)}$$

which is holomorphic since $h(z_0) = a_{-k} \neq 0$

2) Let $f : U \to \bar{\mathbb{C}}$ be a holomorphic map to the Riemann sphere, and let $\Lambda = f^{-1}(\infty)$. By definition, $f : U - \Lambda \to \mathbb{C}$ is holomorphic. For $z \in \Lambda$ we have $\frac{1}{f(z)}$ is holomorphic:

$$\frac{1}{f(z)} = b_k(z - z_0)^k + \ldots = (z - z_0)^k [b_k + \ldots]$$

Hence $f(z) = (z - z_0)^{-k}$[holomorphic]. Hence f has a pole at z_0. □

A **Riemann surface** is a Hausdorff topological space S together with an open covering $\{U_j\}$ of S with homeomorphisms $\phi : U_k \to V_k \subset \mathbb{C}$ onto open sets V_k of $\mathbb{C}$ such that the transition coordinates

$$\phi_{k,j} : \phi_j(U_j \cap U_k) \to \phi_k(U_j \cap U_k)$$

are bijective holomorphic maps.

An example of a Riemann surface is the Riemann sphere $\bar{\mathbb{C}}$.

Concepts which are invariant under biholomorphic maps can be introduced on a Riemann surface by precomposing with the local coordinate charts. For example a **meromorphic function** f on the Riemann surface C is a function $f : C - \Lambda \to \mathbb{C}$, where Λ is a discrete set of points on C such that $f \circ \phi_\alpha^{-1}(z_\alpha)$ is a meromorphic function for $\{(U_\alpha, z_\alpha)\}$ a covering of C by coordinates charts. A **holomorphic map** between Riemann surfaces $f : C \to C'$ is a continuous map such that in local coordinates of domain and codomain it is a holomorphic function. Theorem 1.14 extends to Riemann surfaces, gives a 1-1 correspondence between meromorphic functions on C

Figure 6: Coordinate charts of a Riemann surface

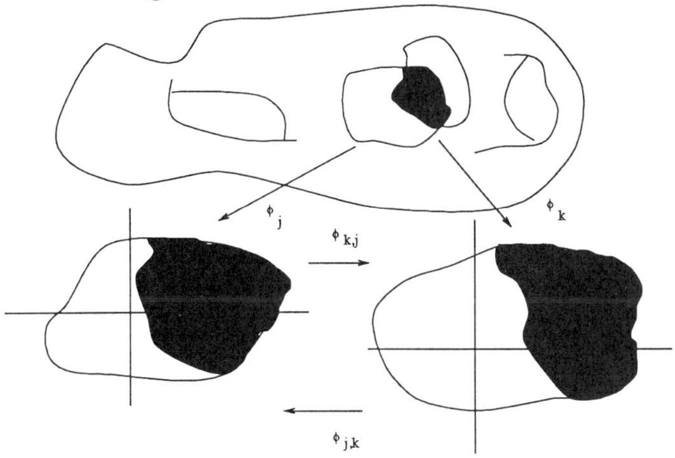

and holomorphic maps from C to the Riemann sphere $\bar{\mathbb{C}}$. A **meromorphic 1-form** ω on C is a collection of meromorphic 1-forms $\omega_\alpha = A_\alpha(z_\alpha)dz_\alpha$ on a covering $\{(U_\alpha, z_\alpha)\}$ such that

$$A_\alpha(z_\alpha) = A_\beta(z_\beta)\frac{dz_\beta}{dz_\alpha}$$

Meromorphic objects defined on compact Riemann surfaces will be called **rational**. Implicitly in this notation, we are using that there is a 1 to 1 correspondence between compact Riemann surfaces and complete smooth complex curves, and this correspondence extends from meromorphic objects defined on the Riemann surface to rational objects defined on the complex curve.

An oriented compact topological surface has a unique topological invariant, the genus g, that can be any non-negative integer $0, 1, , \ldots$. Every such surface is equivalent to a torus with g-handles. Its fundamental group is a free group with 2g generators and such that they satisfy the unique relation

$$[\alpha_1, \alpha_{g+1}] \ldots [\alpha_g, \alpha_{2g}] = id \quad ; \quad [\alpha, \beta] := \alpha\beta\alpha^{-1}\beta^{-1}$$

Recall that the first homology group $H_1(C,\mathbb{Z})$ of the compact surface C of genus g is the abelianization of the fundamental group $\pi_1(C,p)$, and so it is a free abelian group with the same 2g generators as $\pi_1(C,p)$. These generators may be chosen as $\alpha_1,\ldots,\alpha_{2g}$ with the property that α_i intersects α_{g+i} once positively, and do not intersect any other α_k, This basis will be called a **canonical basis**.

Proposition 1.15 *Let $f : C \to \mathbb{C}$ be a holomorphic function defined on the compact and connected Riemann surface C, then f is a constant function.*

Proof: The real valued function $|f| : C \to \mathbb{R}$ attains its maximum value at some point p_0. Take coordinates (U,z) around p_0. By the Maximum Modulus Principle, f is constant on U, but then by analytic continuation, it will be constant on C. □

Recall a classical **Theorem of Euler** that says that if one triangulates a compact orientable surface of genus g, and e_0, e_1, e_2 denotes respectively the number of $0, 1$ and 2 cells then $e_0 - e_1 + e_2 = 2 - 2g$.

Riemann-Hurwitz Formula *Let $f : C \to C'$ be a non-constant holomorphic map between 2 compact and connected Riemann surfaces of genuses g and g', and let B be the ramification points of f. Then $f : C - f^{-1}(f(B)) \to C' - f(B)$ is a finite covering map, say with n sheets, and we have*

$$2 - 2g = n(2 - 2g') - \sum_{q\in B}(\nu(f,q) - 1)$$

where $\nu(f,q)$ is the branching order of f at q.

Proof: Outside the critical values $f(B)$ it follows from Proposition 1.12 that f is a covering map, since it is locally covering at each point and there are only a finite number of points in the inverse image by the compactness hypothesis. The connectedness hypothesis implies that the number of inverse images remains constant, so that f outside of $f^{-1}(B)$ is a covering map with n sheets.

Triangulate C' so that the singular values of f are 0-cells in the

Figure 7: $x^2 + y^2 - 1 = 0, \quad y = \sqrt{1 - x^2}$

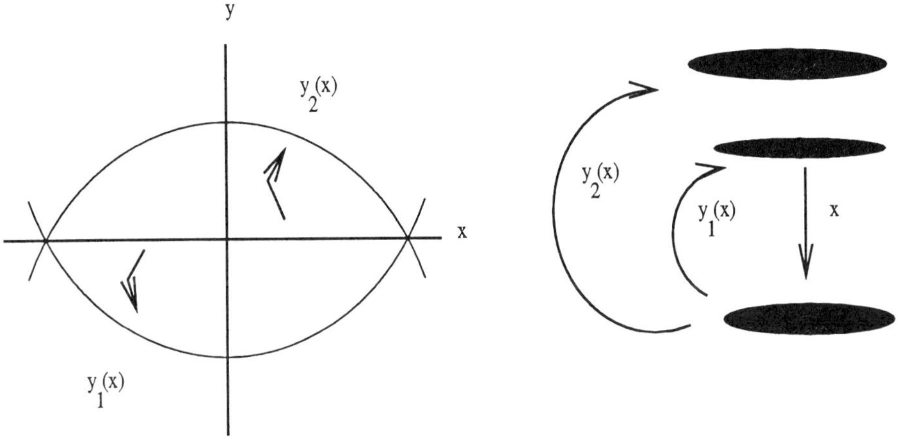

triangulation, and pull back the triangulation to C'. It is immediate that $e'_2 = ne_2$, $e'_1 = ne_1$ and $e'_0 = ne_1 - \sum_{q \in B}(\nu(q, f) - 1)$. By taking the alternating sum and using Euler's Theorem we obtain the formula. □.

1.4 Algebraic Functions

Let $f(x, y) = \sum_{j=1}^{n} a_j(x)y^j$ be an irreducible polynomial in 2 variables x and y with complex coefficients and with positive degree n in y. The height y of a point (x, y) on the Riemann surface C' is by definition an **algebraic function** of the horizontal position.

Algebraic functions are multivalued holomorphic functions, but acquiring only a finite number of distinct values (that is, they have only 'mild' multivaluedness).

Theorem 1.16 *Given an irreducible polynomial* $f(x, y) = \sum_{j=1}^{n} a_j(x)y^j$, $n > 0$, *and* $C' \subset \mathbb{C}^2$ *the complex algebraic curve defined by* $f = 0$, *then there is a unique way to construct a connected compact Riemann surface* C *with 2 rational functions* x *and* y *such that* $(x, y) : C \to C' \subset \mathbb{C}^2$ *is a birational isomorphism; that is, such that any rational function on* C' *may be expressed as a rational*

Figure 8: $x^4 + y^2 - 1 = 0, \quad y = \sqrt{1 - x^4}$

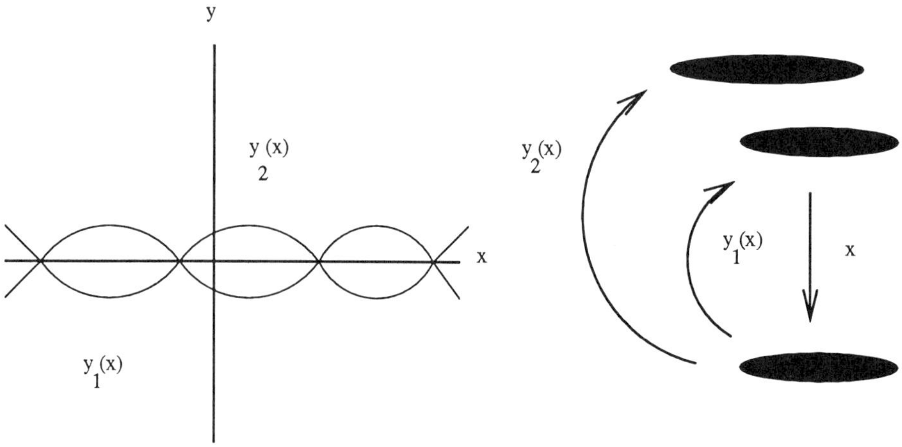

function of x and y.

Proof: 1) There are a finite number of points $B \subset C'$ that satisfy $f = \frac{\partial f}{\partial x} = 0$. Outside of $x(B)$, C' looks locally like the graph of n holomorphic functions by the Implicit Function Theorem (of functions from $\mathbb{R}^4 = \mathbb{C}^2$ to $\mathbb{R}^2 = \mathbb{C}$).

2) If $(x_1, y_1) \in B$ is such that $\frac{\partial f}{\partial y}(x_1, y_1) \neq 0$, we may use y as a local variable of C' in a neighborhood of (x_1, y_1). C' is locally the graph of a function $x(y)$. The map $y \to (x(y), y) \to x(y)$ has a branch point of index $k := \nu(y_1, x(y)) > 1$, so that C' is around (x_1, y_1) the graph of a multivalued function of type $x^{\frac{1}{k}}$.

3) If (x_1, y_1) satisfies $f = \frac{\partial f}{\partial x} = \frac{\partial f}{\partial y} = 0$, then (x_1, y_1) is a singular point of C'. If we intersect C' with a small sphere around (x_1, y_1) we will obtain a finite number of closed loops in S^3 which may be knotted. This implies that locally $C' - \{(x_1, y_1)\}$ has the analytic type of several punctured disc, embedded in a non-trivial way in $\mathbb{C}^2$. We may complete 'abstractly' each punctured disc with a disc to obtain a desingularization of C' at these points.

4) Connectedness follows directly from the irreducibility of f.

5) Carry out the above procedure for the other 3 canonical charts of

$\bar{\mathbb{C}} \times \bar{\mathbb{C}}$ by taking coordinates $x_1 = \frac{1}{x}$ and/or $y_1 = \frac{1}{y}$. Since $\bar{\mathbb{C}} \times \bar{\mathbb{C}}$ is compact and the condition defined by $f = 0$ is closed, we obtain a compact Riemann surface C. □

Given an irreducible polynomial $f(x, y)$ and a rational function $X(x, y)$, let $y(x)$ be the algebraic function defined by $f(x, y) = 0$. The expression $X(x, y(x))$ is a rational function of x and $y(x)$ and it corresponds to a rational function defined on the compact Riemann surface C associated to f in Theorem 1.16. **The integral of a rational function of x and the algebraic function defined by** $f(x, y) = 0$

$$\int X(x, y(x))dx \qquad\qquad f(x, y(x)) = 0$$

is the (multivalued) function obtained by integrating a rational 1-form on the Riemann surface C associated to f.

The procedure of computing the integral $\int X(x, y(x))dx$ of an algebraic function (hence finite multivalued) on $\bar{\mathbb{C}}$ can then be replaced by the integral of a rational 1-form on a compact Riemann surface. In this way we have removed the obstruction of integrating a multivalued function and we can now concentrate our attention on the serious part of the multivaluedness of $\int X(x, y(x))dx$.

2 Abel's Theorem

2.1 The Universal Abelian Integral

Let C be a compact and connected Riemann surface of genus g and ω a meromorphic 1-form on C. The 'function' $\int_p^c \omega : C \to \mathbb{C}$ is in general multivalued, and as we have seen in the previous section, this can happen for 2 different types of reasons:

1) ω has a pole with a non-zero residue (Corollary 1.11).

2) Due to the non-trivial topology of C: i.e. The value of the integral $\int_p^c \omega$ depends on the homotopy class of the path from p to c that we use to carry out the integration, and for any closed curve $\alpha \subset C$ with base point in p, the integral $\int_\alpha \omega$ is the difference between 2 branches of $\int_p^c \omega$, differing one from the other by precomposing the path of integration with α.

In order to concentrate our attention on the second kind of multi-valuedness, we will only integrate special kinds of 1-forms on C:

Definition *An* **Abelian Differential** ω *on the compact and connected Riemann surface C is a holomorphic 1-form on C. Namely, ω may be written locally as $\omega_j dz_j$, with ω_j a holomorphic function, and we have the compatibility relationship $\omega_j = \frac{dz_k}{dz_j}\omega_k$.*

Theorem 2.1 *The space $H^0(C,\Omega_C)$ of Abelian Differentials on a Riemann surface of genus g has dimension g. The De'Rham 1-cohomology $H^1_{Dr}(C) := \frac{closed\ 1-forms}{exact\ -forms}$ is isomorphic to*

$$H^0(C,\Omega_C) \oplus \bar{H}^0(C,\Omega_C)$$

and there is a perfect pairing induced by integration

$$\int : H_1(C,\mathbb{C}) \times H^1_{Dr}(C) \to \mathbb{C}$$

We will accept this Theorem without proof (see [3]). We will just point out its content. The dimension of the space of Abelian differentials is a computation. The description of the De'Rham cohomology group as holomorphic plus antiholomorphic forms is choosing a canonical representative on each cohomology class. This is a consequence of Hodge Theory. The perfect pairing is a consequence of the duality between (singular) homology and cohomology (Poincaré duality). The last assertion uses the above and the isomorphism between singular cohomology and De'Rham cohomology (Differential Topology).

Example 2.2

We want to analyze the integral

$$\int_p^c \frac{dy}{\sqrt{1-x^4}}dx$$

Let $C' \subset \mathbb{C}^2$ be the curve defined by $f(x,y) = x^4+y^2-1 = 0$. The set of points defined by $f = \frac{\partial f}{\partial y} = 0$ are $B := \{(1,0),(i,0),(-1,0),(-i,0)\}$. Outside of these points C' may be defined as the graph of a function $y(x)$. On B we have $\frac{\partial f}{\partial x} \neq 0$, so that C' is a smooth curve. Changing coordinates $x_1 = \frac{1}{x}$ and $y_1 = \frac{1}{y}$ we obtain an ordinary double point on $\bar{\mathbb{C}}' \subset \bar{C} \times \bar{\mathbb{C}}$. This means that the Riemann surface C constructed in Theorem 1.10 is C' union 2 points (over $x = \infty$).

The projection to the x coordinate gives a holomorphic map $x : C \to \bar{\mathbb{C}}$ which is 2 to 1, being branched on B. We claim that C has genus 1. To see this, we apply the Riemann-Hurwitz formula to the above map x: It is 2 to 1, and has 4 branch points of index 2. Since the genus of $\bar{\mathbb{C}}$ is 0 we have $2-2g = 2(2)-4 = 0$, so $g = 1$.

We claim that $\omega = \frac{dx}{y}$ is a holomorphic differential on C (and hence it is a basis for all holomorphic differentials on C). To see this, observe that dx is a meromorphic differential on $\bar{\mathbb{C}}$ that has at ∞ a pole of order 2 (since $d(\frac{1}{x}) = -\frac{1}{x^2}$). Hence $x^*(dx)$ on C is a meromorphic differential on C that has 2 poles of order 2 (at the points over $x = \infty$) and vanishes at the 4 points B of order 1 (the branch points of x). Now y is a meromorphic function on C that vanishes on B and has also poles of order 2 at the 2 points over $x = \infty$. Hence ω is a non-vanishing holomorphic differential on C.

We can write out generators of the fundamental group of C as follows. Consider the arc of the unit circle in the x-plane parametrized by $e^{2\pi it}$, $t \in [0, \frac{1}{4}]$, and its 2 liftings to C': $\tau_1 := (e^{2\pi it}, +\sqrt{1-e^{4\pi it}})$ and $\tau_2 := (e^{2\pi it}, -\sqrt{1-e^{4\pi it}})$. Let α_1 be the loop on C obtained by going first with τ_1 and then returning with τ_2^{-1}. Define α_2 in a similar manner, but with respect to the parameter $t \in [\frac{1}{4}, \frac{1}{2}]$. One checks that this is a basis of $H_1(C,\mathbb{Z})$.

The map $M : \mathbb{C}^2 \to \mathbb{C}^2$ defined by $M(x,y) = (ix,y)$ induces an automorphism of C of order 4, that sends the loop α_1 to α_2 and $M^*\omega = i\omega$. If $\lambda_j := \int_{\alpha_j} \omega$ then by the above we have $\lambda_2 = i\lambda_1$. One can see that $\lambda_j \neq 0$, since otherwise $\int \omega$ would define a holomorphic function on C (uni-valued, since it would be independent of the homotopy class of the curve on which we integrate), but such functions are constant by Proposition 1.15 and so $0 = d(\int \omega) = \omega$, a contradiction.

Denote by $\Gamma = \mathbb{Z}\lambda_1 \oplus \mathbb{Z}\lambda_2 \subset \mathbb{C}$ the additive subgroup generated by λ_1 and $\lambda_2 := i\lambda_1$, which is a discrete subgroup (in the sense of having no accumulation points). The quotient group $\mathbb{C}/\Gamma$ is a compact topological group, which also has the structure of a Riemann surface, with the group structure being holomorphic. Since the integral $\int_p^c \omega$ is well defined modulo the addition of an element of Γ, we obtain a well defined holomorphic map

$$\int_p^c : C \to \mathbb{C}/\Gamma$$

One can show that this map is a biholomorphism. □

A key idea in generalizing the above example to $g > 1$ is to compute the Abelian integrals $\int \omega$ simultaneously for all Abelian differentials.

Let $\omega_1, \ldots, \omega_g$ be a basis for the Abelian differentials on the connected and compact Riemann surface of genus g. Consider the multi-valued map

$$\int_p^c (\omega_1, \ldots, \omega_g) : C \to \mathbb{C}^g$$

Let $\alpha_1, \ldots, \alpha_{2g}$ be a canonical basis of generators of $\pi_1(C, p)$ and Γ be the additive subgroup of $\mathbb{C}^g$ defined by

$$\Gamma := \{\sum_{j=1}^{2g} m_j \Lambda_j \ / \ m_j \in \mathbb{Z}\} \quad , \quad \Lambda_j := \int_{\alpha_j} (\omega_1, \ldots, \omega_g)$$

The elements of Γ are the **periods of the Abelian Integrals on C.**

Theorem 2.3 *The additive subgroup $\Gamma \subset \mathbb{C}^g$ is a discrete subgroup, the quotient $\mathbb{C}^g/\Gamma$ is a compact complex manifold of dimension g with an Abelian holomorphic group structure and the map*

$$\int_p^c (\omega_1, \ldots, \omega_g) : C \to \mathbb{C}^g/\Gamma \tag{2.1}$$

is a holomorphic map (univalued).

Proof: It will suffice to show that the $2g$ periods Λ_j of the Abelian integrals are $\mathbb{R}$-linearly independent in $\mathbb{C}^g$. So assume that they satisfy a linear relation:

$$\sum_{j=1}^{2g} r_j \Lambda_j = 0$$

By conjugating we have that for $k = 1, \ldots, g$:

$$\sum_{j=1}^{2g} r_j \int_{\alpha_j} \omega_k = 0 \quad , \quad \sum_{j=1}^{2g} r_j \int_{\alpha_j} \bar{\omega_k} = 0$$

Since by Theorem 2.1 the holomorphic and antiholomorphic 1-forms are a basis of the de'Rham 1-cohomology group and the 1-homology and the de'Rham 1-cohomology form a perfect pairing, we have that the above equalities imply that $\sum_{j=1}^{2g} r_j[\alpha_j] = 0 \in H_1(C, \mathbb{C})$, which implies that $r_j = 0$ since α_j are basis for $H_1(C, \mathbb{Z})$. □

Definition The Jacobian of the compact and connected Riemann surface *C is the compact complex manifold $J(C) := \mathbb{C}^g/\Gamma$ and the map (2.1) is called the* **universal Abelian Integral**

$$\int_p : C \to J(C)$$

As we mentioned in the introduction, the Abelian differential 1-forms on $J(C)$ may be written as $\sum_{j=1}^{g} a_j dz_j$ where a_j are constants (since the only holomorphic functions on $J(C)$ are constants, and the cotangent bundle of $J(C)$ is a direct sum of trivial bundles). Its integral correspond to the multivalued functions on $J(C)$ induced by the linear function $\sum_{j=1}^{g} a_j z_j$ on $\mathbb{C}^g$. These 'constant' differential forms pull back via the universal Abelian integral to the Abelian differentials on C. Hence the integrals of Abelian differentials on C may be decomposed as the composition of an algebraic map and a 'multivalued linear function' on $J(C)$. The fibres of the integral of an Abelian differential

$$\int \sum_{j=1}^{g} a_j dz_j$$

on $J(C)$ corresponds to parallel hyperplanes in $\mathbb{C}^g$ projected to $J(C)$, and may be understood as the interplay of the hyperplane with the periods. We then want to intersect this codimension 1 foliation in $J(C)$ with the image of the universal Abelian integral, which as we mentioned, is an algebraic map. This reduces the problem of understanding the integrals of Abelian differentials to understanding an algebraic map and a linear foliation of codimension 1 in $J(C)$ and how they intersect.

Another image that one has is to consider the 'lift' of the universal Abelian integral to the universal covers

$$\tilde{\int}_{n\tilde{p}}^{\tilde{*}} (\omega_1, \ldots, \omega_g) : \tilde{C} \to \mathbb{C}^g$$

The image curve is an analytic curve in $\mathbb{C}^g$, defined by the vanishing of a finite number of entire functions. This map is not injective, since it factors through the 'maximal Abelian cover' of C, which is the covering of C corresponding to the subgroup of commutators of the fundamental group of C. The image curve is actually biholomorphic to the maximal Abelian cover of C.

2.2 Abel's Theorem

Abel discovered in 1826 a remarkable property of Abelian integrals.

Let C be the compact Riemann surface associated to the polynomial $f(x,y)$ as in Theorem 1.10, and let $\omega = X(x,y)dx$ be a rational 1-form defined on C. Let $y_1(x),\dots,y_n(x)$ be different branches of the algebraic function defined by $f=0$ on a small open set in $U \subset \bar{\mathbb{C}}$, and consider the meromorphic 1-form on U:

$$[X(x,y_1(x)) + \ldots + X(x,y_n(x))]dx \tag{2.2}$$

If by analytic continuation of $\{y_j(x)\}$ in $\bar{\mathbb{C}}$ we return back to U we will obtain that the $\{y_j(x)\}$ are reshuffled, but in any case the meromorphic 1-form in (2.2) on U remains the same, since addition in $\mathbb{C}$ is commutative ($\mathbb{C}$ after all is an 'Abelian' group). This is also true at the ramification points of $y_j(x)$. Hence the expression (2.2) gives rise to a meromorphic (i.e. rational) 1-form in $\bar{\mathbb{C}}$, but by example 1.1 we know all about the integrals of such differential forms:

$$\int_p^{(x,y_1(x))} X(x,y)dx + \ldots + \int_p^{(x,y_n(x))} X(x,y)dx = Easy\ Function(x)$$

The most dramatic case is when $X(x,y)dx$ is an Abelian differential, since there are no Abelian differentials in $\bar{\mathbb{C}}$ (it's genus is 0), we obtain:

Abel's Theorem *If $X(x,y)dx$ is an Abelian differential on the compact Riemann surface C associated to the irreducible polynomial $f(x,y)$ of degree n in y, and $y_1(x),\dots,y_n(x)$ are the branches of the algebraic function defined by f, then*

$$\int_p^{(x,y_1(x))} X(x,y)dx + \ldots + \int_p^{(x,y_n(x))} X(x,y)dx \quad = Constant$$

Figure 9: Sums of local 1-forms

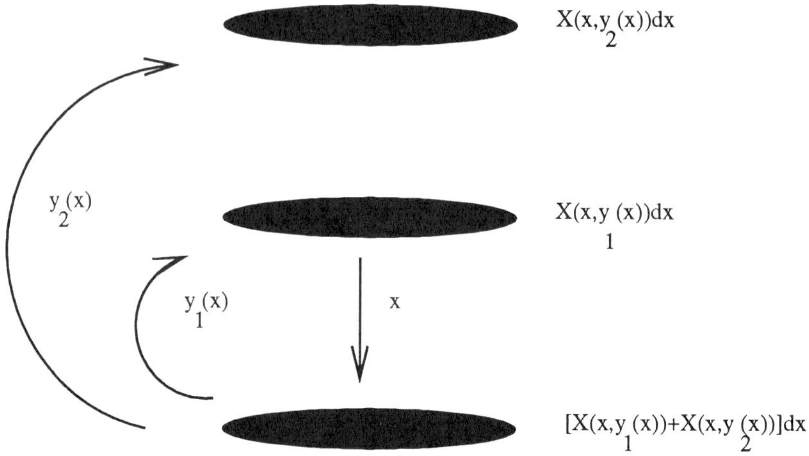

We change slightly our point of view. Let C be a compact and connected Riemann surface and let $f : C \to \mathbb{C}$ be a rational function (or a holomorphic map to $\bar{\mathbb{C}}$). For every $x \in \bar{\mathbb{C}}$ let $\{c_1(x), \ldots, c_n(x)\} = f^{-1}(x)$ be the inverse images of x under f, counting multiplicities. Since we do not have a preferred order for these n-points of C, we use the additive notation $c_1(x)+\ldots+c_n(x)$, where with this notation we are just meaning n-points of C with no preferred order.

Abel's Theorem: *Let ω be an Abelian differential on the compact and connected Riemann surface and f a non-constant rational function on C, with $f^{-1}(x) = c_1(x) + \ldots + c_n(x)$, then*

$$\int_p^{c_1(x)} \omega + \ldots + \int_p^{c_n(x)} \omega = Constant$$

Let's introduce a better language to express Abel's Theorem. For $n \geq 1$ we can define a function

$$\int_{np} (\omega_1, \ldots, \omega_g) : C \times \ldots \times C \to J(C)$$

Figure 10: An illustration of Abelian sums

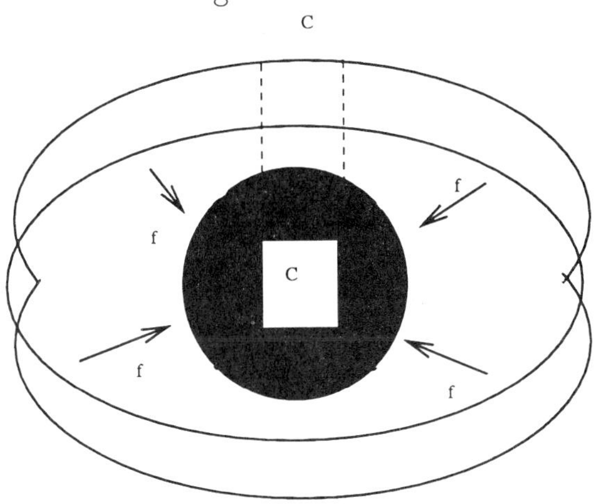

$$(c_1, \ldots, c_n) \mapsto \int_p^{c_1} (\omega_1, \ldots, \omega_g) + \ldots \int_p^{c_n} (\omega_1, \ldots, \omega_g)$$

Actually, if $i : X \to G$ is a function from a set X to an abelian group G, we can define a function $i^n : X \times \ldots \times X \to G$ by $i^n(x_1, \ldots, x_n) = i(x_1) + \ldots + i(x_n)$ by using the group structure in the image. If we denote by Sym_n the group of permutations on n letters, the map i^n is invariant under the action of Sym_n, and hence gives rise to a map defined on the orbit space $i^{(n)} : \frac{X^n}{Sym_n} \to G$. The set $\frac{X^n}{Sym_n}$ is called the symmetric product of X and is denoted by $X^{(n)}$.

For our case of interest, let $C^{(n)}$ be the nth-**symmetric product of the Riemann surface** C.

Lemma 2.4 *Let C be a compact and connected Riemann surface and f a non-constant rational function on C then:*

1) $C^{(n)}$ is a compact complex manifold of dimension n,

2) There is an integer n (called the degree of the rational function f) and a holomorphic map $f^{-1} : \bar{\mathbb{C}} \to C^{(n)}$ defined by $x \in \bar{\mathbb{C}} \to f^{-1}(x)$.

Proof: : 1) We will begin by observing that $\mathbb{C}^{(n)}$ is biholomorphic to the projective space $\mathbb{P}^n$. This is true, since given a polynomial in t with complex coefficients of degree less than or equal to n we can associate to it its roots (with multiplicities), that will be an 'unordered' set of complex numbers. If the degree is less than n, we just complete putting ∞ with multiplicity so that the total number is n. Such polynomials have $n+1$ complex coefficients and they determine the same roots if and only if they differ by a non-zero constant. In all, we see that $\bar{\mathbb{C}}^{(n)}$ may be identified with the projective space of polynomials of degree less than or equal to n. Taking coordinate charts of the Riemann surface C, we can use the above computation locally to conclude that $C^{(n)}$ is a complex manifold of dimension n.

2) It follows from Theorem 1.14 that rational functions are holomorphic maps to the Riemann sphere $\bar{\mathbb{C}}$. In the proof of the Riemann-Hurwitz formula we saw that a holomorphic map may be interpreted as a branched covering, with a finite number of sheets n. Outside of the critical values $f(B)$, one obtains n holomorphic functions $c_1(x), \ldots, c_n(x)$ describing the different branches. So we have a holomorphic map $c_1(x)+\ldots+c_n(x)$. From the description in Proposition 1.12 of the behaviour of f at the branch points (i.e. coordinates were it takes the form w^k) we see that it extends to a holomorphic function at the critical values of f. □

The nth Abelian map is the holomorphic map:

$$\int_{np} : C^{(n)} \to Jac(C)$$

defined by

$$\int_{np} (c_1 + \ldots + c_n) = \int_p^{c_1} (\omega_1, \ldots, \omega_g) + \ldots + \int_p^{c_n} (\omega_1, \ldots, \omega_g)$$

where as before $\omega_1, \ldots, \omega_g$ is a basis of the Abelian differentials.

Abel's Theorem *Let C be a compact and connected Riemann surface of genus g, f a non-constant rational function of C then the map*

$$\int_{np} \circ f^{-1} : \bar{\mathbb{C}} \to C^{(n)} \to Jac(C) \tag{2.3}$$

is a constant map.

Proof: The 1-forms $\sum_{j=1}^{g} a_j dz_j$ on $\mathbb{C}^g$ descend to holomorphic 1-forms on $Jac(C)$ for any value of the constants a_j. Pull back these holomorphic 1-forms to $\bar{\mathbb{C}}$ via the map (2.3). In local coordinates we may parametrize C by $(h_1(w), \ldots, h_g(w))$, to obtain holomorphic 1-forms $\sum_{j=1}^{g} a_j \frac{\partial h_j}{\partial w} dw$ on $\bar{\mathbb{C}}$. But there are no non-zero holomorphic 1-forms on $\bar{\mathbb{C}}$. Since this is true for any values of a_j, we obtain that $\frac{\partial h_j}{\partial w} = 0$, and hence h is constant. □

It follows also from the above proof of Abel's Theorem that if we have a holomorphic map from a projective space $\mathbb{P}^r \to C^{(n)}$ then it is also transformed by the nth Abelian map $\int_{np}$ to a constant: $\mathbb{P}^r$ is full of lines, since through any pair of points there is a line and lines are sent to a point. So $\mathbb{P}^r$ is contracted to a point by the Abelian map $\int_{np}$. In the next section we will organize in a more efficient way the rational functions on C, we will find some projective spaces in $C^{(n)}$ and we will be able to describe the fibers $\int_{np}^{-1}$ of the nth Abelian map.

2.3 Meromorphic Functions on Compact Riemann Surfaces

Let C be a compact and connected Riemann surface. We will call the points of the symmetric product $C^{(n)}$ **positive divisors of degree n**, and we will write them as $D = n_1p_1 + \ldots + n_rp_r$, with $n_j \in \mathbb{Z}_+$ and $p_j \in C$, with $n_1 + \ldots + n_r = n$ to reflect the multiplicity of the point p_j in the divisor D (i.e. the number of times that the point appears repeated in the divisor D). More generally, the **group of divisor** $Div(C)$ **on** C is the free abelian group generated by the points of C and its elements, called **divisors**, are expressions of the form $D = n_1p_1 + \ldots + n_rp_r$, with $n_j \in \mathbb{Z}$ and $p_j \in C$, where $n_1 + \ldots + n_r$ is called the **degree of the divisor**.

Let f be a rational function on C of degree n (recall from Lemma 2.4 that n is the number of sheets of the associated holomorphic map $C \to \bar{\mathbb{C}}$). The **divisor of zeroes** $(f)_0$ of f is the positive divisor $f^{-1}(0)$, the **divisor of poles** $(f)_\infty$ of f is the positive divisor $f^{-1}(\infty)$ and the **divisor** of f is $(f) = (f)_0 - (f)_\infty$. Divisors of rational functions on C are called **principal divisors**.

We say that the positive divisors $D_1 = m_1p_1 + \ldots + m_rp_r$ and $D_2 = n_1q_1 + \ldots + n_sq_s$ are **linearly equivalent** if and only if there is a rational function f on C and a positive divisor D such that

$$(f)_0 + D = m_1p_1 + \ldots + m_rp_r \qquad (f)_\infty + D = n_1q_1 + \ldots + n_sq_s$$

That is, two positive divisors are equivalent if after cancelling the common factors there is a rational function on C that has poles in one and zeroes on the other. Or better, if $D_1 - D_2 = (f)$ is a principal divisor (i.e. we let the cancelling be automatic in $Div(C)$).

Example 2.5

All divisors of degree n on the Riemann sphere are linearly equivalent, since given $D_1 = m_1p_1 + \ldots + m_rp_r$ and $D_2 = n_1q_1 + \ldots + n_sq_s$ two positive divisors of $\bar{\mathbb{C}}$ of degree n we can construct the rational function

$$\frac{\Pi_{p_j \neq \infty}(t - p_j)^{m_j}}{\Pi_{q_j \neq \infty}(t - q_j)^{n_j}}$$

on $\bar{\mathbb{C}}$ that vanishes on D_1 and has poles on D_2. We had already observed in the proof of Lemma 2.4 that $\bar{\mathbb{C}}^{(n)} = \mathbb{P}^n$ and that there are no Abelian differentials on $\bar{\mathbb{C}}$, so that the nth Abelian map degenerates to a constant map. So its fiber is all of $\bar{\mathbb{C}}^{(n)}$: That is, the fibers of the nth Abelian map are the equivalence classes of divisors (one class in this case) and this fiber is a projective space $\mathbb{P}^n$.

Example 2.6

Let C be now a Riemann surface of genus $g > 1$, and we want to analyze when two divisors of degree 1 are linearly equivalent. Let $p, q \in C$, $p \neq q$ be two distinct points in C and assume that they are linearly equivalent. Hence there is a rational function on C that has zeroes only at p with multiplicity 1 and poles only at q with multiplicity 1. Interpreting f as a holomorphic map $f : C \to \bar{\mathbb{C}}$ to the Riemann sphere, then it has degree 1. Applying Lemma 2.4 we see that f is injective, and also surjective; hence a biholomorphism, which contradicts the hypothesis that the genus $g > 0$. Hence, on Riemann surfaces of $g > 0$ divisors of degree one are not linearly equivalent (except to itself).

Example 2.7

Let now C be a curve of genus $g > 0$ and assume that $D_1 = p_1 + p_2$ is linearly equivalent to $D_2 = q_1 + q_2$. By Example 2.6, we have that $p_j \neq q_k$, and let f be the rational function on C that vanishes on D_1 and has poles on D_2. Viewing $f : C \to \bar{\mathbb{C}}$ as a holomorphic map to the Riemann sphere of degree 2, we can apply the Riemann-Hurwitz formula to obtain

$$2 - 2g = 2(2 - 0) - \sum(\nu(q, f) - 1) = 4 - d$$

where f has $d = 2g + 2$ branch points of order 2. Let $\{\alpha_j\}$ be the critical values of f (i.e the images of the branch points) and assume for simplicity that they are all finite (otherwise, postcompose with a holomorphic automorphism of $\bar{\mathbb{C}}$ moving ∞ to a finite value). One checks that C is biholomorphic to the Riemann surface associated to the polynomial

$$F(x, y) = y^2 - \Pi_{j=0}^{2g+2}(x - \alpha_j)$$

So, there are Riemann surfaces of arbitrary genus $g > 0$ such that the 2nd Abelian map $\int_{2p} : C^{(2)} \to Jac(C)$ defined on the complex 2-dimensional manifold $C^{(2)}$ contracts a line. These Riemann surfaces are called **hyperelliptic**, and it is more difficult to see that only a 'few' Riemann surfaces of $g \geq 3$ are hyperelliptic. Hence, the nature of the Abelian maps can depend on the complex structure of the Riemann Surface C. For non-hyperelliptic Riemann surfaces the 2nd Abelian map is injective and $C^{(2)}$ does not have any copy of $\bar{\mathbb{C}}$ (even though the topological structure of $C^{(2)}$ is independent of the complex structure on C).

Given a positive divisor $D = m_1p_1 + \ldots + m_rp_r$ on the compact and connected Riemann surface C define $Rat(D)$ as the set of rational functions on c that have poles at most on the points $p_1, \ldots, p_r$ with order at p_j at most m_j: $\nu(f, p_j) \leq m_j$.

Proposition 2.8 *1) Linear Equivalence of positive divisors of degree n is an equivalence relation on $C^{(n)}$.*

2) For a positive divisor D of degree n, $Rat(D)$ is a vector space of dimension at most $n+1$.

3) For a positive divisor D of degree n, there is a holomorphic map from the projective space of lines in $Rat(D)$ to $C^{(n)}$

$$Proj(Rat(D)) \to C^{(n)} \qquad f \to (f)_0 - (f)_\infty + D$$

whose image is the set $|D|$ of positive divisors linearly equivalent to D.

Proof: 1) D is linearly equivalent to itself through a constant function, and if f reflects the linear equivalence between D and E, then the rational function $\frac{1}{f}$ reflects the equivalence between E and D.

Assume that D_1 is linearly equivalent to D_2 and that D_2 is linearly equivalent to D_3. Let this be expressed by

$$D_1 - D_2 = (f) \qquad D_2 - D_3 = (g)$$

for suitable rational functions f and g on C. Let F_1 be the positive divisor representing the zeroes of (f) which are also poles of (g), and F_2 the positive divisor representing the zeroes of (g) which are also poles of (f), then

$$\begin{aligned}(fg)_0 &= (f)_0 + (g)_0 - (F_1 + F_2)\\ (fg)_\infty &= (f)_\infty + (g)_\infty - (F_1 + F_2)\\ (fg) &= (f) + (g)\end{aligned}$$

Hence

$$D_1 - D_3 = (D_1 - D_2) + (D_2 - D_3) = (f) + (g) = (fg)$$

2) Let $D = \sum_{j=1}^{r} m_j p_j$ and $V(D)$ be the vector space of dimension n formed by the partial Laurent expansions:

$$V(D) = \oplus_{j=1}^{r}[\oplus_{k_j=1}^{m_j}[\frac{\mathbb{C}}{(z_j - p_j)^{k_j}}]],$$

where z_j are local variables around p_j. Consider the map which associates to a rational function in D the negative terms of its Laurent series expansion at its poles:

$$\rho : Rat(D) \to V(D)$$

An element in the kernel of ρ is a rational function with poles at most on D, but such that its Laurent series expansion at these points is actually a power series expansion. That is, a holomorphic function on C. But by Proposition 1.15 it is a constant function. Hence, by linear Algebra, the dimension of $Rat(D)$ is at most $n + 1$.

3) Let $f_1, \ldots, f_s$ be a basis for $Rat(D)$ and consider the holomorphic function

$$F : [\mathbb{C}^s - \{0\}] \times [C - \cup_{j=1}^r \{p_j\}] \to \mathbb{C}, \quad F(t_1, \ldots, t_s, z) = \sum_{j=1}^{s} t_j f(z)$$

and let Z be the algebraic subvariety of $[\mathbb{C}^s - \{0\}] \times [C - \cup j = 1^r \{p_j\}]$ defined by $F = 0$. The projection to the first factor $[\mathbb{C}^s - \{0\}]$ is a finite map, so it will have a behaviour of a 'branched covering'. Using local coordinates and properties of the behaviour of the roots of polynomials with respect to parameters (see [3]) one obtains local holomorphic maps to $C^{(m)}$. To understand the behaviour of Z near the points of D, one considers coordinates around p_j. The family of maps F has a 'general' order of a pole at a point p_j and this drops precisely when there is a point of Z approaching. Analyzing this situation carefully gives the result. (Actually it is easier to prove this part using holomorphic sections of a line bundle by cancelling the poles by multiplying by a holomorphic section of the line bundle. The result will then follow since the similar variety Z will be proper and finite over $\mathbb{C}^s - \{0\}$). Since the holomorphic map to $C^{(n)}$ is independent of multiplying by a non-zero constant, we obtain that it actually is defined on $Proj(Rat(D))$. $\square$.

Our previous observations then gives the final form of:

Abel's Theorem *Let D be a positive divisor of degree n, and let $|D|$ be the projective space of divisors linearly equivalent to D in $C^{(n)}$, then the nth Abelian map $\int_{np}$ sends $|D|$ to a point.*

A proof of the converse of Abel's Theorem may be found in [3] p.235. It asserts that the fibers of the nth Abelian map $\int_{np}$ are the different complete linear series.

3 The Riemann-Kempf Singularity Theorem

3.1 The Canonical Curve

Let C be a compact and connected Riemann surface of genus $g \geq 2$ and let $\omega_1, \ldots, \omega_g$ be a basis of the Abelian differentials on C. The rational functions obtained by dividing the Abelian differentials with ω_1 defines a holomorphic map

$$(\frac{\omega_2}{\omega_1}, \ldots, \frac{\omega_g}{\omega_1}) : C - \{\omega_1 = 0\} \to \mathbb{C}^{g-1}$$

This map extends to a holomorphic map from C to projective space $\mathbb{P}^{g-1}$. In local coordinates (U_α, z_α) the differentials may be written as $\omega_j = A_j^\alpha(z_\alpha)dz_\alpha$ and in the intersection of coordinates we obtain the relations $A_j^\beta = A_j^\alpha \frac{dz_\alpha}{dz_\beta}$. Hence, if we define maps to projective space

$$U_\alpha \to \mathbb{P}^{g-1} \qquad z_\alpha \to (A_1^\alpha(z_\alpha) : \ldots : A_g^\alpha(z_\alpha))$$

then they coincide in the intersection. The map obtained by gluing these local maps

$$i_C : C \to \mathbb{P}^{g-1}$$

is called the **canonical map of** C and the image curve is called the **canonical curve**.

The geometry of the canonical curve $i_C(C)$ contains many secrets of the (rational) function theory of the Riemann surface C. We begin to describe **the linear geometry of the canonical curve**. The canonical map is injective if C is not hyperelliptic, and we identify C with its image $i_C(C)$. The canonical curve C is not contained in any hyperplane in $\mathbb{P}^{g-1}$, since $\omega_1, \ldots, \omega_g$ is a basis of the space of Abelian differentials on C. Let H be a hyperplane in $\mathbb{P}^{g-1}$. The intersection of H with the canonical curve is a divisor on C of degree $2g-2$. The divisors that one obtains by intersecting with all hyperplanes is the complete linear series called the **canonical linear series** $|\kappa|$ and it correspond to the divisors of degree $2g-2$ that are zero-sets of Abelian differentials on C.

Let $D = p_1 + \ldots + p_n$ be a positive divisor in C with n distinct points. The **span** $< D >$ of D is the smallest linear subspace of $\mathbb{P}^{g-1}$ that contains the points $p_1, \ldots, p_n$. If the points are independent, then the dimension of the span is $n-1$, otherwise it is $n-1-r$ for some positive integer r. r measures the number of independent linear relations of the points $\{p_i\}$.

Geometric Riemann-Roch Theorem *If C is a non-hyperelliptic curve and $D = p_1 + \ldots + p_n$ is a positive divisor whose span $< D >$ has dimension $n-1-r$, then the complete linear system $|D|$ has dimension r.*

For a proof see [3], p.248.

Corollary 3.1 *If C is a non-hyperelliptic curve and $D = p_1 + \ldots + p_n$ is a positive divisor then:*

1) If the span $< D >$ of D is $\mathbb{P}^{g-1}$, then the dimension of the complete linear series $|D|$ is $n-g$.

2) If the span $< D >$ of D has dimension less than $g-1$, let E be a positive divisor with $D + E \in |\kappa|$, then

$$dim(|D|) = codim(< E >, \mathbb{P}^{g-1}) - 1$$

3) If $D' \in |D|$ then $dim < D > = dim < D' >$.

Proof: 1) If D spans all $\mathbb{P}^{g-1}$ (necessarily $n \geq g$), the Riemann-Roch Theorem says exactly that the complete linear system $|D|$ has dimension $n-g$.

2) If on the other hand the span of D is not all $\mathbb{P}^{g-1}$ let H be a hyperplane containing D. The intersection of H with the curve C is $C \cap H = D + E$, where E is a positive divisor of degree $2g - 2 - n$. We have to prove that the span of E has codimension $r + 1$, where $r = \dim|D|$. To see this, let H' be a hyperplane containing $< E >$, then. $H' \cap C = D' + E$. By construction, $D' \in |D|$. Conversely, given an element $D' \in |D|$ then $E + D' \in |\kappa|$ and so there is a hyperplane containing E that intersects C on $D' + E$. So there is a one to one correspondence between hyperplanes containing E and elements in $|D|$.

3) Apply 2). □

As 3) shows, if the span of D has dimension $n - 1 - r$, then for any other divisor $D' \in |D|$ we will also have that the span $< D' >$ has dimension $n - 1 - r$. In this manner, we obtain an r-dimensional family of $n-1-r$-linear subspaces of $\mathbb{P}^{g-1}$. The union of these linear subspaces

$$Cone(|D|) = \cup_{D' \in |D|} < D' >$$

gives rise to a variety of $\mathbb{P}^{g-1}$ that can be shown to have dimension $n - 1$ and that we will call the cone of $|D|$.

3.2 The Infinitesimal Geometry of the Abelian maps

Now we turn to show that the canonical curve is related to the infinitesimal geometry of the universal Abelian integral.

Lemma 3.2 *The derivative of the universal Abelian Integral* $\int_p : C \to Jac(C)$ *is the canonical map* i_C.

Proof: Recall first of all that we have chosen (and fixed) a basis $\omega_1, \ldots, \omega_g$ of the Abelian differentials and that $Jac(C) = \mathbb{C}^g/\Lambda$ is a compact Abelian group. This implies that the tangent space of $Jac(C)$ at any point p can be identified with the tangent space $T_0Jac(C)$ at 0 via the derivative of the translation in $Jac(C)$ that sends p to 0. That is, the natural identification that one has of the tangent vector in $\mathbb{C}^g$ descends to give canonical isomorphisms of tangent spaces in $Jac(C)$.

Using local coordinates, then

$$\frac{d}{dz}\int_p^z (A_1(z),\ldots,A_g(z))dz|_{z=z_0} = (A_1(z_0),\ldots,A_g(z_0))$$

The tangent to the curve $\int_p(C)$ at the point $\int_p^c$ can be identified (after translating it to the origin in $Jac(C)$ and taking the point in $Proj(T_0Jac(C))$ which it represents) with $i_C(c)$. □

We can generalize this computation from the universal Abelian integral to any Abelian map $\int_{np} : C^{(n)} \to Jac(C)$

Lemma 3.3 *The derivative of the Abelian map $\int_{np} : C^{(n)} \to Jac(C)$ at the point $D = p_1+\ldots+p_n$ is a linear map $T_DC^{(n)} \to T_{\int_{np} D}Jac(C)$ whose image may be identified to the linear span $< D >$ of D in $\mathbb{P}^{g-1} = ProjT_{\int_{np} D}Jac(C)$.*

Proof: Assuming for simplicity that the points are distinct, then the derivative we want to calculate coincides with the derivative of the map $\int_{np} : C^n \to Jac(C)$ defined on the product of C, instead of the symmetric product, at D. Here, we can take partial derivatives with respect to each factor of C. By Lemma 3.2 each of this partial derivatives corresponds to a point $i_C(p_i)$, and hence the image of the linear map correspond to the span of D. □

3.3 The Riemann-Kempf Singularity Theorem

Define

$$\mathcal{W}_n = \int_{np}(C^{(n)}) \subset Jac(C)$$

as the image variety of the nth Abelian map. It is a compact algebraic variety, possibly with singularities. $\mathcal{W}_n$ has in the structure of its singularities information about the function theory of the curve C:

Riemann-Kempf Singularity Theorem *Let C be a compact Riemann surface of genus $g \geq 2$. For $|D|$ a linear system of degree n and dimension r, the tangent cone to $\mathcal{W}_n$ at $\int_{np}(D)$ is the previously defined cone:*

$$T_{\int_{np}(D)}\mathcal{W}_n = Cone(|D|) = \cup_{D' \in |D|} < D' > .$$

It has degree $\binom{g-n+r}{r}$, *and is swept out once by the planes* $|D'|$.

There are 2 kinds of proofs:

1) Analytic ([1,5]): Lemma 3.3 gives a description of the the derivative of the nth Abelian map, and one continues with this idea.

2) Algebraic: Uses Grothendieck's variational machinery. It is described in full in [4] and we will finish these notes by giving a brief description of its method of proof.

The role of the Jacobian $J(C)$ is taken by the Picard varieties $Pic_n(C)$ which parametrize classes of invertible sheaves on C of degree n (see [2] or [3] for assumed background in what follows). There is a map

$$\int : C^{(n)} \to Pic_n(C) \tag{3.1}$$

which associates to each divisor D the invertible sheaf $\mathcal{O}_C(D)$ of meromorphic functions with poles bounded by D. The image varieties again are denoted by $\mathcal{W}_n$. The fibres of the map (3.1) over $\mathcal{O}_C(D)$ may be identified with the projective space of lines in the space of global sections $H^0(C, \mathcal{O}_C(D))$ of $\mathcal{O}_C(D)$. Hence, the understanding of how the space of global sections $H^0(C, \xi)$ of the invertible sheaf ξ vary with $\xi \in Pic_n(C)$ is related to the problem of the structure of the varieties $\mathcal{W}_n$.

The main reduction is to convert the problem to a problem of 'matrices with parameters'. That is, to show that given $\xi_0 \in Pic_n(C)$ there is an affine neighborhood U of ξ_0 in $Pic_n(C)$ and an $s \times t$ matrix $A = (A_{i,j})$ with coefficients in the ring of regular functions $H^0(U, \mathcal{O}_U)$ such that for every $\xi \in U$ we have

$$dim H^0(C,\xi) = dim Ker A(\xi)$$

$$dim H^1(C,\xi) = dim Coker A(\xi)$$

This in particular shows that the varieties $\mathcal{W}_n$ are determinental varieties, since the condition defining a point in $\mathcal{W}_n$ is that the rank of $A(\xi)$ is one less than maximal, which may be written as the vanishing of all the maximal minors of $A(\xi)$.

The case $n = g - 1$ is the simplest, in the sense that $\mathcal{W}_{g-1}$ has codimension 1, and so is defined by the vanishing of a single function. The above reduction says in this case that A is an $s \times s$ matrix and $det(A)$ is the equation defining $\mathcal{W}_{g-1}$. One can actually find an A such that $s = dim H^0(C, \xi_0)$ and $A(\xi_0) = 0$. Expanding A in terms of regular parameters at ξ_0 we have that the entries of the matrix A begin at least with linear terms, and hence the determinant will begin with terms of degree at least s.

If $s = 1$ then one has to show that A, which is now just 1 function, has a non-trivial linear term in its expansion around ξ_0. The linear terms of A at ξ_0

$$A'(\xi_0) : Tan_{\xi_0}(Pic_{g-1}C) = H^1(C, \mathcal{O}_C) \to Hom_{\mathbb{C}}(H^0(C,\xi_0), H^1(C,\xi_0)) \tag{3.2}$$

can be interpreted as the first order variation of the connected pair of functors (H^0, H^1) ([4] p.81). (3.2) is seen to be the cup product. Now one uses:

*) For any non-zero section $\sigma \in H^0(C,\xi)$, the cup product

$$\sigma\cup\ : H^1(C, \mathcal{O}_C) \to H^1(C, \xi)$$

is surjective.

This is enough to show that the 'first order variation' of A is non-zero, and hence $\mathcal{W}_{g-1}$ is smooth at ξ. And conversely, at any smooth point we must have that $s = dim H^0(C, \xi) = 1$, since we just saw that the degree is at least s. This proves that the non-singular points of $\mathcal{W}_{g-1}$ correspond exactly to invertible sheaves with global sections of dimension 1.

For $s > 1$ one has to show that for the corresponding map (3.2) the subset

$$\{v \in Tan_{\xi_0} Pic_{g-1} C \ \mid \ A'(0)v : H^0(C, \xi_0) \to H^1(C, \xi_0) \\ \text{has nontrivial kernel}\}$$

is not the entire tangent space. This again is carried out using *) above. This shows that $\mathcal{W}_{g-1}$ is defined at ξ_0 by a function whose first non-zero term has degree exactly $s = dim H^0(C, \xi_0)$. This sketches a proof of Riemann's Theorem.

Homological Algebra provide methods to compute 'intrinsic derivatives' using higher order operations, as is shown in (3.2). Note that by using Serre-duality, (3.2) is equivalent to the multiplication map:

$$H^0(C, \xi_0) \otimes H^0(C, \kappa \otimes \xi_0^*) \to H^0(C, \kappa)$$

One of the main advantages of this method, is that it is valid independently if the points of a divisor D are distinct or not.

We finish by writing Riemann's Theta function. Let $\delta_1, \ldots, \delta_{2g}$ be normalized generators of the fundamental group and choose a basis $\omega_1, \ldots, \omega_g$ of the Abelian differentials such that $\int_{\delta_i} \omega_j = \delta_{i,j}$ for $1 \leq$

$i, j \leq g$. If Z denotes $\int_{\delta_i} \omega_j = \delta_{i,j}$ for $g+1 \leq j \leq 2g$ and $1 \leq j \leq g$, then Z is a symmetric matrix with positive definite imaginary part ([3], 232). The Theta function is the function on $\mathbb{C}^g$ defined by the convergent power series

$$\Theta(w) = \sum_{n \in \mathbb{Z}^g} e^{\pi i <n, Zn>} e^{2\pi i <n, w>} \tag{3.3}$$

References

[1] Arbarello, E., Cornalba, M., Griffiths, P., Harris, J.: Geometry of Algebraic Curves, I, Springer Grundlehren series 267, 1985.

[2] Gómez-Mont, X: Meromorphic Functions and Cohomology on a Riemann Surface, in Riemann Surfaces, Proceedings of School on Riemann Surfaces, ICTP, 1987, ed. X. Gómez-Mont, M.Cornalba, A. Verjovsky, World Scientific, Singapur, 1989, 245–301.

[3] Griffiths, Ph., Harris, J.: Principles of Algebraic Geometry, Wiley, 1978, NY.

[4] Kempf, G. Abelian Integrals, Monografias del IMATE-UNAM, 13, 1983, Mexico.

[5] Smith, R. Theta Vanishing in Jacobians in 'Riemann Surfaces', Proceedings of School on Riemann Surfaces, ICTP, 1987 ed. X. Gómez-Mont, M.Cornalba, A. Verjovsky, World Scientific, Singapur, 1989,

MacPherson's Graph Construction

Michał Kwieciński

Uniwersytet Jagielloński,
Instytut Matematyki,
ul. Reymonta 4,
30-059 Kraków, Poland

e-mail: `kwiecins@im.uj.edu.pl`

1 Definition of the Graph Construction.

MacPherson's Grassmannian graph construction is a tool for studying the singularities of a morphism of vector bundles over an algebraic variety. The aim of this lecture is to discuss selected topics concerning this construction and its applications, nb. results contained in the author's thesis [10] (where detailed proofs can be found). We will focus on the definition of the graph construction, its equivalent description using Fitting ideals and some applications to characteristic classes. We will also relate it to the Nash transform using a "modified fibred product".

Consider a morphism $d : E_2 \to E_1$ of algebraic vector bundles over a complex algebraic variety M. Locally d can be represented by a matrix whose entries are regular functions on M. If, at a point of M, the rank of that matrix is smaller than its generic value, we say that that point is in the singular locus of that morphism. Morphisms with non empty singular locus often occur in the study of characteristic classes of singular varieties and of coherent sheaves. The following construction ([11],[1],[5]), has been introduced by MacPherson. Its avatars appear already in [12],[13].

Let e_2 be the rank of E_2 and let us write $G = \mathrm{Grass}_{e_2}(E_2 \oplus E_1)$ for the Grassmann bundle of e_2-dimensional subspaces of the vector bundle $E_2 \oplus E_1$. Consider the embedding ($\mathbf{C}$ is the complex line)

$$\Gamma : M \times \mathbf{C} \to G \times \mathbf{P}^1\mathbf{C},$$

$$\Gamma(x, \lambda) = (graph(\lambda d_x), (1 : \lambda)).$$

Let W be the closure of $\Gamma(M \times \mathbf{C})$. Now W is an algebraic variety, which comes equipped with two natural projections : $\pi : W \to M$, induced by that of the bundle G and $p : W \to \mathbf{P}^1\mathbf{C}$, the restriction of the projection of $G \times \mathbf{P}^1\mathbf{C}$ on its second factor. The scheme $Z_\infty = p^{-1}((0:1)) \subset W$ is canonically embedded in G, by

$$Z_\infty \hookrightarrow G \times \{(0:1)\} \cong G$$

and the restriction of π to Z_∞ coincides with the restriction of the natural projection $G \to M$. The pullback of the tautological bundle on G by the projection $\eta : G \times \mathbf{P}^1\mathbf{C} \to G$ is a bundle on $G \times \mathbf{P}^1\mathbf{C}$ whose restriction to W is denoted Ξ. Set $\xi = \Xi|_{Z_\infty}$.

Definition 1 *The collection of space, maps and bundle (W, π, p, Ξ) will be called the graph construction for the morphism d. The triple $(Z_\infty, \pi|_{Z_\infty}, \xi)$ will be called the scheme at infinity for the morphism d.*

The idea is that the graph construction deforms $M = p^{-1}((1:0))$ to Z_∞, also improving the behaviour of the bundles. We will mainly be interested not in the Z_∞ scheme itself, but in the cycle it induces in the Chow group of M. (See [5], for a description of how a scheme gives rise to a cycle. Its irreducible components are given multiplicities which reflect how "non reduced" they are.)

The data of the graph construction are summarized by the following diagramme.

$$\begin{array}{ccc}
\xi & & \Xi \\
\downarrow & & \downarrow \\
Z_\infty & \hookrightarrow & W \\
\downarrow & & \downarrow \\
M \cong M \times \{(0:1)\} & \hookrightarrow & M \times \mathbf{P}^1\mathbf{C}
\end{array}$$

First, what does the graph construction do to morphisms which do not have any singularity ? This is easily answered (see the proof of lemma 1.1 in Chapter I of [1]). If $d : E_2 \to E_1$ is a morphism of vector bundles, of constant rank, then :

- The embedding $\Gamma : M \times \mathbf{C} \to G \times \mathbf{P}^1\mathbf{C}$ extends to an isomorphism
$$\overline{\Gamma} : M \times \mathbf{P}^1\mathbf{C} \stackrel{\cong}{\to} W .$$
- In particular, the restriction $\overline{\Gamma}|_{M\times\{(0:1)\}}$ gives an isomorphism
$$M \times \{(0:1)\} \stackrel{\cong}{\to} Z_\infty .$$
- Identifying M and $M \times \{(0:1)\}$ one gets
$$(\overline{\Gamma}|_{M\times\{(0:1)\}})^*\xi \cong (\ker d \oplus \operatorname{Im} d).$$

Before examining an example of a singular morphism of vector bundles, we need to be able to calculate the graph construction more explicitly. The following section shows how to do this.

2 Fitting Ideals.

It turns out that the graph construction for a morphism of vector bundles is isomorphic to a blowup of an ideal constructed from the Fitting ideals of the cokernel sheaf of the morphism. This ideal was introduced independently in [16] and [10].

Let $\mathcal{G}$ be a coherent sheaf on M. Its j-th Fitting ideal - $\mathrm{Fitt}_j(\mathcal{G})$ (see [3]) is a coherent sheaf of ideals in $\mathcal{O}_M$, defined locally using a finite presentation ϕ of the sheaf $\mathcal{G}$ on an open subset U of M :

$$\mathcal{O}_U^q \xrightarrow{\phi} \mathcal{O}_U^p \to \mathcal{G}|_U \to 0 .$$

For $j \geq p$ one defines $\mathrm{Fitt}_j(\mathcal{G})|_U = \mathcal{O}_U$ and if $q < p$ then for $0 \leq j < p-q$, on sets $\mathrm{Fitt}_j(\mathcal{G})|_U = (0)$. For $\max\{0, p-q\} \leq j < p$, $\mathrm{Fitt}_j(\mathcal{G})|_U$ is defined as the sheaf of ideals generated by the $(p-j)$-minors of the matrix of ϕ in some chosen bases of the two free modules $\mathcal{O}_U^q$ and $\mathcal{O}_U^p$. It is well known that this definition is independent of the choice of the finite presentation and of the bases.

Let us now consider the variety $M \times \mathbf{P}^1\mathbf{C}$. We will define an ideal $\mathcal{J}(\mathcal{G}) \subset \mathcal{O}_{M\times\mathbf{P}^1\mathbf{C}}$, using the Fitting ideals of $\mathcal{G}$. Let $(\mu : \nu)$ be homogeneous coordinates on $\mathbf{P}^1\mathbf{C}$. Denote $U_0 = \mathbf{P}^1\mathbf{C} - \{(0:1)\}$ and $U_\infty = \mathbf{P}^1\mathbf{C} - \{(1:0)\}$.

Definition 2 *Let $\mathcal{G}$ be a coherent sheaf on M. The ideal $\mathcal{J}(\mathcal{G}) \subset \mathcal{O}_{M\times\mathbf{P}^1\mathbf{C}}$ is defined by*

$$\begin{aligned} \mathcal{J}(\mathcal{G})|_{M\times U_0} &= \mathcal{O}_{M\times U_0} \\ \mathcal{J}(\mathcal{G})|_{M\times U_\infty} &= \sum_{j=0}^{\infty} \left(\left(\frac{\mu}{\nu}\right)^j \right) \mathrm{Fitt}_{j+\mathrm{rk}\,\mathcal{G}}(\mathcal{G}) \mathcal{O}_{M\times U_\infty} \end{aligned}$$

The sum above is just the ideal of sums of elements of the summand ideals. Remark that in a neighbourhood of $x \in M$, the Fitting ideals of index greater than $\dim_{\mathbf{C}} (\mathcal{G}_x/\mathbf{m}_x\mathcal{G}_x) - 1$ are trivial. One deduces, that the sum defining $\mathcal{J}(\mathcal{G})$ is in fact locally finite and that the two expressions of the above definition coincide on $M \times (U_0 \cap U_\infty)$.

Theorem 3 *Let $d : E_2 \to E_1$ be a morphism of vector bundles on M, $\mathcal{G} =$ CoKer d - the cokernel sheaf of this morphism. Then the graph construction W for the morphism d is canonically isomorphic to the blowup of $M \times \mathbf{P}^1\mathbf{C}$ in the ideal $\mathcal{J}(\mathcal{G})$ defined above :*

$$W \cong \mathrm{Bl}_{\mathcal{J}(\mathcal{G})}\left(M \times \mathbf{P}^1\mathbf{C}\right) .$$

By this we understand that also the natural projections of these two spaces on $M \times \mathbf{P}^1\mathbf{C}$ coincide. In particular W depends only on CoKer d.

The proof is done essentially by computing the graph construction in Plücker coordinates and using standard facts about blowups and the vanishing of Fitting ideals.

The above result gives us a straightforward way of computing the scheme at infinity :

Corollary 4 *Under the same assumptions as before, Z_∞ is isomorphic to the total transform of $M \times \{\infty\}$ in the blowup of $M \times \mathbf{P}^1\mathbf{C}$ in the ideal $\mathcal{J}(\mathcal{G})$.*

Describing the graph construction by a blowup has several advantages. First of all, in any concrete example, it allows us to calculate the graph construction effectively. Indeed, the computation of a blowup is the computation of a closure of an image of an algebraic set in a regular map, or equivalently the calculation of a kernel of a morphism of algebras. It can thus be done using Gröbner basis based algorithms.

3 An Example.

The results of the preceding section allow us to study the following example of a singular map of (trivial) vector bundles over $\mathbf{C}$. In fact this example is quite general in the sense that the study of the graph construction for any morphism of vector bundles over a smooth curve can be easily brought down to the study of this example.

Example 5 *We will study a morphism of trivial bundles of the same rank over the affine 1-space* $\mathbf{C}$*, with isolated singularity:*

$$\mathbf{C} \times \mathbf{C}^k \to \mathbf{C} \times \mathbf{C}^k$$

defined by the matrix

$$A = \begin{pmatrix} x^{n_k} & 0 & \cdots & 0 \\ 0 & x^{n_{k-1}} & \cdots & 0 \\ \vdots & \vdots & \ddots & \vdots \\ 0 & 0 & \cdots & x^{n_1} \end{pmatrix}$$

Here x *is the variable of the base* $\mathbf{C}$ *and the* n_i *are positive integers, such that* $n_k > \cdots > n_1 > 0$*. In this case, the cycle* $[Z_\infty]$ *can be expressed as follows :*

$$[Z_\infty] = [V_{k+1}] + n_k[V_k] + \cdots + n_1[V_1] ,$$

where V_{k+1} *projects isomorphically to the base* $\mathbf{C}$ *and* $V_i \cong \mathbf{P}^1\mathbf{C}$ *for* $i = 1, \ldots, k$*. For all* i, j *the intersection* $V_i \cap V_j$ *is empty if* $|i-j| > 1$ *and is transversal in one single isolated point if* $|i-j| = 1$*. In particular* $V_{k+1} \cong \mathbf{C}$ *intersects* V_k *in* 0*. All copies of* $\mathbf{P}^1\mathbf{C}$ *project to* 0*.*

Here is a sketch of the computation. Use Theorem 3 to calculate the Grassmannian graph construction as a blowup. The blowup in question is (over $\mathbf{C} \times U_\infty$) the Proj of the Rees algebra :

$$\begin{aligned} A &= \mathbf{C}[x, \mu, Tx^{n_k+\cdots+n_1}, T\mu x^{n_{k-1}+\cdots+n_1}, \ldots, T\mu^{k-1}x^{n_1}, T\mu^k] \\ &\subset \mathbf{C}[x, \mu, T] \end{aligned}$$

(graded by the formal variable T). The affine open subschemes of that Proj are the Spec of the algebras A_j, each of which is defined as the subalgebra of elements of degree 0 of the localization of A in the element $T\mu^j x^{n_{k-j}+\cdots+n_1}$. One easily shows that A_j is isomorphic to a sub-algebra of $\mathbf{C}[x,\mu,\frac{1}{x},\frac{1}{\mu}]$:

$$A_j \cong \mathbf{C}\left[x,\mu,\frac{x^{n_{k-j+1}}}{\mu},\frac{\mu}{x^{n_{k-j}}}\right] .$$

Expressing A_j as a quotient of a polynomial ring will allow us to see the components of Z_∞ and their multiplicities. This is done as follows : if n, m are positive integers with $n > m \geq 1$, then there is an isomorphism of $\mathbf{C}[x,\mu]$ - algebras :

$$\mathbf{C}[x,\mu,a,b]\big/(\mu - bx^m, ab - x^{n-m}) \xrightarrow{\cong} B = \mathbf{C}\left[x,\mu,\frac{x^n}{\mu},\frac{\mu}{x^m}\right] \quad (*)$$

which sends a to $\frac{x^n}{\mu}$ and b to $\frac{\mu}{x^m}$. Hint:

$$B \cong \mathbf{C}[x,\mu,a,b]\big/(J \cap \mathbf{C}[x,\mu,a,b]) ,$$

where $J = (x\tilde{x}-1,\ \mu\tilde{\mu}-1,\ a-\tilde{\mu}x^n,\ b-\mu\tilde{x}^m) \subset \mathbf{C}[x,\mu,a,b,\tilde{x},\tilde{\mu}]$ and

$$\left\{\tilde{\mu}-\tilde{x}^n a,\ \tilde{x}x-1,\ \mu-bx^m,\ ab-x^{n-m}\right\}$$

is a reduced Gröbner basis of J in the pure lexicographic order such that

$$\tilde{\mu} > \tilde{x} > \mu > a > b > x$$

(it generates J and all the S-polynomials of Buchberger vanish).

Clearly one can apply the isomorphism $(*)$ to $B = A_j$ and in our case the corresponding affine part of Z_∞ is computable by letting $\mu = 0$ in the quotient algebra in $(*)$. A primary decomposition and simple computation of lengths ends the calculation.

4 Analyticity.

Another consequence of using Fitting ideals is the analyticity of the graph construction in the case of an analytic morphism of analytic vector bundles over an analytic space. The blowup of an analytic coherent sheaf of ideals being analytic, from theorem 3 (from its proof to be precise) we deduce the following result

Theorem 6 *For an analytic morphism $d : E_2 \to E_1$, of analytic vector bundles on an analytic space M,* the set *"graph construction" - W defined as at the beginning of this paragraph is an analytic subset of* $\mathrm{Grass}_{e_2}(E_2 \oplus E_1) \times \mathbf{P}^1\mathbf{C}$. *In particular Z_∞, defined also as at the beginning, is a well defined analytic space.*

This generalizes results of Sinan Sertöz ([20], 4.Theorem 1.), ([21], Theorem III.1.) who proved the analyticity of the graph construction in the case, where the base space is compact kählerian, using the Białynicki-Birula decomposition for a $\mathbf{C}^*$ action. Other results of Sertöz about residues of singular foliations [21] are thus immediately generalized.

5 The Modified Fibred Product.

To pursue our study of the graph construction, we need yet another tool - the "modified fibred product". A blowup is an example of a modification. A modification is a proper morphism for which there exists a nowhere dense subset of the target space (called the discriminant set) such that this morphism is an isomorphism outside this set and its inverse image. The term modification is used more frequently in analytic than in algebraic geometry. The results developed in this section are valid in both categories. We will also say that a space is a modification of another space X, if it comes equipped with a morphism to X, which is a modification. The modifications of a fixed variety X form a category (morphisms between them must commute with the modifications).

It is well known, that the fibred product of two modifications is usually not a modification (consider the product of two copies of the same blowup of an algebraic variety). The remedy is to pick some of the components of the fibred product and form a new space. The theorem below explains the precise meaning of this statement.

Theorem 7 *Let Y_1 and Y_2 be two reduced modifications of an algebraic variety X, with discriminant sets D_1 and D_2. Denote by*

π, π_1, π_2 the canonical projections of $Y_1 \times_X Y_2$ on X, Y_1, Y_2 respectively. Define $Y_1 \overset{m}{\times}_X Y_2$ by

$$Y_1 \overset{m}{\times}_X Y_2 = \overline{\left(Y_1 \times_X Y_2 - \pi^{-1}(D_1 \cup D_2)\right)}$$

and the maps ρ, ρ_1, ρ_2 as the restrictions to $Y_1 \overset{m}{\times}_X Y_2$ of π, π_1, π_2 respectively. Then $\left((Y_1 \overset{m}{\times}_X Y_2, \rho), \rho_1, \rho_2\right)$ is the (categorical) product of Y_1 and Y_2 in the category of reduced modifications of X. We will call it the **modified fibred product** *of Y_1 and Y_2 over X.*

Several similar constructions exist in the literature: the *junction* of two morphisms which are compositions of blowups [8],2.9, the *join* of two morphisms of which one is a modification [11], the *strict transform* of a morphism by a blowup [9] and the *meromorphic junction* [22] (applied to "inverses" of modifications).

The two properties below of the modified fibred product, show that it is essentially different from the usual fibred product and allow it to be computed in certain important cases.

- The modified fibred product is idempotent, ie. $Y \overset{m}{\times}_X Y \cong Y$.

- More generally, if there exists a morphism $Y_1 \to Y_2$ of modifications of X, then $Y_1 \overset{m}{\times}_X Y_2 \cong Y_1$.

Theorem 7 would not be true if we had not limited ourselves to considering reduced spaces. Indeed one can show that in the category of all modifications of $\mathbf{C}^2$ the product (as a categorical product) of two copies of the blowup of $\mathbf{C}^2$ at the origin does not exist (show that it would be a subscheme of the usual fibred product and would have to contain a "double", "triple", "quadruple" etc. exceptional divisor).

6 The Nash Transform.

An earlier notion akin to the graph construction needs to be mentioned here. It is the so called Nash transform with respect to a coherent sheaf. For a coherent sheaf $\mathcal{F}$ on X, we will denote by $\mathrm{T}(\mathcal{F})$ its $\mathcal{O}_X$ - torsion. For a morphism $f : Y \to X$, (after [17]) we will denote by $\mathcal{F} \circ f$ the "torsion free analytic inverse image" of $\mathcal{F}$ by the morphism f (the usual analytic inverse image modulo its torsion over $\mathcal{O}_Y$).

In the sequel, all the spaces considered will be irreducible and reduced. Consider all the reduced modifications $\phi : Y \to X$ of X such that $\mathcal{F} \circ \phi$ is a locally free sheaf (it is then of the same rank as $\mathcal{F}$). By [17] or [15], there is a minimal such modification (i.e. every other such modification is a modification of this minimal one), which is called **the Nash transform** of X with respect to $\mathcal{F}$.

Locally, the Nash transform is constructed in [17] as follows. One represents $\mathcal{F}$ as the cokernel of a morphism of vector bundles $d : E_1 \to E_0$. Then d is of constant rank over an open dense set U of the base space. One defines $X_{\mathcal{F}}$ as the closure, in the total space of the the bundle $H = \mathrm{Grass}_{\mathrm{rk}\,\mathcal{F}}(E_0^\vee)$, of the image of the section "kernel of the dual morphism ". In [17] it is shown that $\mathcal{F} \circ \phi_{\mathcal{F}}$ corresponds to the restriction of the dual bundle to the tautological bundle on H.

7 The Distinguished Component.

We now take a closer look at the scheme at infinity. It turns out that it has a distinguished irreducible component. We will describe this component using the Nash transform and the modified fibred product.

One of the components of Z_∞ for a morphism $d : E_2 \to E_1$ is a modification of the base space M. In [1] this distinguished component, which will be denoted $\widehat{M}$ is described as the closure of the section "kernel $\oplus$ image" of the Grassmann bundle:

$$\widehat{M} \hookrightarrow \mathrm{Grass}_{e_2-s}E_2 \times_{\widehat{M}} \mathrm{Grass}_s E_1 \hookrightarrow \mathrm{Grass}_{e_2}(E_2 \oplus E_1) ,$$

$$\widehat{M} = \overline{\ker d|_U \oplus \mathrm{Im}\, d|_U} .$$

Above, s is the generic rank of the morphism d and U is the open dense subset of M where the rank of the morphism is equal to the generic rank. Below, we prove different characterizations of the distinguished component, establishing in particular that it is a modification of the Nash transform of the cokernel sheaf considered. In [1] it is also proved that the union of the irreducible components of Z_∞ other than $\widehat{M}$ projects by $\pi|_{Z_\infty}$ to the singular locus of the map d. In fact, a stronger scheme-theoretic result (below) is true.

Theorem 8 *Let $d : E_2 \to E_1$ be a morphism of vector bundles. Consider the Z_∞ scheme for d. Then*

- *Z_∞ has a distinguished primary component $\widehat{M}$, which is isomorphic to the blowup of M in the ideal* $\mathrm{Fitt}_{\mathrm{rk}\,(\mathrm{CoKer}\, d)}(\mathrm{CoKer}\, d)$,
- *the restriction of the projection $\pi|_{Z_\infty}$ to an isolated primary component of Z_∞, other than $\widehat{M}$, factorizes through the inclusion in M of the zero - subscheme of the ideal* $\mathrm{Fitt}_{\mathrm{rk}\,(\mathrm{CoKer}\, d)}(\mathrm{CoKer}\, d)$,
- *$\widehat{M}$ is isomorphic to the modified fibred product of Nash transforms :*

  $$M_{\mathrm{CoIm}\, d} \overset{m}{\times}_M M_{\mathrm{CoKer}\, d} ,$$

  and the restriction to $\widehat{M}$ of the tautological bundle ξ is linked to the tautological bundles on the Nash transforms by the formula:

  $$(\pi|_{\widehat{M}})^*(E_2 \oplus E_1)\big/(\xi|_{\widehat{M}}) \cong (\mathrm{CoIm}\, d \oplus \mathrm{CoKer}\, d) \circ \pi|_{\widehat{M}} .$$

The first two statements of this theorem are proved by examining the ideal in theorem 3. This theorem implies in particular that the Z_∞ scheme has a distinguished primary component $\widehat{M}$, which is the strict transform (in the sense of [7],II.7.15) of $M \times \{\infty\}$ in the blowup of $M \times \mathbf{P}^1\mathbf{C}$ in the ideal $\mathcal{J}(\mathrm{CoKer}\, d)$. Remark that the restriction of the ideal $\mathcal{J}(\mathrm{CoKer}\, d)$ to $M \times \{\infty\}$ is equal to $\mathrm{Fitt}_{\mathrm{rk}\,(\mathrm{CoKer}\, d)}(\mathrm{CoKer}\, d)$ (and in particular is non zero). To prove the third statement, one would have to delve more deeply into the construction of the Nash transform.

8 The Graph Construction for Sheaves.

A natural habitat of singular morphisms of vector bundles is the locally free resolution of a coherent sheaf. Suppose that M is smooth and let $\mathcal{F}$ be a coherent algebraic sheaf on M. After [2] or [7] (exercise III.6.9), $\mathcal{F}$ has a finite locally free resolution:

$$0 \to \mathcal{E}_k \stackrel{d_k}{\to} \mathcal{E}_{k-1} \stackrel{d_{k-1}}{\to} \ldots \stackrel{d_2}{\to} \mathcal{E}_1 \stackrel{d_1}{\to} \mathcal{E}_0 \stackrel{d_0}{\to} \mathcal{E}_{-1} = \mathcal{F} \to 0 \ . \ (*)$$

This resolution gives a complex of vector bundles (corresponding to the locally free sheaves in the resolution).

$$0 \to E_k \stackrel{d_k}{\to} E_{k-1} \stackrel{d_{k-1}}{\to} \ldots \stackrel{d_2}{\to} E_1 \stackrel{d_1}{\to} E_0 \stackrel{d_0}{\to} E_{-1} = 0 \ , \ (**)$$

This complex is exact over the open dense subset of M on which the sheaf $\mathcal{F}$ is locally free. We will keep the symbols $d_k, \ldots, d_1$ for the induced morphisms of vector bundles. Furthermore, taking some liberties with the notation, $d_0 : E_0 \to E_{-1} = 0$ will denote the zero morphism. Any time an ambiguity is possible we will state precisely whether we have in mind a morphism of sheafs or of bundles. For example there is no ambiguity in the definition of $\mathrm{CoKer}\, d_i$.

Consider the graph constructions (W_i, π_i, p_i, Ξ_i) and the schemes at infinity $(Z_{\infty,i},\ \pi_i|_{Z_{\infty,i}},\ \xi_i)$ for each one of the morphisms d_i of the complex $(**)$.

Definition 9 *The* **graph construction for the sheaf** $\mathcal{F}$ *will be the collection* (W, π, p, Ξ) *defined as follows*

$$W \ = \ W_k \overset{m}{\times}_{M \times \mathbf{P}^1\mathbf{C}} \cdots \overset{m}{\times}_{M \times \mathbf{P}^1\mathbf{C}} W_0 \ ,$$

the map (π, p) will be the natural projection of the modified fibred product on $M \times \mathbf{P}^1\mathbf{C}$. Further, Ξ will be a virtual bundle defined as the formal alternating sum

$$\Xi = \sum_{i=0}^{k} (-1)^i \rho_i^* \Xi_i ,$$

where ρ_i is the projection of the modified fibred product on its i-th factor. The **scheme at infinity for the sheaf** *$\mathcal{F}$ is the triple $(Z_\infty, \pi|_{Z_\infty}, \xi)$, defined by $Z_\infty = p^{-1}((0:1))$ and $\xi = \Xi|_{Z_\infty}$.*

Most interesting for us will be the class of the virtual bundle in the Grothendieck group of algebraic vector bundles (locally free sheaves) K^0. Over a manifold, this is isomorphic to the Grothendieck group of coherent sheaves K_0 (see [2] or [7] exercise III.6.9).

The use of Fitting ideals allow us to see that in fact the graph construction does not depend on the chosen resolution.

Theorem 10 *For a coherent sheaf $\mathcal{F}$ on an algebraic manifold M, the following objects of the graph construction for $\mathcal{F}$ do not depend on the choice of a locally free resolution :*

- *the spaces (W_i, π_i, p_i) of the graph construction for the i-th boundary morphism,*
- *the space (W, π, p) of the graph construction for the sheaf $\mathcal{F}$,*
- *the scheme $(Z_\infty, \pi|_{Z_\infty})$ and the class of the virtual tautological bundle ξ in coherent K-theory $K^0(Z_\infty)$.*

The original definition of the graph construction for a sheaf did not use the modified fibred product, but a simultaneous construction for all morphisms concerned. The above independence theorem first appeared [6] where a slight modification of the graph construction is presented (the bundles of the resolution are twisted). That paper too, uses neither modified fibred products nor Fitting ideals, the use of which, except for the part concerning bundles, allows us to sketch a quick proof of the above theorem :

By theorem 3, W_i is the blowup of $M \times \mathbf{P}^1\mathbf{C}$ in the ideal $\mathcal{J}(\text{CoKer } d_i)$ of the definition 2. Now, by ([3], theorem 20.2), a resolution is locally uniquely determined up to a trivial component. Hence, $\mathcal{J}(\text{CoKer } d_i)$ does not depend on the locally free resolution chosen. This ends the proof.

The Z_∞ scheme for a sheaf also has a distinguished irreducible component $\widehat{M}$ which is a modification of M. The union of the other components of Z_∞ projects by $\pi|_{Z_\infty}$ to the singular locus of the sheaf. In [1] it is described as the closure of a section of the Grassmann bundle made up of simultaneous sums of the kernel and image sections of all morphisms involved in the resolution of the sheaf. The proposition below gives a different description of this component, similar to the one given in the preceding section for the single morphism case. In particular it compares the distinguished component to the Nash transform $M_\mathcal{F}$ of the sheaf and the Nash transforms of its syzygies, comparing also the tautological bundles. This part generalizes the result of [1] (proof of II.2.1), which says that if the support of a sheaf $\mathcal{F}$ is nowhere dense in M, then $\xi_{|\widehat{M}} = 0$.

Proposition 11 *Let $\widehat{M}_i$ be the distinguished component of $Z_{\infty,i}$, for a coherent sheaf $\mathcal{F}$, $\widehat{M}$ the distinguished component of Z_∞. Then*

- $\widehat{M} \;\cong\; \widehat{M}_k \overset{m}{\times}_M \cdots \overset{m}{\times}_M \widehat{M}_0$,
- $\widehat{M} \;\cong\; M_{\text{CoKer } d_k} \overset{m}{\times}_M \ldots \overset{m}{\times}_M M_{\text{CoKer } d_0}$,
- *There exists a canonical modification $\Phi : \widehat{M} \to M_\mathcal{F}$ and in $K^0(\widehat{M})$ we have :*

$$\xi|_{\widehat{M}} \;=\; \Phi^*(\mathcal{F} \circ \Phi_\mathcal{F}) \ .$$

- *The sheaf $\mathcal{F} \circ (\pi|_{\widehat{M}})$ is locally free and, in $K^0(\widehat{M})$, we have : $\xi|_{\widehat{M}} = \mathcal{F} \circ (\pi|_{\widehat{M}})$.*

9 Mather Classes.

As is usual in the study of this kind of problems, by "cohomology", we shall understand one of the classical functorial theories, together with the corresponding homology theory, equipped with the usual formal properties: cup and cap product, projection formula, fundamental class (in homology) of an irreducible algebraic set (such that for a modification, the direct image of the fundamental class of the source space is the fundamental class of the target space). We shall suppose further, that there is a homomorphism from the Chow group (of algebraic cycles modulo rational equivalence) to homology. To a subscheme of an algebraic variety, we associate a fundamental algebraic cycle, as in ([5],1.5) and so a fundamental class in homology, which coincides with the above mentioned fundamental class in the case of an irreducible and reduced subvariety. For example, if we restrict our attention to compact varieties, then singular homology and cohomology will do.

A characteristic class in full generality is a "way to associate" a cohomology class cl^* to a vector bundle, such that for a morphism $f : X \to Y$ and for a vector bundle E on Y, we have the naturality formula $\mathrm{cl}^* f^* E = f^* \mathrm{cl}^* E$. We will also suppose, that this natural transformation takes its values only in even dimensional cohomology. For example the total Chern class in singular cohomology with integer or rational coefficients is a characteristic class in this sense. Likewise, any polynomial in the Chern classes is a characteristic class. Another important example of a characteristic class is the Chern character, see for example ([5], 3.2) or ([14], Problem 16B). which appears in different Riemann-Roch - type theorems [2],[1].

Having a characteristic class defined on vector bundles, we would like to obtain one defined on all coherent sheaves. One way of achieving that is the Mather class construction, so called because it resembles the construction of Chern - Mather classes for singular varieties. This construction works over all varieties (smooth or singular), but has the disadvantage of producing a homology class instead of a cohomology class.

Definition 12 *Let* cl^* *be any characteristic class. It induces a (homology)* **Mather class** cl_**, defined for coherent sheaf* $\mathcal{F}$ *on a variety* X*, by:*

$$\mathrm{cl}_*\mathcal{F} = (\phi_\mathcal{F})_*(\mathrm{cl}^*(\mathcal{F}\circ\phi_\mathcal{F})\cap[X_\mathcal{F}]) .$$

Here $(X_\mathcal{F},\phi_\mathcal{F})$ is the Nash transform of X relative to $\mathcal{F}$ and hence $\mathcal{F}\circ\phi_\mathcal{F}$ is a locally free sheaf (so it makes sense to take its characteristic class). Further $[X_\mathcal{F}]$ is the fundamental (homology) class of $X_\mathcal{F}$ and $(\phi_\mathcal{F})_*$ is the morphism induced in homology by the map $\phi_\mathcal{F}$.

The universal property of the Nash transform allows us to characterize the Mather class in an axiomatic way. For a coherent sheaf $\mathcal{F}$ on a variety X, we have :

- if $\mathcal{F}$ is locally free, then $\mathrm{cl}_*\mathcal{F} = \mathrm{cl}^*\mathcal{F}\cap[X]$,
- $\mathrm{cl}_*\mathcal{F} = \mathrm{cl}_*(\mathcal{F}/\mathrm{T}(\mathcal{F}))$,
- for a modification $f : Y \to X$ $\mathrm{cl}_*\mathcal{F} = f_*\mathrm{cl}_*f^*\mathcal{F}$.

These properties characterize the Mather class cl_* completely.

The above definition concerning Chern - Mather classes applies in particular to the Chern character ch^* and Chern class c^*. inducing a *Chern-Mather character* ch_* and a (total) *Chern-Mather class* c_* - both in homology. These objects have some nice natural properties, which one can prove using the universal property of the Nash transform and the modified fibred product. The key to most proofs is the observation that

$$X_{\mathcal{F}_1\oplus\mathcal{F}_2} \cong X_{\mathcal{F}_1}\overset{m}{\times}_X X_{\mathcal{F}_2} .$$

Here is a sample of properties of Mather classes : For two coherent sheaves $\mathcal{F}_1$ and $\mathcal{F}_2$ on X we have:

$$\mathrm{ch}_*(\mathcal{F}_1\oplus\mathcal{F}_2) = \mathrm{ch}_*\mathcal{F}_1 + \mathrm{ch}_*\mathcal{F}_2.$$

If $\mathcal{F}$ is a coherent sheaf on X and $\mathcal{E}$ a locally free sheaf on X, then

$$\mathrm{ch}_*(\mathcal{E}\otimes\mathcal{F}) = \mathrm{ch}^*\mathcal{E}\cap\mathrm{ch}_*\mathcal{F} .$$

And if we have a short exact sequence of coherent sheaves

$$0 \to \mathcal{F}' \to \mathcal{F} \to \mathcal{E} \to 0 \ ,$$

the sheaf $\mathcal{E}$ being locally free, then

- $\mathrm{ch}_*\mathcal{F} \ = \ \mathrm{ch}_*\mathcal{F}' + \mathrm{ch}_*\mathcal{E}$,
- $\mathrm{c}_*\mathcal{F} = \mathrm{c}^*\mathcal{E} \cap \mathrm{c}_*\mathcal{F}'$.

The big drawback of Chern - Mather classes is that they are not well defined in K-theory. For example the Chern - Mather character of the middle term of a short exact sequence of coherent sheaves need not be equal to the sum of the Chern characters of the lateral terms.

Chern - Mather classes and characters were studied in a slightly different setup by Marie - Hélène Schwartz in [18] and [19]. There, a Nash transform relative to a *linear space* (see [4]) is constructed and is subsequently used to develop the theory. This Nash transform corresponds to the one mentioned above by the anti-equivalence of coherent sheaves and linear spaces and so also these classes also correspond to those defined here.

10 Mather Classes and K-theory Classes.

On a manifold, the Chern class and the Chern character can be extended to natural transformations of K_0 to cohomology, considered as functors to the category of sets. For the Chern character, this transformation can be considered also as a transformation of functors from K-theory to the category of abelian (additive) groups. The Chern class (resp. Chern character) of a sheaf is defined as the alternating product (resp. alternating sum) of the Chern classes (resp. Chern characters) of locally free sheaves appearing in a resolution of the sheaf concerned. We will show how to compare a part of them with corresponding Mather classes.

In fact, any characteristic class cl^* defined on K_0, is in particular a characteristic class on vector bundles only, which we shall also denote cl^*. Hence it induces a homological Mather class cl_*. The theorem below shows that parts of the two classes are equal. It generalizes theorem II.2.1.b) of [1], which states that the dual of the Chern character of a coherent sheaf on an algebraic manifold has a representative with support included in the support of the sheaf (if this support is nowhere dense in the base manifold, then on one hand it is equal to the singular locus of the sheaf and on the other hand its Chern Mather character is zero).

Theorem 13 *Let M be a compact complex manifold. Let cl^* be a characteristic class on the Grothendieck group and cl_* - the corresponding homological Mather class. Then for a coherent sheaf $\mathcal{F}$ on M, the homology class*

$$\mathrm{cl}^*(\mathcal{F}) \cap [M] \; - \; \mathrm{cl}_*(\mathcal{F})$$

has a representative supported in the singular locus of the sheaf $\mathcal{F}$. In particular, if $n = \dim_{\mathbf{C}} M$ and s is the dimension of the singular locus of the sheaf $\mathcal{F}$ then, for $i = s+1, \ldots, n$ we have

$$\mathrm{cl}_i(\mathcal{F}) \; = \; \mathrm{cl}^{(n-i)}(\mathcal{F}) \cap [M].$$

The proof of this theorem makes use of the graph construction (W, π, p, Ξ) for the sheaf $\mathcal{F}$. Like in [1], one remarks that the cycles associated to the schemes $Z_0 = p^{-1}((1:0))$ and Z_∞ are rationally equivalent and so, that

$$\mathrm{cl}^*\Xi \cap [Z_0] \; = \; \mathrm{cl}^*\Xi \cap [Z_\infty] \; ,$$

Then, one studies the two terms of the above equality, after composing with π_*. The left hand side becomes $\mathrm{cl}^*(\mathcal{F}) \cap [M]$ - this can be proved directly. The right hand side becomes $\mathrm{cl}_*(\mathcal{F})$ plus a cycle supported in the singular locus. Essential in the proof of this last statement is the description of the distinguished component of Z_∞ in proposition 11. The distinguished component provides $\mathrm{cl}_*(\mathcal{F})$ and the other components only contribute a cycle supported in the singular locus. Apart from that, the proof makes repetitive use of all the naturality properties involved, projection formulas and universal properties of the Nash transform and modified fibred products.

11 A Vanishing Theorem.

The Chern classes of a vector bundle, of index greater than the rank of the vector bundle vanish. This is no longer true for coherent sheaves. However, Chern - Mather classes are pushforwards of Chern classes of vector bundles and so they do vanish for certain indices. This remark, coupled with Poincaré duality and theorem 13 implies the following theorem on the vanishing of the usual Chern classes of sheaves.

Theorem 14 *Let $\mathcal{F}$ be a coherent sheaf on a compact complex manifold M of (complex) dimension n. Let s be the dimension of the singular locus of $\mathcal{F}$. Suppose $i > \mathrm{rk}\ \mathcal{F}$ and $i < n - s$. Then the i-th Chern class of $\mathcal{F}$ vanishes :*

$$c^i \mathcal{F} = 0 \ .$$

References

[1] A.Baum, W.Fulton, R.MacPherson, *Riemann-Roch for singular varieties,* Publ. Math. I.H.E.S. 45 , pp. 101-145, 1975.

[2] A.Borel et J.-P.Serre, *Le théorème de Riemann-Roch (d'après Grothendieck),* Bull. Soc. Math. France 86 , pp. 97-136, 1958.

[3] D.Eisenbud, *Commutative Algebra with a view toward Algebraic Geometry*, Springer GTM vol. 150, 1995.

[4] G.Fischer, *Complex Analytic Geometry*, Springer LN 538, 1976.

[5] W.Fulton, *Intersection Theory*, Springer Ergebnisse, 3 Folge, Band 2, 1984.

[6] H.Gillet and C.Soulé, An arithmetic Riemann-Roch theorem, Inventiones Math. 110 , pp. 473-543, 1992.

[7] R.Hartshorne, *Algebraic Geometry*, Springer GTM vol. 52, 1993.

[8] H.Hironaka, *La Voute Etoilée,* dans Singularités à Cargèse, Astérisque 7 et 8, 1973 , pp. 415-440.

[9] H.Hironaka, M.Lejeune-Jalabert, B.Teissier, *Platificateur Local en Géométrie Analytique et Aplatissement Local,* dans Singularités à Cargèse, Astérisque 7 et 8, 1973, pp. 441-449.

[10] M.Kwieciński, *Sur le transformé de Nash et la construction du graphe de MacPherson*, doctoral thesis, Université de Provence, Marseille, 1994.

[11] R.MacPherson, *Chern classes for singular algebraic varieties,* Ann. of Math. 100, 1974, pp. 423-432.

[12] R.MacPherson, *Singularities of vector bundle maps* , in *Proceedings of Liverpool Singularities Symposium I (1970)*, ed. C.T.C. Wall, Springer L.N. 192, 1971, pp. 316-318.

[13] R.MacPherson, *Generic vector bundle maps* , in *Dynamical Systems (Proc. Univ. Bahia, Salvador 1971)*, Academic Press, New York, pp. 165-175, 1973.

[14] J.Milnor, *Characteristic Classes*, Annals of Math. Studies No. 76, Princeton Univ. Press.

[15] V.Navarro-Aznar, *Sur les multiplicités de Schubert locales des faisceaux algébriques cohérents,* Compositio Math. 48, 1983, pp. 311-326.

[16] Roberts *Chern classes of matrices over Noetherian rings*, Math. Zeitschrift 216, 1994, pp. 157-168.

[17] O.Riemenschneider, *Characterizing Moišezon Spaces by Almost Positive Coherent Analytic Sheaves,* Mathematische Zeitschrift 123, pp. 263-284, 1971.

[18] M.H.Schwartz, *Classes et Caractères de Chern des Espaces Linéaires,* Pub. Int. Univ. Lille 2 , Fasc. 3, 1980.

[19] M.H.Schwartz, *Classes et Caractères de Chern-Mather des Espaces Linéaires*, CRAS t.295, 1982 , pp. 399-402.

[20] S.Sertöz, *C^*-actions on Grassmann bundles and the cycle at infinity,* Mathematica Scandinavica 62, pp. 5-18, 1988.

[21] S.Sertöz, *Residues of singular holomorphic foliations,* Compositio Mathematica 70, pp. 227-243, 1989.

[22] K.Stein, *Meromorphic Mappings,* L'Enseignement Mathématique, 14, 1968, pp. 29-46.

[23] B.Teissier, *The hunting of invariants in the geometry of discriminants*, in *Real and Complex Singularities, Proc. Oslo*, ed. P.Holm, 1976, Stijthof and Noordhof Int. Publ., 1977.

An Overview of Algebraic Surfaces

Rick Miranda

Department of Mathematics
Colorado State University
Ft. Collins, CO 80523, USA

e-mail: `miranda@riemann.math.colostate.edu`

Preface

These notes are a faithful record of five lectures given at the Summer School on Algebraic Geometry, held at the Bilkent International Center for Advanced Studies of Bilkent University, in August of 1995. The intention of the lectures was to give a quick overview of the classification of algebraic surfaces. In the first lecture a brief digest of the general theory is given, concentrating on the major invariants for algebraic surfaces and on the standard theorems relating them. In the next three lectures the standard constructions of algebraic surfaces of special type are presented; rational and ruled surfaces in the second lecture, abelian, $K3$, Enriques, and hyperelliptic surfaces in the third lecture, and elliptic surfaces in the fourth lecture. In the last lecture the geography of surfaces of general type is discussed, along with some examples of particular constructions which are the most well-known to illustrate the situation. In general the approach is not at all systematic, with hardly any proofs given; my intention is only to display some of the highlights of the classification of algebraic surfaces, hoping to whet the reader's appetite for a more detailed study.

It gives me great pleasure to thank TÜBITAK, the British Council in Ankara, and Bilkent University for their generous contributions to the summer school, as well as the local organizing committee from Bilkent and METU for giving me this wonderful opportunity to present this material. I especially want to thank Dr. Ihsan Dogramaci, Prof. Mefharet Kocatepe, and Prof. Sinan Sertöz and his family for their warm hospitality during the school.

Contents

1 Algebraic Surfaces: Basic Invariants

In this first section we present a quick digest of the general theory of algebraic surfaces, simply to fix notation and to set the stage. The basic references for algebraic surfaces would include [Z], [E], [S2], [Be], and [BPV]. There are also sections of [S1], [GH], and [Ha] which deal specifically with surfaces.

1.1 The Definition of an Algebraic Curve

We will work over the field of complex numbers $\mathbf{C}$. With this point of view an algebraic curve is, first of all, a Riemann surface, that is, a complex manifold of dimension one. Being a complex manifold means that at every point there is a local complex coordinate z, and the change-of-coordinate functions are holomorphic. Hence if both z and w are local complex coordinates in the neighborhood of a point p, then $z = T(w)$ and $w = S(z)$ near p, where T and S are (inverse) holomorphic functions.

For a compact Riemann surface X to be an algebraic curve, its field of global meromorphic functions $\mathcal{M}(X)$ must be sufficiently rich. To be precise:

Definition 1.1 *A compact Riemann surface X is an* algebraic curve *if and only if*

(a) for every two points p, q on X there is a global meromorphic function f on X having different values at p and q, and

(b) for every point p on X there is a global meromorphic function f on X such that f is a local coordinate at p.

The first condition is usually referred to as "separating points"; the second as "separating tangents".

It is a basic theorem that *every* compact Riemann surface is an algebraic curve. The field of functions $\mathcal{M}(X)$ is an extension field of $\mathbf{C}$ of transcendence degree one.

1.2 The Definition of an Algebraic Surface

We take our definition of an algebraic surface directly from that of an algebraic curve. Let X be a compact connected complex manifold of dimension two; this means that at each point of X there are two local complex coordinates (z, w) and change-of-coordinate maps are holomorphic. Let $\mathcal{M}(X)$ be the field of global meromorphic functions on X.

Definition 1.2 *X is an* algebraic surface *if and only if the field $\mathcal{M}(X)$ separates points and tangents, i.e.,*

(a) for every two points p, q on X there is a global meromorphic function f on X having different values at p and q, and

(b) for every point p on X there are global meromorphic functions f and g on X such that (f, g) are local coordinates at p.

This implies rather readily that the field of functions $\mathcal{M}(X)$ has transcendence degree two; the converse is true, but only for surfaces: there are counterexamples in dimension three and higher.

The compactness of X implies that in fact there are finitely many meromorphic functions $f_1, \ldots, f_n$ which separate points and tangents. We may use these functions to map X to projective space $\mathbf{P}^n$, via the mapping $x \to [1 : f_1(x) : \cdots : f_n(x)]$; then X becomes embedded in $\mathbf{P}^n$. By Chow's theorem X is an algebraic subset of $\mathbf{P}^n$, i.e., defined by the vanishing of a set of homogeneous polynomials. Therefore an algebraic surface in the above sense is indeed a projective algebraic surface.

It is trivial that every smooth projective algebraic surface is algebraic in the above sense: the ratios of the homogeneous coordinates on $\mathbf{P}^n$ restrict to meromorphic functions on X which separate points and tangents.

One of the basic examples of projective algebraic surfaces are the *hypersurfaces* in $\mathbf{P}^3$, defined by the vanishing of a single homogenous polynomial in the four projective coordinates. More generally, one has *complete intersections* in $\mathbf{P}^n$, by taking the common zeroes of a set of $n-2$ homogeneous polynomials, whose zero locus is a surface. If the polynomials have degrees $(d_1, \ldots, d_{n-2})$, then such a surface is said to be a *complete intersection of type* $(d_1, \ldots, d_{n-2})$.

1.3 Functions and Forms

The concept of a differentiable function or a meromorphic function on an algebraic surface is straightforward. That of a form may be less familiar; let us briefly review it.

Suppose that (z, w) are local coordinates on X. If we write $z = x+iy$ and $w = u + iv$ then (x, y, u, v) become real coordinates on X. A differentiable 1-form on X is, locally, an expression of the form

$$f_1(x,y,u,v)dx + f_2(x,y,u,v)dy + f_3(x,y,u,v)du + f_4(x,y,u,v)dv$$

where the f_i are differentiable functions. Upon changes of coordinates such an expression transforms to another such, and a global 1-form can be thought of as a collection of such expressions, one for every choice of local coordinates, all of which transform to one another. It is sufficient to give a set of such expressions, at least one of which are valid at each point of X.

It is more useful to substitute the variables z, $\overline{z}$, w, and $\overline{w}$ for the 4 real variables; the obvious relationships between these four and x, y, u, v make the concepts completely equivalent. In this notation then a 1-form is a collection of compatible expressions of the form

$$g_1(z, \overline{z}, w, \overline{w})dz + g_2 d\overline{z} + g_3 dw + g_4 d\overline{w}$$

with the g_i's differentiable.

Note that under holomorphic changes of coordinates, say to new complex coordinates (s,t), the dz and dw terms transport to the ds and the dt terms, while the $d\overline{z}$ and $d\overline{w}$ terms transport to the $d\overline{s}$ and the $d\overline{t}$ terms. Therefore the 1-form written above naturally decomposes *globally* into the dz, dw part (called the $(1,0)$ part) and the $d\overline{z}, d\overline{w}$ part (called the $(0,1)$ part).

Higher forms are defined as sets of compatible expressions which are linear combinations of terms like

$$f(z,\overline{z},w,\overline{w})da \wedge \ldots db$$

where f is differentiable and the expressions $da,\ldots,db$ stand for any one of the differentials dz, $d\overline{z}$, dw, and $d\overline{w}$. The basic exterior algebra laws that $da \wedge da = 0$ and $da \wedge db = -db \wedge da$ hold. A term like this has a *type* (p,q), where there are p dz and dw's, and q $d\overline{z}$ and $d\overline{w}$'s. The possible types are restricted by $0 \leq p,q, \leq 2$ of course; the form above is then called a $(p+q)$-form, and for the same reasons as above any n-form decomposes naturally into its (p,q) parts, for all p, q with $p+q=n$.

For example, a 2-form is locally an expression of the form

$$\begin{gathered} f(dz \wedge dw) \\ +g_1(dz \wedge d\overline{z}) + g_2(dz \wedge d\overline{w}) + g_3(dw \wedge d\overline{z}) + g_4(dw \wedge d\overline{w}) \\ +h(d\overline{z} \wedge d\overline{w}) \end{gathered}$$

and the first line is the $(2,0)$ part, the second line is the $(1,1)$ part, and the third line is the $(0,2)$ part.

Let $\mathcal{E}^n$ denote the sheaf of differentiable n-forms on X $(0 \leq n \leq 4)$; denote by $\mathcal{E}^{p,q}$ the sheaf of $(p+q)$-forms of type (p,q). The global sections H^0 of these sheaves are the global forms.

In addition to these differentiable forms, there are the holomorphic and meromorphic forms. These forms, always of type $(p,0)$, have holomorphic (respectively meromorphic) functions as coefficients.

The sheaf of holomorphic p-forms is denoted by Ω^p; Ω^0 is more commonly denoted by $\mathcal{O}$, the sheaf of holomorphic functions. The sheaf of meromorphic two-forms is denoted $\mathcal{M}^2$.

To be more explicit, a section of $\mathcal{O} = \Omega^0$ is a holomorphic function; a section of Ω^1 is a 1-form of the form $f(z,w)dz + g(z,w)dw$ with f and g holomorphic; and a section of Ω^2 is a 2-form of the form $f(z,w)dz \wedge dw$ with f holomorphic. A section of clM^2 is a 2-form of the form $f(z,w)dz \wedge dw$ with f meromorphic.

1.4 Divisors

A divisor on X is locally defined by a single nonzero meromorphic function f; one thinks of the divisor as the locus $f = 0$, but when f has multiple roots this is taken into account also. Two such meromorphic functions locally define the same divisor when their ratio has no zero or poles.

This then is how divisors are globalized: one takes a meromorphic function $f(z,w)$ for every choice of local coordinates (z,w), and requires that in an open set where two choices of local coordinates exist, the ratio of the given functions has no zeros or poles.

Of course it is sufficient to give a meromorphic function on each open set of an open covering of the surface, which satisfy the ratio condition.

With this definition, two collections of local meromorphic functions form the same divisor if and only if their union is a valid divisor, that is, each function of one set agrees (up to multiplication by a nowhere zero holomorphic function) with each function of the other set, where they are both defined.

Divisors form an abelian group $\mathrm{Div}(X)$, by multiplying the corresponding functions; the group operation on divisors is usually denoted by addition however.

Divisors are used to organize meromorphic functions and forms, by serving as bounds on the possible poles. If D is a divisor on X, locally defined by f, then a term $g(z,w)$ or $g(z,w)dz$ or $g(z,w)dz \wedge dw$ of a p-form is said to *have poles bounded by* D if the product fg is holomorphic.

We may then introduce the sheaves $\Omega^p[D]$ of meromorphic p-forms with all terms having poles bounded by D; as above, it is more common to write $\mathcal{O}[D]$ for $\Omega^0[D]$.

The zeroes and poles of a meromorphic function always have codimension one; therefore on a surface they form curves. A divisor D then gives rise to a finite formal sum $\sum_i n_i C_i$ where the C_i are curves on X and the n_i are integers, representing the order of the local function along the curve C_i. This representation is called a *Weil divisor*; on a smooth surface, there is a 1-1 correspondence between Weil divisors and divisors. When there is any possibility of confusion, ordinary divisors are called *Cartier divisors.*

If f is a *global* meromorphic function, then in every local coordinate one can take the function f; this gives *divisor of* f, denoted by $\text{div}(f)$. We note that $\text{div}(fg) = \text{div}(f) + \text{div}(g)$ and $\text{div}(1/f) = -\text{div}(f)$. Such divisors are called *principal divisors*, and form a subgroup of the group $\text{Div}(X)$ of all divisors.

Two divisors are said to be *linearly equivalent* if their difference is principal; this equivalence relation is simply congruence modulo the subgroup of principal divisors.

Cohomology does not "see" linear equivalence, in the sense that if $D_1 = D_2 + \text{div}(f)$, then multiplication by f induces natural isomorphisms between $H^i(\Omega^p[D_1])$ and $H^i(\Omega^p[D_2])$.

The *Picard group* $\text{Pic}(X)$ is the group of divisors modulo linear equivalence. It may be identified with the first cohomology group $H^1(\mathcal{O}^*)$ of the sheaf of nowhere zero holomorphic functions; given a divisor, the ratios of the local functions give a 1-cocycle for this sheaf. Changing local coordinates on an open set changes the cocycle by a coboundary.

If X is a projective surface, lying in $\mathbf{P}^n$, then if one takes a hyperplane H defined by a linear polynomial, this polynomial may be taken as the numerator of a local defining equation for a divisor on X, where any nonzero denominator is chosen. If one changes the hyperplane one obtains a linearly equivalent divisor, so that the linear equivalence class is unique; this class is called the *hyperplane class.*

Divisors are rather well-behaved for maps $\pi : X \to Y$ of algebraic surfaces; if D is a divisor on Y, then one defines $\pi^*(D)$ as a divisor on X by taking as the local function at a point $p \in X$ to be the composition $f \circ \pi$, where f is the local function for D at the image $\pi(p)$. The divisor $\pi^*(D)$ is called the *pullback* of D.

1.5 The Canonical Class and the Plurigenera

Let ω be a meromorphic 2-form, which is therefore locally of the form $f(z,w)dz \wedge dw$. Suppose that one changes coordinates, to (s,t), where $z = z(s,t)$ and $w = w(s,t)$. Then

$$dz = (\partial z/\partial s)ds + (\partial z/\partial t)dt$$

and

$$dw = (\partial w/\partial s)ds + (\partial w/\partial t)dt$$

so that

$$dz \wedge dw = (\frac{\partial z}{\partial s}\frac{\partial w}{\partial t} - \frac{\partial z}{\partial t}\frac{\partial w}{\partial s})ds \wedge dt$$

and this coefficient, being the Jacobian J of the change of coordinates map, is nowhere zero and holomorphic. Therefore the coefficient function $f(z,w)$ is transformed to fJ, and these two meromorphic functions define the same divisor locally. Hence we may define the *divisor* K of ω by taking this coefficient function in any local expression for ω, and since this is well-defined up to multiplying by a nowhere zero holomorphic function, the divisor is well-defined; it is denoted by $\text{div}(\omega)$. We have

$$\text{div}(f\omega) = \text{div}(f) + \text{div}(\omega)$$

for a meromorphic function f and a meromorphic 2-form ω. Such a divisor is called a *canonical divisor* on X.

Note that if ω_1 and ω_2 are nonzero meromorphic 2-forms on X, then there is always a global meromorphic function f on X such that $\omega_1 = f\omega_2$. Therefore

$$\text{div}(\omega_1) = \text{div}(f\omega_2) = \text{div}(f) + \text{div}(\omega_2)$$

so that the two canonical divisors $\text{div}(\omega_1)$ and $\text{div}(\omega_2)$ are linearly equivalent. Therefore the linear equivalence class of canonical divisors is well-defined; this is called the *canonical class* of divisors, and is usually denoted by K (as is any particular canonical divisor).

The *geometric genus* of the surface X is the dimension of the space $H^0(\Omega^2)$ of global holomorphic 2-forms on X. If $K = \text{div}(\omega)$ is a canonical divisor, and f is any global meromorphic function with poles bounded by K, then $f\omega$ is holomorphic. This gives an isomorphism between $H^0(\mathcal{O}[K])$ and $H^0(\Omega^2)$. The geometric genus is commonly denoted by p_g.

Generalizing this, we may define the *plurigenera* of X, to be the dimensions of the spaces $H^0(\mathcal{O}[nK])$ as n varies; precisely,

$$P_n = \dim H^0(\mathcal{O}[nK])$$

for $n \geq 0$. Then $P_0 = 1$, $P_1 = p_g$, etc.

1.6 Kodaira Dimension

The growth rate κ of the plurigenera is an important invariant of X, called the *Kodaira dimension*. Precisely, if the plurigenera are all 0, then we set $\kappa = -\infty$ (some authors set $\kappa = -1$ for this case). Otherwise κ is the integer such that the plurigenera sequence P_n grows like n^κ, in the sense that κ is minimal such that P_n/n^κ is bounded. For algebraic surfaces, using the Riemann-Roch Theorem or otherwise, we have that $\kappa \leq 2$, so that the four possible values of κ are $-\infty$, 0, 1, or 2.

The situation is illustrated by briefly studying the corresponding invariant for curves. There the canonical class has degree $2g-2$, where g is the genus of the curve. Therefore if $g = 0$, the canonical divisor is negative, and all plurigenera are 0: we have Kodaira dimension equal to $-\infty$. If $g = 1$, then 0 is a canonical divisor, so all plurigenera equal 1 and $\kappa = 0$. Finally if $g \geq 2$ then the plurigenera grow: by Riemann-Roch one has $P_n = n(2g-2) + 1 - g$ for $n \geq 2$, which is linear in n, so $\kappa = 1$.

The Kodaira dimension is obviously a rather coarse invariant, but it serves well as a starting tool to classify surfaces. As the case of curves indicates, the "general" situation is that κ is maximal, equal to 2. The surfaces with $\kappa = 2$ are therefore said to be *of general type*, while those with $\kappa \leq 1$ are *of special type.* The classification of algebraic surfaces begins by first classifying those of special type; then attacking those of general type.

1.7 Numerical Invariants: q, $h^{p,q}$, b_i, e, χ

There are several other important numerical invariants which it is well to be aware of. Most involve some form of cohomology.

We begin with the *Betti numbers* b_i, which are the ranks of the simplicial (or singular) homology groups:

$$b_i = \text{rank} H^i(X, \mathbf{Z}) = \dim_{\mathbf{R}} H^1(X, \mathbf{R}) = \dim_{\mathbf{C}} H^i(X, \mathbf{C}).$$

The Betti numbers b_i are defined for $0 \leq i \leq 4$.

The *Euler number* e is the alternating sum of the Betti numbers:

$$e = \sum_{i=0}^{4} (-1)^i b_i.$$

It is also the alternating sum of the numbers of i-simplices in any triangulation of X.

Next we have the *irregularity* q, defined as the dimension of the space of global holomorphic 1-forms:

$$q = \dim H^0(\Omega^1).$$

More generally we have the *Hodge numbers*, which we may define as the dimensions of the cohomology groups of the sheaves of p-forms:

$$h^{p,q} = \dim H^q(\Omega^p).$$

Note that the irregularity $q = h^{1,0}$ and the geometric genus $p_g = h^{2,0}$.

Another definition avoiding cohomology is to consider the $\overline{\partial}$ operator on (p,q)-forms, which (locally) sends a term $f(z,w)da \wedge \dots db$ to

$$(\partial f/\partial \overline{z})da \wedge \dots db \wedge d\overline{z} + (\partial f/\partial \overline{w})da \wedge \dots db \wedge d\overline{w}.$$

This maps (p,q)-forms to $(p,q+1)$-forms, and $\overline{\partial} \circ \overline{\partial} = 0$. Then one defines

$$H^{p,q} = \ker(\overline{\partial} : H^0(\mathcal{E}^{p,q}) \to H^0(\mathcal{E}^{p,q+1}))/\text{image}(\overline{\partial} : H^0(\mathcal{E}^{p,q-1}) \to H^0(\mathcal{E}^{p,q}))$$

and $h^{p,q} = \dim H^{p,q}$. These numbers are defined for $0 \leq p, q \leq 2$.

These invariants have various relationships, which form the heart of Hodge theory for algebraic surfaces; these are summed up as follows:

(a) $h^{p,q} = h^{q,p} = h^{2-p,2-q} = h^{2-q,2-p}$

(b) $b_i = \sum_{p+q=i} h^{p,q}$

(c) $b_0 = b_4 = 1$

The second property comes from the Hodge decomposition

$$H^n(X, \mathbf{C}) = \bigoplus_{p+q=n} H^{p,q}$$

where $H^n(X, \mathbf{C})$ has its DeRham interpretation. The symmetry properties (a) are induced by conjugation and the $*$ operator.

One often arranges the Hodge numbers in a "Hodge diamond"

$$\begin{array}{ccccc} & & h^{0,0} & & \\ & h^{1,0} & & h^{0,1} & \\ h^{2,0} & & h^{1,1} & & h^{0,2} \\ & h^{2,1} & & h^{1,2} & \\ & & h^{2,2} & & \end{array}$$

with 5 rows; because of the above identifications we see that the Hodge diamond looks like

$$\begin{array}{ccccc} & & 1 & & \\ & q & & q & \\ p_g & & h^{1,1} & & p_g \\ & q & & q & \\ & & 1 & & \end{array}$$

whose rows sum to the Betti numbers. Therefore we have that

$$b_1 = b_3 = 2q \quad \text{and} \quad b_2 = 2p_g + h^{1,1}$$

while the alternating sum of the rows gives the Euler number

$$e = 2 + 2p_g + h^{1,1} - 4q.$$

Finally we have the holomorphic Euler-Poincare characteristic $\chi = \chi(\mathcal{O})$, the alternating sum of the dimensions of the cohomology groups of the sheaf $\mathcal{O}$ of holomorphic functions:

$$\chi = \dim H^0(\mathcal{O}) - \dim H^1(\mathcal{O}) + \dim H^2(\mathcal{O}) = 1 - q + p_g.$$

1.8 The Neron-Severi Group and the Lefschetz (1,1) Theorem

On any algebraic surface we have the exponential short exact sequence of sheaves

$$0 \to \underline{\mathbf{Z}} \to \mathcal{O} \to \mathcal{O}^* \to 0$$

where the map on the right is $\exp(2\pi i -)$. This is also exact on global sections, so we have an exact sequence starting at the H^1 level which begins

$$0 \to H^1(\mathbf{Z}) \to H^1(\mathcal{O}) \to H^1(\mathcal{O}^*) \to H^2(\mathbf{Z}) \to H^2(\mathcal{O}).$$

Using the standard names for the invariants we see that this is

$$0 \to \mathbf{Z}^{2q} \to \mathbf{C}^q \to \mathrm{Pic}(X) \to H^2(\mathbf{Z}) \to H^2(\mathcal{O}).$$

We see therefore that the Picard group $\mathrm{Pic}(X)$ of divisors modulo linear equivalence has a continuous part $\mathbf{C}^q/\mathrm{image}(\mathbf{Z}^{2q})$ (which is a complex torus of dimension q) and its image in $H^2(\mathbf{Z})$ is its discrete part. This map to $H^2(\mathbf{Z})$ is called the *Chern class map* on divisors. The image, modulo torsion, is called the *Neron-Severi group* of X, denoted by $\mathrm{NS}(X)$. It is a finitely generated free abelian group; its rank ρ is called the *Picard number* of X, and tells us how many discrete parameters divisors on X depend on.

Note that any class in $H^2(\mathbf{Z})$ which comes from a divisor must then go to zero in $H^2(\mathcal{O})$. This map factors as

$$H^2(\mathbf{Z}) \subset H^2(\mathbf{C}) \to H^2(\mathcal{O})$$

which shows that such a class, when considered in $H^2(\mathbf{C})$, is in the kernel of the map to $H^2(\mathcal{O}) \cong H^{0,2}$. A conjugation argument shows that it is also zero in $H^{2,0}$, and so must lie in $H^{1,1}$.

The converse of this is the Lefschetz $(1,1)$ Theorem: a class in $H^2(\mathbf{Z})$ is the class of a divisor if and only if, when considered in $H^2(\mathbf{C})$, it lies in the $H^{1,1}$ part.

Thus the divisors on X are controlled by how the complex subspace $H^{1,1}$ of $H^2(\mathbf{C})$ intersects the discrete subgroup $H^2(\mathbf{Z})$. For surfaces with $p_g = 0$, for which therefore $H^2(\mathbf{C}) = H^{1,1}$, this is no condition, but for surfaces of positive geometric genus this can be a subtle arithmetic problem.

1.9 Intersection Theory, Riemann-Roch, Index Theorems

On the middle-dimensional cohomology groups $H^2(\mathbf{R})$ and $H^2(\mathbf{C})$ the cup-product induces a nondegenerate bilinear form, called the *intersection form* on X. The group of divisors, through the Chern class map, inherits this bilinear form, as does the Picard group and the Neron-Severi group $\mathrm{NS}(X)$.

The intersection of divisors can be defined algebraically. It suffices to define $(C_1 \cdot C_2)$ for two curves on X, by linearity. If the curves are different, then they intersect at only finitely many points $\{p_i\}$. At each $p = p_i$, let C_j be defined by the local function f_j in the local ring $\mathcal{O}_p$. Then the local intersection number

$$(C_1 \cdot C_2)_p := \mathrm{length}(\mathcal{O}_p/(f_1, f_2))$$

is well-defined and finite, and

$$(C_1 \cdot C_2) = \sum_i (C_1 \cdot C_2)_{p_i}.$$

The case of $(C \cdot C)$ can be handled in two ways: either replace one of the C's by a linearly equivalent divisor not containing C as a component, or in an ad hoc way set $(C \cdot C)$ to be the degree of the normal bundle of C on X. In particular note that $(C \cdot C)$ may be negative.

An important invariant is $K^2 = (K \cdot K)$ where K is a canonical divisor.

We are finally in a position to state some of the most important results in surface theory.

Theorem 1.3 *Let X be an algebraic surface.*

(a) (Thom-Hirzebruch Index (or Signature) Theorem:) *The signature τ of the intersection form on $H^2(\mathbf{R})$ satisfies*

$$\tau = (K^2 - 2e)/3.$$

(b) (Hodge Index Theorem:) *The intersection form on* $H^{1,1} \cap H^2(\mathbf{R})$ *and hence on the Neron-Severi group* NS(X) *has one positive eigenvalue and* $h^{1,1} - 1$ *negative eigenvalues.*

(c) (Noether's Formula:)

$$12\chi = K^2 + e.$$

(d) (Riemann-Roch Theorem:)

$$\chi(\mathcal{O}[D]) = \frac{(D \cdot (D - K))}{2} + \chi.$$

(e) (The Genus Formula:) *If* C *is an irreducible curve on* X, *then its arithmetic genus* $p_a(C)$ *is*

$$p_a(C) = \frac{(C \cdot (C + K))}{2} + 1.$$

Here $\chi(\mathcal{F}) = h^0(\mathcal{F}) - h^1(\mathcal{F}) + h^2(\mathcal{F})$ is the Euler-Poincare characteristic of the sheaf in question; without specifying the sheaf one takes $\chi = \chi(\mathcal{O})$. (Here we write h^i for the dimension of a cohomology group H^i.)

1.10 Projective Invariants: Degree, Class

Suppose that $X \subset \mathbf{P}^n$ is a projective surface. Then as noted above we have the hyperplane divisor class H on X. A typical hyperplane divisor should be thought of, as a Weil divisor, as the intersection of X with a hyperplane. If we intersect X with two hyperplanes we obtain the *degree* of X. On the other hand intersecting first with X and then intersecting again, we see that this degree is equal to the self-intersection number $H^2 = (H \cdot H)$. This is a fundamental invariant of a projective surface, but note that it depends very much on how X is embedded in projective space; the same surface can be re-embedded in different ways having different degrees.

A hypersurface in $\mathbf{P}^3$ defined by taking the zeroes of a homogeneous polynomial of degree d has degree d. More generally, if X is a smooth complete intersection surface of type $(d_1, \ldots, d_n)$ in $\mathbf{P}^{n+2}$, then X has degree $d_1 d_2 \cdots d_n$.

Other projective invariants abound; one of the most common is the *class*, defined to be the degree of the dual variety of tangent hyperplanes.

1.11 Birational Maps, Blowups and Minimal Models

A holomorphic map f between algebraic surfaces is a map which, locally, is defined by two holomorphic functions. In the case of surfaces (and any varieties of higher dimension) there are always interesting maps which are not defined everywhere however. The prototype is the map from $\mathbf{C}^2$ to $\mathbf{P}^1$ sending a point (z, w) to the line $[z : w]$ through 0 and (z, w). This is not well-defined at the origin, and is not definable there as a continuous function.

So the concept of a *rational* map exists, given locally by meromorphic functions, but which may not be defined at all points. Rational maps $X \dashrightarrow Y$ are in correspondence with function field inclusions $\mathcal{M}(Y) \to \mathcal{M}(X)$. A *birational* map is a rational map with a rational inverse. A birational map is therefore determined by an isomorphism between function fields; birational equivalence is essentially this isomorphism property.

The fundamental construction in the birational theory of surfaces is the *blowup* at a point p. Suppose that one chooses local coordinates (z, w) on X at p, so that p is $z = w = 0$. Choose an open set around p, giving an open set $U \subset \mathbf{C}^2$. Replace that open set by two open sets V_1 and V_2, where V_1 has coordinates (z, u) and V_2 has coordinates (w, v). Glue these into the surface X by the change of coordinates

$$w = zu, \quad uv = 1, \quad z = wv.$$

Note that in particular these two open sets are glued together, and map to the open set U; the inverse image of p (which is $z = w = 0$) is the line $z = 0$ (with coordinate u) in V_1 and the line $w = 0$ (having coordinate v) in V_2. Since $u = 1/v$, these lines are glued to form a $\mathbf{P}^1$ which we will denote by E. The inverse image of any other point is simply one point. Therefore we have constructed another complex manifold B, which maps to X. If we call the map π, we have the diagram

$$\begin{array}{ccc} E & \subset & B \\ \downarrow & & \downarrow \pi \\ p & \in & X \end{array}$$

It is easy to see that B is an algebraic surface if X is, and has an isomorphic field of meromorphic functions: B is birational to X.

The following basic facts are easy consequences of local computations:

Lemma 1.4 *Consider $\pi : B \to X$ as above, with exceptional curve E.*

(a) $(E \cdot E) = -1$.

(b) $(E \cdot \pi^*(D)) = 0$ *for any divisor D on X.*

(c) π is an isomorphism between $B - E$ and $X - p$.

(d) If C is a curve on X having multiplicity m at p, then the proper transform C' of C on B (which is defined to be the closure of $\pi^{-1}(C' - p)$ in B) satisfies

$$(C' \cdot E) = m \quad \text{and} \quad (C')^2 = C^2 - m^2.$$

(e) $K_B = \pi^*(K_X) + E$, *hence* $K_B^2 = K_X^2 - 1$.

(f) q, p_g, χ, and the plurigenera P_n are the same for B as for X.

(g) $e(B) = e(X) + 1,\ h^{1,1}(B) = h^{1,1}(X) + 1,\ b_2(B) = b_2(X) + 1,$ $\rho(B) = \rho(X) + 1.$

(*h*) $\mathrm{Pic}(B) = \pi^*(\mathrm{Pic}(X)) \oplus \mathbf{Z} \cdot E$.

Reversing this construction is *Castelnuovo's Contraction Criterion*: if on an algebraic surface B one finds a curve E with $E \cong \mathbf{P}^1$ and $E^2 = -1$, then B is the blowup of an algebraic surface X and E is the exceptional curve. This contraction of E to a point p is unique. Such a curve E on a surface is called a (-1)-*curve*.

A surface is therefore said to be *minimal* if it has no (-1)-curves; it does not arise as the blowup of any other surface. Most of the detailed work in algebraic surfaces concentrates on minimal ones.

Blowups are so important because *any* birational map between algebraic surfaces factors through blowups in the following sense:

Proposition 1.5 *Let* $f : X \dashrightarrow Y$ *be a birational map between algebraic surfaces. Then there is an algebraic surface* Z *and maps* $\pi_X : Z \to X$ *and* $\pi_Y : Z \to Y$*, which are compositions of blowup maps, such that the diagram*

$$\begin{array}{ccc} & Z & \\ \swarrow & & \searrow \\ X & \overset{f}{\dashrightarrow} & Y \end{array}$$

commutes.

As a corollary, we obtain that q, p_g, χ, and the plurigenera P_n are birational invariants, since they are invariant under blowups.

1.12 Homeomorphism and Diffeomorphism

To close this first lecture I want to briefly mention the great strides that have been made in the last decade on understanding the topology and differential topology of algebraic surfaces. Firstly, when are two algebraic surfaces homeomorphic? At least in the simply connected case, we have an answer, given by Freedman (see [F]):

Theorem 1.6 *Two simply connected surfaces are homeomorphic if and only if their cohomology groups $H^2(\mathbf{Z})$ are isomorphic, as groups with bilinear form.*

This group, modulo torsion, is free abelian of rank b_2, and the form is unimodular with $2p_g + 1$ positive eigenvalues. There are at most two such forms, depending on whether the form is even or odd, and this depends on whether the canonical class is even or not. Thus a complete answer is at hand.

The diffeomorphism question is more subtle, and there are examples of simply connected surfaces which are homeomorphic but not diffeomorphic.

There has been a flurry of activity in just this past year on the diffeomorphism question; it is now known that, for minimal surfaces of general type, the canonical class is a diffeomorphism invariant up to sign, and that the plurigenera are diffeomorphism invariants, for example. The reader may consult the recent article [FM] for more complete information.

1.13 The Mori Point of View

Mori theory is now seen as the best way to understand algebraic threefolds, and its point of view also lends some insight into the classification of algebraic surfaces. The Kodaira Dimension κ recedes somewhat in importance, and the canonical class K itself comes into the foreground. Specifically, Mori's approach says to consider the intersection properties of the canonical class K, and in particular we say that K is *nef* (meaning numerically eventually free), if $(K \cdot C) \geq 0$ for all curves C on X. Mori's first main theorem is that if K is *not* nef, then either there is a (-1)-curve on X to blow down (which is the curve meeting K negatively) or K is either rational or ruled. The latter case is the $\kappa = -\infty$ situation. In the former case one then blows down until K does become nef. Then there are three possibilities: either K is numerically trivial (which corresponds to the $\kappa = 0$ case) or K is not numerically trivial but $K^2 = 0$ (which corresponds to the $\kappa = 1$ case) or $K^2 > 0$ (which corresponds to the $\kappa = 2$ case). This all then leads to the same basic finer classification, but with a slight alteration of attitude. The reader would do well to consult [CKM] for a gentle introduction to Mori's ideas.

2 Surfaces with Negative Kodaira Dimension

2.1 The Projective Plane

Let $\mathbf{P}^2$ be the projective plane, with homogeneous coordinates $[x : y : z]$. We have that $\mathbf{P}^2 = \mathbf{C}^2 \cup \mathbf{C} \cup \{p\}$, which is a cell decomposition with one cell of dimension 0, 2, and 4. Therefore in the cell chain complex all boundary maps must be zero and

$$\begin{aligned} H^0(\mathbf{P}^2, \mathbf{Z}) &\cong \mathbf{Z} \\ H^1(\mathbf{P}^2, \mathbf{Z}) &= 0 \\ H^2(\mathbf{P}^2, \mathbf{Z}) &\cong \mathbf{Z} \\ H^3(\mathbf{P}^2, \mathbf{Z}) &= 0 \\ H^4(\mathbf{P}^2, \mathbf{Z}) &\cong \mathbf{Z}. \end{aligned}$$

Hence

$$b_1 = b_3 = 0, \quad b_2 = 1, \quad e = 3.$$

Therefore also we must have

$$q = p_g = 0, \quad \chi = h^{1,1} = 1.$$

The intersection form on $H^2(\mathbf{Z})$ must be simply (1). If we denote by H the class of a line, then H is a divisor which generates the Picard group of $\mathbf{P}^2$. The intersection formula $H^2 = 1$ means that two lines intersect in one point!

If $u = x/z$ and $v = y/z$, then (u, v) are local coordinates on the $\mathbf{C}^2$ chart of $\mathbf{P}^2$ where $z \neq 0$. The 2-form $du \wedge dv$ is holomorphic on this chart, but has a triple pole at infinity (that is, on the line $z = 0$). To check this explicitly, use coordinates $s = z/x$ and $t = y/x$ at infinity ($s = 0$ defines the line at infinity in this chart). Then $u = 1/s$ and $v = t/s$, so that $du = (-1/s^2)ds$ and $dv = (-t/s^2)ds + (1/s)dt$. Therefore

$$du \wedge dv = [(-1/s^2)ds] \wedge [(-t/s^2)ds + (1/s)dt] = (-1/s^3)ds \wedge dt$$

exhibiting a triple pole at the line H at infinity. Therefore $K = -3H$ is the canonical class, and $K^2 = 9$.

Since K is a negative divisor, so is nK for every $n \geq 1$, and hence all plurigenera P_n for the plane are zero. Therefore $\kappa = -\infty$ for $\mathbf{P}^2$.

The Riemann-Roch Theorem for $\mathbf{P}^2$ says that

$$h^0(\mathcal{O}[dH]) = (dH \cdot (d+3)H)/2 + \chi = d(d+3)/2 + 1$$

which represents the number of coefficients of a form of degree d in 3 homogeneous variables; the χ equals the h^0 since the higher cohomology is all zero. The Genus Formula for a curve C of degree d (and hence linearly equivalent to dH) is

$$p_a(C) = (dH \cdot (d-3)H)/2 + 1 = (d-1)(d-2)/2$$

which is the classical Plücker formula.

An algebraic surface is said to be *rational* if it is birational to the plane.

2.2 Quadric Surfaces

Just as for $\mathbf{P}^2$, the canonical class of any projective space is a multiple of the hyperplane class; indeed, for $\mathbf{P}^n$ we have $K = -(n+1)H$. The adjunction formula then states that if X is a complete intersection surface in $\mathbf{P}^n$ of type $(d_1, \ldots, d_{n-2})$, then $K_X = (\sum_i d_i - n - 1)H$. The first case to apply this is for a smooth quadric surface X of degree 2 in $\mathbf{P}^3$. Then $K = -2H$, and since this is clearly negative, so are all multiples, and we have that $P_n = 0$ for all n. Therefore again X has $\kappa = -\infty$.

Since $H^2 = 2$ (the surface has degree 2), we see that $K^2 = 8$.

Any two smooth quadrics are isomorphic (there is only one nondegenerate quadratic form up to isomorphism over $\mathbf{C}$) and we may take $xy = zw$ (in the homogeneous coordinates $[x : y : z : w]$ of 3-space) as a defining equation for X. We see easily that $X \cong \mathbf{P}^1 \times \mathbf{P}^1$ by the map sending $([a : b], [c : d])$ to $[x = ac : y = bd : z = ad : w = bc]$. Therefore the Künneth formulas give that

$$
\begin{aligned}
H^0(\mathbf{Z}) &\cong \mathbf{Z} \\
H^1(\mathbf{Z}) &= 0 \\
H^2(\mathbf{Z}) &\cong \mathbf{Z}^2 \\
H^3(\mathbf{Z}) &= 0 \\
H^4(\mathbf{Z}) &\cong \mathbf{Z}.
\end{aligned}
$$

Hence

$$b_1 = b_3 = 0, \quad b_2 = 2, \quad e = 4.$$

Again $q = p_g = 0$ and $\chi = 1$; this time $h^{1,1} = 2$. In the $\mathbf{P}^1 \times \mathbf{P}^1$ representation, we have the obvious horizontal curve G and the vertical curve F; these generate $H^2(\mathbf{Z})$ and the Picard group, with $F^2 = G^2 = 0$ and $(F \cdot G) = 1$. The hyperplane class (as a quadric surface) is $H = F + G$, hence $K = -2F - 2G$.

Riemann-Roch now gives that

$$\chi(rF + sG) = ((rF + sG) \cdot (r+2)F + (s+2)G)/2 + \chi = (r+1)(s+1)$$

which represents the number of monomials of a bihomogenous polynomial of bidegree (r, s). The Genus Formula for a curve C linearly equivalent to $rF + sG$ is

$$p_a(rF + sG) = (r-1)(s-1).$$

Projection of the quadric X from a point p on X gives a birational map from X to $\mathbf{P}^2$; this map blows down the two lines through p and blows up p. In particular, X is a rational surface.

2.3 Ruled Surfaces

A *ruled surface* is a surface birational to $X = C \times \mathbf{P}^1$ for some curve C of genus g. This decomposition implies that we can construct a meromorphic 2-form on X by taking a 1-form on C and wedging with a 1-form on $\mathbf{P}^1$. Therefore a canonical divisor on X is gotten by taking a canonical divisor on C (which therefore has degree $2g-2$) and pulling back via the first projection, plus taking a canonical divisor on $\mathbf{P}^1$ (which is -2 points) and pulling back via the second projection. If we denote by F the $\mathbf{P}^1$ fiber and by C the C fiber, then *numerically* (that is, in $H^2(\mathbf{Z})$) we have that $K = (2g-2)F - 2C$. No multiple of this divisor can be linearly equivalent to a positive divisor; its intersection with the curve F is strictly negative. Therefore again $\kappa = -\infty$ for this surface, and hence for any ruled surface.

The Künneth decomposition gives that

$$\begin{aligned} H^0(\mathbf{Z}) &\cong \mathbf{Z} \\ H^1(\mathbf{Z}) &= 2g \\ H^2(\mathbf{Z}) &\cong \mathbf{Z}^2 \\ H^3(\mathbf{Z}) &= 2g \\ H^4(\mathbf{Z}) &\cong \mathbf{Z}. \end{aligned}$$

Hence

$$b_1 = b_3 = 2g, \quad b_2 = 2, \quad e = 4 - 4g.$$

Now $q = g$, $p_g = 0$, $\chi = 1 - g$, and $h^{1,1} = 2$; the Neron-Severi group is generated by F and C, with $F^2 = C^2 = 0$ and $(F \cdot C) = 1$. Also $K^2 = 8 - 8g$. Note that now if $g > 0$ there is a continuous part to the Picard group, coming from the Jacobian of C. The 1-forms on X are all pulled back from the g linearly independent 1-forms on C, which is why $q = g$.

2.4 Elementary Transformations

Suppose that X is ruled surface, birational to $C \times \mathbf{P}^1$, with $g(C) > 0$. Then as noted in the previous lecture X may be obtained by blowing up $C \times \mathbf{P}^1$ a finite number of times, then blowing down some (-1)-curves. Now every (-1)-curve is a rational curve, and therefore cannot dominate C if $g > 0$. Therefore these (-1)-curves must all map to points under the first projection to C; in other words, they must lie in fibers of this first projection $\pi : C \times \mathbf{P}^1 \to C$. We conclude that this map π to C is also defined on X, and so X comes also with a projection to C, which is hence natural.

The prototype for this phenomenon is to choose any point of X, contained in a smooth fiber F, and blow that point up, obtaining the exceptional curve E. Then the proper transform of F on the blowup also is a (-1)-curve, and can therefore be blown down. This produces a new surface, and this operation is called an *elementary transformation* of the ruled surface. We note that elementary transformations preserve essentially all discrete invariants.

Again assuming $g > 0$, any (-1)-curve on X must live in a fiber of the map π to C. Therefore if every fiber is a smooth curve (the so-called *geometrically ruled* case) then since fibers have self-intersection 0 (not -1) there can be no (-1)-curves on X, and X is a minimal surface. The converse is also true: if X is minimal, then every fiber of π is smooth.

The basic theorem relating minimal ruled surfaces with $q > 0$ is the following.

Theorem 2.7 *Let X be a minimal ruled surface over a curve C of genus $g > 0$. Then X is obtained from the product surface $C \times \mathbf{P}^1$ by a finite number of elementary transformations.*

Since elementary transformations preserve all discrete invariants, we therefore know these invariants for all minimal ruled surfaces over positive genus curves.

We remark that there is not a unique minimal surface birational to any ruled surface, but infinitely many: just perform an elementary transformation on one to get another.

2.5 Hirzebruch Surfaces

If X is a ruled surface with $q = 0$, then X is a rational surface. In this case the ruling on X may not be unique, especially if X is not minimal. However the concept of the elementary transformation still exists; let us apply it iteratively to $\mathbf{P}^1 \times \mathbf{P}^1$ and see what we get.

Define $\mathbf{F}_0 = \mathbf{P}^1 \times \mathbf{P}^1$. Choose a horizontal section G_0 (here $G_0^2 = 0$) and perform an elementary transformation, blowing up a point on G_0 and blowing down the proper transform of the fiber through p. The formulas give that the proper transform of G_0 is a curve G_1 with self-intersection -1; the ruling as noted above is still preserved. Call this surface $\mathbf{F}_1$. The section G_1 and the fiber F of the ruling generate tho Picard group of $\mathbf{F}_1$.

Recursively, we will have a surface $\mathbf{F}_n$, ruled over $\mathbf{P}^1$, with a section G_n of self-intersection $-n$ which, together with the fiber F, generates the Picard group. Perform an elementary transformation, blowing up a point on G_n, and blowing down the fiber; the result is a ruled surface $\mathbf{F}_{n+1}$. The proper transform of G_n is a curve G_{n+1} of self-intersection $-n-1$, which is a section of the ruling; it and the fiber generate the Picard group.

The classification of minimal ruled surfaces with $q = 0$ is now at hand:

Theorem 2.8 *Let X be a minimal ruled surface with $q = 0$. Then X is either $\mathbf{P}^2$ or $\mathbf{F}_n$ for some $n \geq 0$, $n \neq 1$. The surface $\mathbf{F}_1$ is the blowup of $\mathbf{P}^2$ at one point.*

These surfaces $\mathbf{F}_n$ are commonly called the *Hirzebruch* ruled surfaces.

2.6 The Classification Theorem

The fundamental classification theorem can now be stated.

Theorem 2.9 *Every algebraic surface with $\kappa = -\infty$ is ruled.*

Every rational surface (birational to $\mathbf{P}^2$) is automatically ruled, since $\mathbf{P}^2$ is birational itself to $\mathbf{P}^1 \times \mathbf{P}^1$. Sometimes this is explicitly stated: every surface with $\kappa = -\infty$ is rational or ruled.

There is a wonderful refinement of this in the case of rational surfaces, namely the Castelnuovo criterion:

Theorem 2.10 *An algebraic surface is rational if and only if $q = P_2 = 0$.*

The Castelnuovo criterion says that if there are no holomorphic 1-forms on X ($q = 0$) and no holomorphic 2-fold 2-forms (forms locally expressible as $f(z,w)(dz \wedge dw)^{\otimes 2}$) then the surface is rational. Of course if $P_2 = 0$ then certainly the geometric genus $P_1 = p_g = 0$ (which indicates that there are no holomorphic 2-forms). It was an open question in the latter part of the last century whether the criterion could be weakened to $q = p_g = 0$; this was shown to be false by Enriques, and we will see an example in the next lecture.

One also has the following criterion of Enriques:

Theorem 2.11 *An algebraic surface has $\kappa = -\infty$ if and only if $P_4 = P_6 = 0$.*

2.7 Scrolls and Rational Normal Scrolls

A ruled surface embedded in projective space such that the fibers of the ruling are straight lines is called a *scroll*. No doubt the most important family of scrolls are the *rational normal scrolls*, which we now describe.

Recall the construction of a *rational normal curve*, which is $\mathbf{P}^1$ embedded into $\mathbf{P}^d$ by all monomials of degree d: $[x : y] \mapsto [x^d : x^{d-1}y :$

$\cdots : xy^{d-1} : y^d]$. Call this map ν_d; it is simply the Veronese map of degree d for $\mathbf{P}^1$. Its image is the rational normal curve, of degree d in $\mathbf{P}^d$. Note that this is the minimal degree for a nondegenerate curve in $\mathbf{P}^d$; if one cuts a curve of degree e with a general hyperplane, one will obtain e points spanning $\mathbf{P}^{d-1}$, so that $e \geq d$.

To build a *rational normal scroll* in $\mathbf{P}^{a+b+1}$, choose two complementary linear spaces of dimension a and b, and in each put a rational normal curve, the image of the line under ν_a and ν_b respectively. Form a surface $S_{a,b}$ by taking the union of all of the lines joining $\nu_a(p)$ to $\nu_b(p)$ as p varies in $\mathbf{P}^1$.

Proposition 2.12 *Assume that $1 \leq a \leq b$. The surface $S_{a,b}$ is nondegenerate, and is isomorphic to $\mathbf{F}_n$, where $n = b - a$. The rational normal curve of degree a is the curve G_{b-a}. The hyperplane divisor on $S_{a,b}$ is $G_n + bF$; the degree of $S_{a,b}$ is $a + b$.*

The case of $S_{1,1}$ is that of a smooth quadric in $\mathbf{P}^3$.

The limiting case where we take $a = 0$ is simply a cone over a rational normal curve of degree b.

2.8 Surfaces of Minimal Degree

Suppose that $C \subset \mathbf{P}^n$ is a nondegenerate curve. As we noted above, cutting C with a general hyperplane H we will find d points in H which span H. Therefore $d \geq n$: any nondegenerate curve in $\mathbf{P}^n$ has degree at least n. The curves of minimal degree are exactly the rational normal curves.

We can iterate this argument for surfaces, and we find that the minimal degree for a nondegenerate surface in $\mathbf{P}^n$ is $n - 1$: the hyperplane section will be a nondegenerate curve in $\mathbf{P}^{n-1}$, and will therefore have degree at least $n - 1$. This is the case for the rational normal scrolls $S_{a,b}$ introduced above: they have degree $a + b$ and lie in $\mathbf{P}^{a+b+1}$.

Another example of a surface of minimal degree is the double Veronese, defined to be the image of $\mathbf{P}^2$ in $\mathbf{P}^5$ under the map sending $[x : y : z]$ to $[x^2 : y^2 : z^2 : xy : xz : yz]$. This has degree 4 (two conics intersect in four points).

These turn out to be all the examples!

Proposition 2.13 *Let X be a surface of degree $n-1$ in $\mathbf{P}^n$. Then X is either a rational normal scroll $S_{a,b}$ or the Veronese surface in $\mathbf{P}^5$*

2.9 The Cubic Surface

Let X be a smooth cubic surface in $\mathbf{P}^3$. By using the exact sequence

$$0 \to \mathcal{O}_{\mathbf{P}^3}[-3H] \to \mathcal{O}_{\mathbf{P}^3} \to \mathcal{O}_X \to 0$$

we see that $q = p_g = 0$. By adjunction we have that $K_X = -H$, which is negative, so all plurigenera are zero and X is a rational surface. $K^2 = H^2 = 3$, so X is either a 6-fold blowup of the plane or a 5-fold blowup of an $\mathbf{F}_n$. In fact the former is true:

Proposition 2.14 *Let X be a smooth cubic surface in $\mathbf{P}^3$. Then X is a 6-fold blowup of $\mathbf{P}^2$, and the embedding of X into $\mathbf{P}^3$ is obtained by the linear system of cubic curves through the 6 blown-up points. There are 27 lines on X, coming from the 6 exceptional curves, the 15 planar lines through 2 of the blown-up points, and the 6 planar conics through 5 of the blown-up points.*

There are whole books written about the 27 lines on the cubic surface: each meets 10 others in a beautiful configuration.

2.10 Del Pezzo Surfaces

The cubic surface does not have minimal degree, but has degree one more than minimal. This is true of all the surfaces obtained by choosing 6 or fewer points in the plane, and embedding the plane into $\mathbf{P}^n$ by the linear systems of cubic curves through the points. If there are r points, the resulting surface is a surface of degree $9-r$ in $\mathbf{P}^{9-r}$, called a *Del Pezzo surface.*

Classifying surfaces of low degree has been fun for centuries.

Theorem 2.15 *Let X be a smooth surface of degree k in $\mathbf{P}^k$. Then X is either a Del Pezzo surface or is the projection into $\mathbf{P}^4$ of the Veronese surface.*

If we blow up $r=0$ points, then the Del Pezzo surface of degree 9 in $\mathbf{P}^9$ is simply the triple Veronese image of the plane. The $r=1$ case is the Hirzebruch surface $\mathbf{F}_1$ embedded in $\mathbf{P}^8$ via the linear system $2G_1+3F$. The $r=6$ case is, as we noted above, the cubic surface in $\mathbf{P}^3$. The $r=5$ case is a complete intersection in $\mathbf{P}^4$ of type $(2,2)$, that is, the intersection of two quadrics. The cases $r \leq 4$ are not complete intersection surfaces.

2.11 Curves on Rational Surfaces

Suppose we choose points $p_1, \ldots, p_n$ on $\mathbf{P}^2$ and consider the linear system of all curves of degree d having a point of multiplicity at least m_i at p_i for each i. The dimension of the linear system of curves of degree d is $d(d+3)/2$; imposing a point of multiplicity m gives $m(m+1)/2$ conditions. Therefore the expected dimension of the linear system is

$$d(d+3)/2 - \sum_i m_i(m_i+1)/2$$

when this quantity is at least -1.

Of course when the points are in special position the conditions imposed by the multiplicities may not be independent; the first example is when all the points lie on a line, forcing the line to be a component of every curve in the system if the multiplicities are high enough. However one can hope that the expected dimension is the correct dimension if one chooses the points generically; but this is an open problem.

Finally let me mention a related question; when does such a linear system embed the blowup of the plane at the n points into $\mathbf{P}^4$? As an exercise, check the case of quartics double at one point, and passing through 7 others; or quartics through 10 points. Can you find all such linear systems which work for generic points?

Smooth surfaces in $\mathbf{P}^4$ are difficult to construct in general, and have only been classified for very low degrees (up to about 10). It is known that the degree of a smooth rational surface in $\mathbf{P}^4$ is bounded (Ellingsrud, Peskine), but the current bound (about 250, proved by Braun and Floystad) is much higher than the highest degrees known to exist (15, after work by Alexander, Aure, Okonek, Ranestad, Serrano, Decker, Ein, Schreyer, Popescu, ...).

3 Surfaces with Kodaira Dimension Zero

3.1 Abelian Surfaces

Let X be a complex torus, obtained by choosing a rank four $\mathbf{Z}$-lattice L inside $\mathbf{C}^2$ and forming the quotient surface $X = \mathbf{C}^2/L$. If z and w are coordinates on $\mathbf{C}^2$, they descend to give local coordinates on X, which are well-defined up to the translations in L. Therefore the 1-forms dz and dw are well-defined, as is the 2-form $dz \wedge dw$, globally. These generate the spaces of holomorphic 1- and 2-forms, and we see that therefore $q = 2$ and $p_g = 1$; $\chi = 0$. The Euler number $e = 0$, since topologically X is a product of four circles; indeed, the simplicial cohomology is

$$\begin{aligned} H^0(\mathbf{Z}) &\cong \mathbf{Z} \\ H^1(\mathbf{Z}) &\cong \mathbf{Z}^4 \\ H^2(\mathbf{Z}) &\cong \mathbf{Z}^6 \\ H^3(\mathbf{Z}) &\cong \mathbf{Z}^4 \\ H^4(\mathbf{Z}) &\cong \mathbf{Z} \end{aligned}$$

by the Künneth formula, and so the Hodge diamond is

$$\begin{array}{ccccc} & & 1 & & \\ & 2 & & 2 & \\ 1 & & 4 & & 1 \\ & 2 & & 2 & \\ & & 1 & & \end{array}$$

We see here our first example of a delicate Neron-Severi group analysis; the lattice $H^2(\mathbf{Z})$ may intersect the subspace $H^{1,1} \subset H^2(\mathbf{C})$ in any rank, from 0 to 4. If the rank is zero, then there are no ample divisors on X, and X is not an algebraic surface. Here's a challenge for your understanding: construct lattices L for which the Picard number $\rho(X)$ is a given integer between 0 and 4 (I believe all are possible).

In any case, the 2-form $dz \wedge dw$ has no zeroes or poles, so that the canonical divisor K is trivial. Therefore $P_n = 1$ for every n, and $\kappa = 0$. A complex torus X which is algebraic is called an *abelian surface.*

3.2 $K3$ Surfaces

Let X be a quartic surface in $\mathbf{P}^3$. By adjunction we have $K = 0$, and by analyzing the ideal sheaf sequence one easily sees that $q = 0$ and $p_g = 1$, so that $\chi = 2$. Since $K = 0$, surely $K^2 = 0$, so that

$e = 24$ and the Hodge diamond is

$$\begin{array}{ccccc} & & 1 & & \\ & 0 & & 0 & \\ 1 & & 20 & & 1 \\ & 0 & & 0 & \\ & & 1 & & \end{array}$$

Again there is a subtle interplay between $H^{1,1}$ and $H^2(\mathbf{Z})$, and the Neron-Severi group can have any rank between 1 and 20. The quartic is simply connected.

If X is a complete intersection of type $(2,3)$ in $\mathbf{P}^4$, or of type $(2,2,2)$ in $\mathbf{P}^5$, similar analyses lead to the same invariants as above.

Definition 3.16 *A* K3 surface *is a (minimal) compact complex surface X with $q = 0$ and $K = 0$.*

With these assumptions, $p_g = 1$, and X is simply connected, with the above Hodge numbers. A $K3$ surface is an algebraic surface only if its Picard number is at least one; there are non-algebraic $K3$ surfaces (with Picard number 0). There are even non-algebraic $K3$ surfaces with Picard number 1, but all of whose divisor classes have $D^2 = 0$. If a $K3$ surface has a divisor D with $D^2 > 0$, then it is algebraic.

3.3 Kummer Surfaces

Another easy family of examples of $K3$ surfaces are obtained as quotients. If A is a complex torus of dimension two, then consider the quotient $X = A/\{\pm I\}$, called a *Kummer surface.* (There are sixteen ordinary double points on X, coming from the 16 points fixed under $\{\pm I\}$, but these singularities resolve in a nice way.) Note that no 1-form on A is invariant under $\{\pm I\}$, but the 2-form is, so certainly $q = 0$ and $p_g = 1$ for X; moreover the 2-form (which has no zeroes or poles on A) descends to one with no zeroes or poles on X, so $K = 0$ too. Thus X is a $K3$ surface.

One finds Kummer surfaces in all of the families of $K3$ surfaces of genus g. For example, in the $g = 2$ case of double covers of the plane branched along a sextic curve, one obtains a Kummer surface if one takes the sextic curve to be six lines all tangent to a given conic. The complex torus which this surface is a quotient of is the Jacobian of the hyperelliptic curve of genus two obtained by taking the double cover of the conic branched at the six points of tangency.

3.4 K3 Surfaces in Projective Space

Counting parameters, one sees that quartics depend on 19 moduli (there are 35 coefficients, with the 16-dimensional group of 4×4 matrices acting). Similarly the complete intersections (of type $(2, 3)$ and of type $(2, 2, 2)$) mentioned above all depend on 19 moduli. However the space of first-order deformations $H^1(\Theta)$ is 20-dimensional for any $K3$ surface; one sees the non-algebraic $K3$ surfaces lurking around every corner. In fact, the moduli space of all $K3$ surfaces is 20-dimensional, with a countable number of 19-dimensional families representing algebraic $K3$ surfaces.

What are these 19-dimensional families? If a $K3$ surface X contains a curve C of genus $g \geq 3$, then C tends to be very ample on X, and by Riemann-Roch the linear system $|C|$ will embed X into projective space as a smooth surface of degree $2g-2$ in $\mathbf{P}^g$. The condition that X contain such a curve is simply a condition on how the lattice $H^2(\mathbf{Z})$ intersects $H^{1,1}$, and so in the 20-dimensional moduli space there are these 19-dimensional families, one for each g, of projective algebraic surfaces of degree $2g-2$ in $\mathbf{P}^g$. Such a $K3$ surface is called a "$K3$ surface of genus g".

There is even a degenerate case of genus 2: a $K3$ surface of degree 2 in $\mathbf{P}^2$ is a double cover of the plane, branched along a smooth sextic curve. Note that sextics depend on 28 coefficients, and subtracting the 9 dimensions for the group of 3×3 matrices acting on the plane, we see the 19 moduli again for these surfaces.

3.5 Curves on $K3$ Surfaces

Given a $K3$ surface of genus g, hence of degree $2g-2$ in $\mathbf{P}^g$, its general hyperplane section C is a smooth curve of degree $2g-2$ in $\mathbf{P}^{g-1}$; indeed, by adjunction $H|_C$ is the canonical class on C, so C is a canonical curve (embedded by its canonical system).

The number of moduli of such curves is clearly $19+g$; 19 for the $K3$ surface and g for the choice of hyperplane. This is less than the number $3g-3$ of moduli for canonical curves in general, so when $g \geq 12$ there is no chance that every canonical curve could be the hyperplane section of a $K3$ surface.

In fact there is an obstruction, whose analysis is due to J. Wahl (see [W]). His idea is to assume that C lies on a $K3$ surface, and to then degenerate the $K3$ surface X to the cone Y over C. This would be a non-trivial deformation of the cone; however the relevant deformation space of the cone is the cokernel of the Gaussian map

$$\phi : \bigwedge^2 H^0(C, K_C) \longrightarrow H^0(C, 3K_C)$$

defined by $\phi(f dz \wedge g dz) = (f'g - fg')(dz)^3$. If the Gaussian map for C is surjective, then the cokernel is trivial, the deformation space is trivial, and the deformation of the cone Y to X cannot exist; so C could not lie on any such X.

It is known that ϕ is surjective for the general curve of genus $g = 11$ and $g \geq 13$, using a degeneration argument; see [CHM]. It is conjectured that C lies on a $K3$ surface if and only if the Gaussian map ϕ for C is not surjective.

3.6 Enriques Surfaces

In the latter part of the last century it had been conjectured that a surface with $q = p_g = 0$ must be rational; no counterexamples were known. Castelnuovo's rationality criterion (that $q = P_2 = 0$) is stronger, since $P_2 = 0$ implies $P_1 = p_g = 0$. It was Enriques who finally settled this question and constructed non-rational surfaces with $q = p_g = 0$, which are named after him.

One of the simplest constructions is given in [GH]. Let S be the quartic $K3$ surface in $\mathbf{P}^3$ defined by $x^4 + y^4 = z^4 + w^4$. Let σ be the automorphism sending $[x : y : z : w]$ to $[x : iy : -z : -iw]$, which has order four, and acts on S. The quotient of S by the cyclic group generated by σ is an algebraic surface X with invariants $p_g = q = 0$ and $2K = 0$, but $K \neq 0$; K is torsion in $H^2(\mathbf{Z})$. The Hodge diamond for X is

$$\begin{array}{ccccc} & & 1 & & \\ & 0 & & 0 & \\ 0 & & 10 & & 0 \\ & 0 & & 0 & \\ & & 1 & & \end{array}$$

and so $h^{1,1} = b_2 = 10$, and the Neron-Severi group has rank $\rho = 10$.

Definition 3.17 *An algebraic surface with $q = p_g = 0$ and $2K = 0$ is an* Enriques surface.

In fact the torsion in the cohomology is reflected by torsion in the fundamental group: $\pi_1(X) \cong \mathbf{Z}/2$. The universal (double) cover of X is a $K3$ surface; in our example above, the universal cover is the quotient of the original quartic S by the involution σ^2, which is a $K3$ surface again; the further quotient by σ gives the Enriques surface.

Since $2K = 0$ but $K \neq 0$, we have that the plurigenera P_n are either 0 or 1, depending on the parity of n. Hence $\kappa = 0$ for an Enriques surface.

Projective models of Enriques surfaces form a fascinating subject; the reader may wish to consult [CD] for lots of information. Probably the most famous is the realization of an Enriques surface as a sextic surface in $\mathbf{P}^3$ which has 6 double lines, arranged as in the edges of a tetrahedron!

The generic Enriques surface can be constructed as follows. Take six quadratic forms Q_1, Q_2, Q_3, R_1 ,R_2, R_3 each in three variables.

Form the complete intersection surface Y of type $(2,2,2)$ in $\mathbf{P}^5$ as the zeroes of the three polynomials

$$Q_i(x,y,z) + R_i(u,v,w) \quad (i = 1,2,3)$$

where $[x : y : z : u : v : w]$ are the homogeneous coordinates in $\mathbf{P}^5$. For general choices of the forms, Y will be a $K3$ surface, and the involution σ which changes the sign of u, v, and w (leaving x, y, and z alone) acts on Y without fixed points. The quotient Y/σ is an Enriques surface.

Every Enriques surface contains elliptic fibrations, which we will discuss in the next lecture; this is an alternative avenue to pursue in constructing them.

3.7 Hyperelliptic/Bielliptic Surfaces

Enriques surfaces come naturally as fixed-point-free quotients of $K3$ surfaces, and are the only such surfaces. One might ask whether similar quotients can be taken of abelian surfaces, to obtain other surfaces with a torsion canonical class. This is the case; let us briefly go over the construction.

It turns out that the abelian surface, in order to support the right kind of automorphism, must be a product of elliptic curves. So take $S = E \times F$, with E and F elliptic. E may always be taken to be arbitrary, $E = \mathbf{C}/(\mathbf{Z} + \mathbf{Z}\tau_1)$. Write $F = \mathbf{C}/(\mathbf{Z} + \mathbf{Z}\tau_2)$. We have the following seven group actions:

1. τ_2 arbitrary, $\phi(e,f) = (e + 1/2, -f)$; $G =< \phi >\cong \mathbf{Z}/2$.
2. $\tau_2 = \omega$ where $\omega^3 = 1$, $\phi(e,f) = (e+1/3, \omega f)$; $G =< \phi >\cong \mathbf{Z}/3$.
3. $\tau_2 = i$, where $i^2 = -1$, $\phi(e,f) = (e+1/4, if)$; $G =< \phi >\cong \mathbf{Z}/4$.
4. $\tau_2 = \omega$, where $\omega^3 = 1$, $\phi(e,f) = (e + 1/6, -\omega^2 f)$; $G =< \phi >\cong \mathbf{Z}/6$.

5. τ_2 arbitrary, $\phi(e,f) = (e + 1/2, -f)$, $\psi(e,f) = (e + \tau_1/2, f + 1/2)$; $G = <\phi, \psi> \cong \mathbf{Z}/2 \times \mathbf{Z}/2$.

6. $\tau_2 = \omega$, $\phi(e,f) = (e + 1/3, \omega f)$, $\psi(e,f) = (e + \tau_1/3, f + (1 - \omega^2)/2)$; $G = <\phi, \psi> \cong \mathbf{Z}/3 \times \mathbf{Z}/3$.

7. $\tau_2 = i$; $\phi(e,f) = (e+1/4, if)$, $\psi(e,f) = (e+\tau_1/2, f+(1+i)/2)$; $G = <\phi, \psi> \cong \mathbf{Z}/4 \times \mathbf{Z}/2$.

This classification goes back to Bagnera and DeFranchis from 1907.

In each of these cases the quotient $X = (E \times F)/G$ is an algebraic surface with $p_g = 0$ and $q = 1$. In every case the canonical class is torsion, of order 2, 3, 4, 6, 2, 3, and 4 respectively. Therefore the plurigenera are all either 0 or 1, and are 1 for all multiples of 12 in every case, so $\kappa = 0$. Since they are covered by abelian surfaces without branching, the Euler numbers are all 0. Therefore the Hodge diamond must be

$$\begin{array}{ccccc} & & 1 & & \\ & 1 & & 1 & \\ 0 & & 2 & & 0 \\ & 1 & & 1 & \\ & & 1 & & \end{array}$$

and so $b_2 = h^{1,1} = 2$. The Neron-Severi group has rank two, with elements easily described by descending the elliptic curves.

As is the case with the Enriques surfaces, all hyperelliptic surfaces have a pencil of elliptic curves (with the above description this pencil is inherited from E).

These surfaces are beginning to be called "bielliptic" as in Beauville's book [Be].

3.8 The Classification Theorem

Theorem 3.18 *Suppose that X is an algebraic surface with $\kappa = 0$. Then X is either abelian, $K3$, Enriques, or hyperelliptic.*

So we have seen them all!

4 Surfaces with Kodaira Dimension One

4.1 Elliptic Surfaces

Suppose that the Kodaira dimension κ of a surface X is one, and consider an n-canonical map for X; this map cannot be onto a surface, else the growth rate of the plurigenera would be two, by applying Riemann-Roch to the multiples of the hyperplane class of the image (which pull back to multiples of nK). Indeed, the map for large n will be onto a curve only, and the fibers of the map will be curves on X which do not meet the canonical class. If F is such a fiber, then $F^2 = 0$ (since it is the fiber of a map) and $(F \cdot K) = 0$ (since it is the fiber of the canonical map). Hence by the Genus Formula we see that F has genus one. We are thereby led to consider the following situation.

Definition 4.19 *Let X be an algebraic surface. An* elliptic fibration *on X is a holomorphic map $f : X \to C$ where C is a curve, such that the general fiber of f is a smooth curve of genus one. An* elliptic surface *is a surface with a given elliptic fibration.*

The main classification result for surfaces with $\kappa = 1$ is:

Theorem 4.20 *Every algebraic surface with $\kappa = 1$ is elliptic.*

The converse is not true; there are elliptic surfaces which are rational, ruled, abelian, and $K3$; and every Enriques and hyperelliptic surface is elliptic. However every elliptic surface has $\kappa \leq 1$.

4.2 Jacobian Surfaces and Weierstrass Equations

If F is a curve of genus one, then F is isomorphic to its Jacobian, once a point (which serves as the origin of the group law on F) has been chosen. A curve of genus one together with a chosen point is called an *elliptic curve*, or a *Jacobian elliptic curve.*

Every elliptic curve can be written in the Weierstrass form

$$y^2 = x^3 + Ax + B$$

where A and B are numbers such that $4A^3 + 27B^2 \neq 0$ (this is a smoothness condition). The chosen point is the point at infinity.

Taking our cue from this, we can try to form a family of elliptic curves $\{F_t\}$ by letting A and B vary with t, obtaining

$$y^2 = x^3 + A(t)x + B(t).$$

In this way one obtains an elliptic surface, with the property that the chosen point also varies with t, forming a *section* S_0 (called the zero-section of the fibration). If t varies on an algebraic curve, and A and B are rational functions, then this describes an algebraic elliptic surface.

There is no problem when A and/or B has a pole: one can "clear denominators" at will, since the same family of curves is obtained using the equation

$$y^2 = x^3 + Af^4x + Bf^6$$

where f is any holomorphic function. (Just change coordinates, replacing y by yf^3 and x by xf^2, then divide through by f^6.) So locally the family is always definable with regular coefficients.

One sees readily now that there is actually a line bundle L (the conormal bundle of the zero-section S_0) such that A and B are global sections of L^4 and L^6 respectively; the discriminant $D = 4A^3 + 27B^2$ is a section of L^{12}, and vanishes at a point p of the curve exactly when the fiber over p is not smooth.

An elliptic surface defined in this way with a Weierstrass equation is called a *Jacobian* elliptic surface; moreover an elliptic surface is a Jacobian surface if and only if the elliptic fibration has a section.

4.3 Kodaira's Classification of Singular Fibers

In the 1950's, in his fundamental papers [K] on surfaces, Kodaira realized the importance of elliptic surfaces in the whole picture and proved the first classification theorems. One of the most fascinating and useful results was his classification of the possible singular fibers of an elliptic fibration; to this day it represents the only complete classification of degenerations which one can remember (degenerations of curves of genus two are also classified, but the list is quite long).

Kodaira's list of singular fibers:

Notation	Description
I_0	smooth elliptic curve
I_1	rational curve with one node
I_2	two lines meeting at two points
I_n, $n \geq 3$	n lines meeting in a cycle
II	rational curve with one cusp
III	two lines simply tangent at one point
IV	three concurrent lines
I_0^*	five lines, one meeting the other four (as in $\tilde{D}_4$)
I_n^*	$n+5$ lines, meeting as in $\tilde{D}_{n+4}$
IV^*	7 lines, meeting as in $\tilde{E}_6$
III^*	8 lines, meeting as in $\tilde{E}_7$
II^*	9 lines, meeting as in $\tilde{E}_8$
${}_mI_n$	like I_n, but all components with multiplicity m

By the word "line" in the above table, I mean a smooth rational curve. These diagrams (the extended Dynkin diagrams) mentioned above are

$\tilde{D}_4$:

```
        1
        |
    1 - 2 - 1
        |
        1
```

$\tilde{D}_{n+4}$:

```
1 - 2 - ... - 2 - 1
    |         |
    1         1
```

$\tilde{E}_6$:

```
1 - 2 - 3 - 2 - 1
        |
        2
        |
        1
```

$\tilde{E}_7$:

```
1 - 2 - 3 - 4 - 3 - 2 - 1
            |
            2
```

$\tilde{E}_8$:

```
1 - 2 - 3 - 4 - 5 - 6 - 4 - 2
                    |
                    3
```

The numbers each stand for a single $\mathbf{P}^1$ component, of self-intersection -2; the value of the number is the multiplicity of that component in the fiber. So for example in the I_0^* case of the $\tilde{D}_4$ diagram, the central component C_0 has multiplicity two, while the other components $C_1, \ldots, C_4$ have multiplicity one, and the fiber (as a divisor on the surface) is $2C_0 + C_1 + C_2 + C_3 + C_4$.

A smooth elliptic curve may always be written in the plane as a smooth cubic curve, and many of these degenerate fibers can also be seen there. For example, I_1 is a nodal cubic curve, II is a cuspidal cubic curve, I_2 is a conic plus non-tangent line, III is a conic plus a tangent line, I_3 are three lines in a triangle, and IV is three concurrent lines.

4.4 The J-function

Two elliptic curves in Weierstrass form $y^2 = x^3 + Ax + B$ are isomorphic if and only if their J-invariants

$$J = \frac{4A^3}{4A^3 + 27B^2}$$

are equal. For an elliptic surface $f : X \to C$, the J-function becomes a function on the base curve C. When there are singular fibers J may have a pole, and so one usually considers it as a map $J : C \to \mathbf{P}^1$. However J may not have a pole, even at singular fibers; for surfaces in Weierstrass form the local behaviour of J is determined by the order of A, B, and $D = 4A^3 + 27B^2$. We have the following table (where $a = \text{order}(A)$, $b = \text{order}(B)$, and $d = \text{order}(D)$):

Name	a	b	d	J	mult(J)
I_0	0	0	0	$\neq 0, 1, \infty$	-
I_0	$a \geq 1$	0	0	0	$3a$
I_0	0	$b \geq 1$	0	1	$2b$
$I_n, n \geq 1$	0	0	n	∞	n
I_0^*	2	3	6	$\neq 0, 1, \infty$	-
I_0	$a \geq 3$	3	6	0	$3a - 6$
I_0	2	$b \geq 4$	6	1	$2b - 6$
$I_n, n \geq 1$	2	3	n+6	∞	n
II	$a \geq 1$	1	2	0	$3a - 2$
III	1	$b \geq 2$	3	1	$2b - 3$
IV	$a \geq 2$	2	4	0	$3a - 4$
IV^*	$a \geq 3$	4	8	0	$3a - 8$
III^*	3	$b \geq 5$	9	1	$2b - 9$
II^*	$a \geq 4$	5	10	0	$3a - 10$

Note that in the above table, we do not consider $a \geq 4$ and $b \geq 6$ together; if this happens, we can divide (locally) by t^4 and t^6 respectively and get an isomorphic family, and continue doing this until either $a \leq 3$ or $b \leq 5$.

Note that if one has a local parameter t, and one replaces A by At^2 and B by Bt^3, the J-function is left invariant. However this operation (called a "quadratic twist") does change the nature of the singular fiber: for example, it switches an I_n fiber to an I_N^* fiber, and vice-versa, and a glance at the table shows that it also switches the types II and IV^*, III and III^*, and IV and II^*. Therefore knowledge of the J-function does not determine the elliptic surface locally: not even the type of the singular fiber is determined completely.

However this quadratic twist is the *only* way that two Jacobian elliptic surfaces can have the same J-function, at least locally: the precise result is that a Jacobian elliptic fibration is determined by the J-function up to quadratic twists.

For more on the basic theory of elliptic surfaces, the reader may consult [Mi], which is an elementary introduction concentrating on the Jacobian elliptic surfaces.

4.5 Multiple Fibers and Logarithmic Transforms

The existence of the multiple fibers is somewhat counter-intuitive: how can an elliptic curve degenerate to a double one? There is no real contradiction, and actually examples are not too hard to come by. Take a plane sextic curve S with 9 double points; this has geometric genus 1, and can be constructed by embedding any elliptic curve in $\mathbf{P}^5$ via a linear system of degree 6 and generically projecting to a plane. Through these nine points there also lies a cubic curve C. Consider the pencil generated by $2C$ (which is a sextic) and S. This pencil has nine base points (the nine double points) since these are the only points at which C and S meet. Blowing these nine points up one obtains an elliptic surface over $\mathbf{P}^1$, and the fiber corresponding to $2C$ is a fiber of multiplicity two (of type ${}_2I_0$ if C is smooth).

Indeed, this pattern continues: one may construct fibers of multiplicity m by taking a S to be a curve of degree $3m$ with 9 points of multiplicity m, and the cubic C through the nine points, and looking at the pencil generated by S and mC.

There can only be a finite set of singular fibers, and in particular only finitely many multiple fibers. Therefore in a punctured neighborhood of a multiple fiber (that is, take a neighborhood and delete the fiber itself), the surface looks like a surface without any multiple fiber, with a well-defined J-function.

Now take an elliptic fibration $f : X \to C$, and a point $0 \in C$, and a small neighborhood Δ of 0 in C. Consider the pre-image $f^{-1}(\Delta - \{0\})$; for small Δ, f restricts to this pre-image and gives an elliptic fibration with no singular fibers.

The punctured disc is unfortunately not a small neighborhood of any point, so the J-function does not quite determine the elliptic fibration over $\Delta - \{0\}$. To be brief, there is also the monodromy of the two real generators of H_1 going around 0. The theorem for punctured discs is that the elliptic fibration is determined by the J-function and the monodromy.

Therefore one can do "surgery" at will: take out this neighborhood and glue in another, with the same J-function and monodromy on the punctured disc of course, but with a different singular fiber.

Now comes the surprise: a multiple fiber has the same J-function and monodromy as a multiplicity one fiber! Therefore this surgery is possible, and we can replace any ${}_mI_n$ (where J has a pole of order n) with any other ${}_kI_n$ (where J also has a pole of order n). This operation is called a "logarithmic transform", and can radically change the nature of the elliptic surface.

4.6 Numerical Invariants

Let $f : X \to C$ be a minimal elliptic surface, with no (-1)-curves lying in the fibers of f. The canonical class K does not meet the general fiber, and is therefore supported at fibers. In fact, there is a canonical bundle formula

$$K = f^*(K_C + D) + \sum_i (m_i - 1)F_i$$

where the F_i are the multiple fibers (F_i having multiplicity m_i) and D is some non-negative divisor on C. (If f is a Jacobian fibration then D is the conormal bundle of the zero-section of f.) This formula implies in particular that $K^2 = 0$ for an elliptic surface.

All one-forms are pulled back from C, unless X is the product surface $C \times E$ for some elliptic curve E. Therefore $q = g(C)$ (unless X is a product, in which case $q = g(C) + 1$). The value of p_g depends on the multiplicities and the divisor D of course.

Suppose for example that we start with the product surface $X = \mathbf{P}^1 \times E$. This is ruled, and has Kodaira dimension $\kappa = -\infty$. We have $K = -2F$, numerically, where F is a fiber of the elliptic structure, namely the projection to $\mathbf{P}^1$. Now start making logarithmic transforms; this will begin to add terms of the form $(m-1)F_i$, and for enough terms, we can switch K to a positive divisor eventually. In particular, we can change the Kodaira dimension!

Suppose that we take the previous example of $\mathbf{P}^1 \times E$ and only perform a *single* logarithmic transform. A relatively straightforward computation of the homology of this space shows that it has $b_1 = 1$; thus it is not even an algebraic surface! (For an algebraic surface $b_1 = 2q$ and is always even.)

As an example of a rational elliptic surface, take two cubic curves in the plane and blow up the nine base points of the pencil they generate; this blowup maps to the parameter line $\mathbf{P}^1$ of the pencil, with general fiber a general cubic curve in the pencil, exhibiting its elliptic structure.

Of course, if E is an elliptic curve, then $E \times \mathbf{P}^1$ is both ruled (over E) and elliptic (over $\mathbf{P}^1$).

Any product $E \times E'$ with both E and E' elliptic is an elliptic abelian surface.

A quartic surface in $\mathbf{P}^3$ containing a line is a $K3$ surface which is elliptic; the pencil of hyperplanes through the line cuts the surface in the line plus a residual cubic curve, which exhibits the elliptic fibration.

If one takes a rational elliptic surface and performs two logarithmic transformations, creating two double fibers, one gets an Enriques surface. Moreover *every* Enriques surface can be obtained in this way; every Enriques surface is elliptic, with infinitely many elliptic fibrations.

The hyperelliptic surfaces are obviously elliptic by their definition, and in fact carry two elliptic structures.

To finally see an example of an elliptic surface with $\kappa = 1$, consider the Weierstrass equation

$$y^2 = x^3 + A_{4n}(t)x + B_{6n}(t),$$

where A and B are polynomials of degree $4n$ and $6n$ respectively. This is the general Weierstrass form for a Jacobian elliptic surface over $\mathbf{P}^1$. For $n = 0$ we have the product surface $E \times \mathbf{P}^1$. For $n = 1$ the surface is a rational elliptic surface, and can be described by a pencil of cubic curves as mentioned above. For $n = 2$ the surface is a $K3$ surface. For $n \geq 3$ the surface has $\kappa = 1$; the canonical class $K = (n-2)F$.

4.7 Classification Questions

The classification problem for elliptic surfaces is considered to be "solved". Let me explain the sense in which this is true.

The first step is to classify all Jacobian elliptic surfaces. As noted above, these are all described by a line bundle L on the base curve C, and two sections A of L^4 and B of L^6; then the Weierstrass equation

$$y^2 = x^3 + Ax + B$$

defines the surface. So up to understanding curves, line bundles, and sections of powers of the bundle, we consider this to be manageable.

The second step is to realize that, using logarithmic transformations, one can add or subtract multiple fibers on a given elliptic surface at will. Therefore one needs only to classify the elliptic surfaces without any multiple fibers; all others are obtained from these by performing logarithmic transformations.

Finally there is the problem of classifying elliptic surfaces with no multiple fibers, but without any section (so that it is not a Jacobian surface). A glance at the list of possible singular fibers shows that for any non-multiple fiber, there is at least one component of multiplicity one, and hence a *local* section. So every such surface is *locally* isomorphic to a Jacobian surface, and the classification problem becomes a cohomological one, as does every classification of objects locally like a given one. For more details, the reader may consult the Shafarevich seminar notes [S2].

5 Surfaces of General Type

5.1 Positivity of K^2, e, χ

We have now introduced all of the surfaces of special type, and finally we turn to the surfaces of general type, that is, surfaces with Kodaira dimension $\kappa = 2$. By Riemann-Roch, one has a formula for the higher plurigenera

$$P_n = \dim H^0(nK) = (nK)(nK - K)/2 + \chi = \chi + n(n-1)K^2/2$$

for $n \geq 2$, using Kodaira vanishing for $H^1(nK)$ (this is assuming the surface is minimal). Therefore since for $\kappa = 2$ the plurigenera grow quadratically, we must have

$$K^2 > 0$$

for a minimal surface of general type. In the Mori minimal model point of view this property, if K is nef, becomes the definition of general type.

Slightly more technical is the positivity of the Euler number

$$e > 0.$$

Therefore by Noether's inequality $12\chi = K^2 + e$ we have that

$$\chi = 1 - q + p_g > 0$$

or, in other words, $p_g \geq q$.

5.2 Noether's Inequality

About 100 year ago Max Noether proved the inequality

$$p_g \leq 2 + K^2/2$$

for a minimal surface of general type. The basic idea is as follows; assume that a general member $C \in |K|$ is irreducible, and recall that by adjunction we have

$$(K + C)|_C = K_C.$$

On the one hand we know that

$$h^0(K_C) = g(C) = (C^2 + CK)/2 + 1 = K^2 + 1;$$

but a general fact says that in this situation we have $h^0((K+C)|_C) \geq h^0(K|_C) + h^0(C|_C) - 1 = 2h^0(K|_C) - 1$. Now from the sequence

$$0 \to \mathcal{O} \to K \to K|_C \to 0$$

we see that $h^0(K|_C) \geq p_g - 1$, so that $h^0((K + C)|_C) \geq 2p_g - 3$. Hence $K^2 + 1 \geq 2p_g - 3$, leading to Noether's inequality.

More care is necessary in the case that the linear system $|K|$ does not have an irreducible member, but this is the main line of the argument.

5.3 The Bogomolov-Miyaoka-Yau Inequality

The positivity of the Euler number e leads immediately to the inequality

$$K^2 \leq 12\chi$$

by Noether's formula. This has been sharpened to the Bogomolov-Miyaoka-Yau inequality

$$K^2 \leq 9\chi$$

which is much deeper and 100 years younger than the previous inequalities.

5.4 Geography

The four inequalities

$$K^2 \geq 1, \quad \chi \geq 1, \quad K^2 \geq 2\chi - 6, \quad \text{and } K^2 \leq 9\chi$$

give a region in the first quadrant of the (χ, K^2) plane, and the classification questions concerning surfaces of general type center primarily on understanding the particular surfaces with K^2 and χ in this region. It is a theorem of Ulf Persson [P] that all values with $K^2 \leq 8\chi$ actually occur; there exist surfaces of general type in this sub-region of the region of invariants.

Note that the signature τ of the intersection form on $H^2(\mathbf{R})$ is

$$\tau = (K^2 - 2e)/3 = (K^2 - 2(12\chi - K^2))/3 = K^2 - 8\chi$$

so that these surfaces constructed by Persson have negative signature. Surfaces with positive signature are in general much harder to construct.

5.5 Complete Intersections

Suppose that X is a complete intersection in $\mathbf{P}^n$ of type $(d_1, d_2, \ldots, d_{n-2})$. Then

$$K = (\sum_{i=1}^{n-2} d_i - n - 1)H$$

so that as soon as $\sum d_i \geq n + 2$ the surface is of general type. Since we should always take $d_i \geq 2$, the only exceptions to this are the complete intersections of the following types:

$n = 3:$	(2)	quadric surface, rational
$n = 3:$	(3)	cubic surface, rational
$n = 3:$	(4)	quartic surface, $K3$
$n = 4:$	$(2,2)$	quartic Del Pezzo surface, rational
$n = 4:$	$(2,3)$	$K3$ surface of genus 4
$n = 5:$	$(2,2,2)$	$K3$ surface of genus 5.

All other complete intersections are of general type.

For hypersurfaces, that is, surfaces of degree d in $\mathbf{P}^3$, one has

$$\chi = 1 + \frac{(d-1)(d-2)(d-3)}{6}$$

while $K = (d-4)H$ so that

$$K^2 = (d-4)^2 d.$$

Hence, asymptotically for large d, one has $K^2 \approx 6\chi$.

A similar but more complicated analysis for general complete intersections yields the following:

Proposition 5.21 *If X is a complete intersection surface of general type, then $K^2 \leq 8\chi$. Moreover asymptotically, $K^2 \approx 8\chi$ as the degrees grow.*

5.6 Double Covers

A *double cover* of the plane is a surface defined affinely by an equation

$$z^2 = f(x,y)$$

where f is a polynomial of even degree (otherwise the covering is branched also at the line at infinity). Note that z is a square root of f, and therefore numerically, if f has degree $2d$, then z has degree d. Note that

$$2z dz = df$$

so that $dz = df/z$.

Globalizing this is a straight-forward matter. Passing to a general divisor B on a general surface Y, suppose that $B \equiv 2L$ for some divisor L (this is the analogue of the "even degree" assumption). Then a double cover $\pi : X \to Y$ exists, branched over the curve B; the local formula that $dz = df/z$ as above, where f is a local equation for B, leads to the global formula that

$$K_X = \pi^*(K_Y + L).$$

The Euler number of the covering is easy to calculate; let us illustrate this for $Y = \mathbf{P}^2$. Assume that the branch curve B is smooth of degree $2d$, so that also its preimage $R \subset X$ is smooth. Then

$$\begin{aligned} e(X) &= e(X - R) + e(R) \\ &= 2e(\mathbf{P}^2 - B) + e(B) \\ &= 2e(\mathbf{P}^2) - e(B) \\ &= 6 - (2 - 2g(B)) \\ &= 4 + (2d-1)(2d-2) = 4d^2 - 6d + 6. \end{aligned}$$

Since $K^2 = \pi^*((d-3)H)$, and since $H^2 = 1$ and π^* doubles intersections, then

$$K^2 = 2(d-3)^2 = 2d^2 - 12d + 18.$$

Therefore

$$\chi = \frac{1}{12}(e + K^2) = \frac{1}{12}(6d^2 - 18d + 24) = \frac{1}{2}(d^2 - 3d + 4).$$

Therefore we see that, asymptotically, double covers of $\mathbf{P}^2$ only gives surfaces with $K^2 \approx 4\chi$.

5.7 Horikawa's Analysis of Surfaces on the Noether Line

Let us consider minimal surfaces of general type whose invariants lie on the Noether line $K^2 = 2\chi - 6$. For these surfaces $q = 0$, and

$p_g = \chi - 1$. The canonical map has total degree K^2 of course, but Horikawa in the 1970s (see [Ho2]) found that in fact, in almost all cases this map is a double cover onto a surface of degree $\chi - 3$ in $\mathbf{P}^{\chi-2}$. This is a surface of minimal degree, and therefore is either a rational normal scroll or the Veronese surface in $\mathbf{P}^5$. Therefore these surfaces are constructed rather easily as double covers of very well-understood surfaces.

If one takes the double cover of $\mathbf{P}^2$ branched over a curve of degree 8 or 10, then the formulas above give invariants

$$B \text{ of degree } 8: \quad \chi = 4; \quad K^2 = 2$$
$$B \text{ of degree } 10: \quad \chi = 7; \quad K^2 = 8$$

which are on the Noether line; for degrees 12 or larger, they are not.

Let us consider also the double covers of $\mathbf{F}_0 \cong \mathbf{P}^1 \times \mathbf{P}^1$ branched over a curve of bidegree $(6, 2d)$, that is, a curve B linearly equivalent to $6G + (2d)F$, where G and F are the horizontal and vertical rulings. Then $L = 3G + dF$, and $K_{\mathbf{F}_0} = -2G - 2F$, so that on the double cover X we have

$$K^2 = 2(G + (d-2)F)^2 = 4d - 8.$$

A similar computation as the above one for $\mathbf{P}^2$ leads to

$$\chi = 2d - 1$$

so that for any $d \geq 3$ we have a minimal surface of general type on the Noether line.

These examples, and others built in the same way from the higher $\mathbf{F}_n$'s, classify all surfaces on the Noether line, except for a few surfaces with low invariants. These surfaces are being called the *Horikawa surfaces.*

5.8 Surfaces on the BMY Line

As indicated above, it is quite difficult to construct surfaces with positive index, and especially difficult to construct surfaces whose invariants satisfy the Bogomolov-Miyaoka-Yau equality $K^2 = 9\chi$. About 25 years ago Mumford in [Mu] made a complicated construction involving lifting a singular configuration of surfaces over a characteristic two field to characteristic zero, and produced an example of a surface with $\chi = 1$ and $K^2 = 9$. Hirzebruch, about 13 years ago, produced relatively simple constructions involving covers of the plane, of degree larger than two, branched over unions of lines (see [Hi]). That is, these examples were surfaces defined by equations of the form

$$z^n = \prod_{i=1}^{k} L_i(x, y)$$

where the L_i's are linear. One needs the lines in a very special position for the resulting surface to lie on the BMY line. The first example is to take four general points, then take the six lines through any two; with $n = 5$, one obtains (as a quintic cover of the plane) a surface with $K^2 = 9 \cdot 5^4$ and $e = 3 \cdot 5^4$ so $\chi = 5^4$ and we are on the BMY line.

Another example is to take a smooth cubic curve, and consider the nine flex points; take the corresponding nine lines in the dual plane, and take the quintic cover. Here $K^2 = 333 \cdot 5^6$!

There are only finitely many such constructions known, though. Constructions of surfaces with $K^2 = 9\chi$ is still an active and interesting area of current research. It is known that every such surface is a quotient of the 2-ball by a group acting freely and properly discontinuously. For each value of χ, there are only finitely many minimal surfaces of general type up to isomorphism with $K^2 = 9\chi$; these surfaces are rigid (i.e., no deformations).

5.9 Quintic Surfaces

A quintic surface in $\mathbf{P}^3$ has $K = H$, so $K^2 = H^2 = 5$; $q = 0$ and $p_g = 4$ so $\chi = 5$ too.

Another example of a surface with these invariants is the double cover of $\mathbf{F}_0$ branched along a curve of bidegree $(6,8)$, with two quadruple points lying on a single fiber of the ruling on $\mathbf{F}_0$. Without the singularities, we saw above that this surface has $\chi = 7$ and $K^2 = 8$. However the singularities drop the invariants; in this case Horikawa shows that χ drops by 2 and K^2 by three. Therefore the resulting surface actually has $\chi = K^2 = 5$ as does the quintic.

Horikawa (see [Ho1]) proved that every surface with $K^2 = \chi = 5$ was either a quintic or a double cover of this type, or a degeneration of one of these; there are two components to the moduli space of surfaces with these invariants. These two components meet; there are common degenerations of these surfaces.

5.10 Godeaux Surfaces

Very few other surfaces are completely classified. As an embarrassing example, let us consider surfaces with $K^2 = 1$ and $p_g = 0$, the smallest possible set of invariants for minimal surfaces of general type. These surfaces are historically important; after Enriques' example of the Enriques surface, disproving the conjecture that if $p_g = q = 0$ then the surface was rational, it was not known that a surface of general type could be constructed with $p_g = q = 0$ until Godeaux in the 1930's gave the following example.

Take the Fermat quintic surface $x^5 + y^5 + z^5 + w^5 = 0$ in $\mathbf{P}^3$. Let G be the cyclic group of order 5, acting on $\mathbf{P}^3$ by letting the generator send a point $[x : y : z : w]$ to $[x : \zeta y : \zeta^2 z : \zeta^3 w]$ (where ζ is a primitive fifth root of unity). This acts on the quintic, freely. The quotient X is a compact surface of general type, and it is easy to see that there are no G-invariant holomorphic 1-forms or 2-forms on the quintic; hence $p_g = q = 0$ for the quotient X. The canonical class on X lifts to that on the quintic, and since $K^2 = 5$ for the quintic, we have $K^2 = 1$ for X. The fundamental group is $\mathbf{Z}/5$.

A minimal surface of general type with $p_g = q = 0$ and $K^2 = 1$ is called a *Godeaux surface* honoring this example. Godeaux surfaces are not classified; it is not known how many components the moduli space has.

Miles Reid in [Rd] has shown that Godeaux surfaces have at most an order five torsion subgroup T of the Picard group, and has classified those with T of order at least 3.

5.11 Rational Double Points (DuVal Singularities) and Canonical Models

Suppose that a surface contains a (-2)-curve, that is, a smooth rational curve C isomorphic to $\mathbf{P}^1$, with self-intersection -2. By the adjunction formula, we have

$$-2 = 2g(C) - 2 = C^2 + (C \cdot K) = -2 + (C \cdot K)$$

so that $(C \cdot K) = 0$.

Suppose that X is a minimal surface of general type, and consider a pluricanonical map for X, given by the sections of nK for some large n. Since X is of general type, the image of this map will be a surface; however since $(C \cdot nK) = 0$, any (-2)-curve C will be contracted by this map.

The contraction of a (-2)-curve creates a singular point on the image. More generally, if one has any connected collection of (-2)-curves on X, they are all contracted, to a single singular point on the image. Any connected configuration of curves which is contracted must be have a negative definite intersection form by the Hodge Index Theorem, and this restricts greatly the possible configurations; they are

A_n:

$$1 - 1 - \dots - 1 - 1$$

D_n:

$$\begin{array}{ccccccccc} 1 & - & 2 & - & \dots & - & 2 & - & 1 \\ & & | & & & & & & \\ & & 1 & & & & & & \end{array}$$

E_6:

$$\begin{array}{ccccccccc} 1 & - & 2 & - & 3 & - & 2 & - & 1 \\ & & & & | & & & & \\ & & & & 2 & & & & \end{array}$$

E_7:

$$\begin{array}{ccccccccccc} 1 & - & 2 & - & 3 & - & 4 & - & 3 & - & 2 \\ & & & & & & | & & & & \\ & & & & & & 2 & & & & \end{array}$$

E_8:

$$\begin{array}{ccccccccccccc} 2 & - & 3 & - & 4 & - & 5 & - & 6 & - & 4 & - & 2 \\ & & & & & & & & | & & & & \\ & & & & & & & & 3 & & & & \end{array}$$

The singular points which are created are the most important in the singularity theory of surfaces; they are called the *rational double points*, or the *DuVal singularities*. They are relatively ubiquitous in the theory of surfaces, although we have not mentioned them up until this point. The reader can see their relationship with the singular fibers of elliptic surfaces; the numbers which appear above are the multiplicities of the corresponding curve in the inverse image of the maximal ideal of the singular point in the smooth resolution X. An interesting account may be found in [D].

These singularities, as we have seen, cannot be avoided in the pluricanonical images of a surface of general type. However these are the only singularities which appear.

Theorem 5.22 *Let X be a minimal surface of general type. Then for any $n \geq 5$, the pluricanonical map given by the linear system*

$|nK|$ maps X birationally to a surface Y, contracting all configurations of (-2)-curves to rational double points. This map is an isomorphism away from the (-2)-curves.

The statement that $5K$ is enough to embed any surface X is due to Bombieri [Bo], and there is much more detailed information concerning surfaces which can be embedded by lower multiples of K. An alternate approach to this was given by Reider in [Rr].

The image Y is independent of n, and is called the *canonical model* of X. It can be described rather succinctly as

$$Y = \text{Proj} \bigoplus_{n=0}^{\infty} H^0(nK_X).$$

References

[BPV] W. Barth, C. Peters, and A. Van de Ven: *Compact Complex Surfaces.* Ergebnisse der Mathematik und ihrer Grenzgebiete, 3. Folge, Band 4, Springer-Verlag (1984).

[Be] A. Beauville: *Complex Algebraic Surfaces.* London Mathematical Society Lecture Note Series # 68. Cambridge University press (1983).

[Bo] E. Bombieri: "Canonical Models of Surfaces of General Type". Publ. Math. Inst. Hautes Etud. Sci., Vol. 42 (1973), 171-219.

[CHM] C. Ciliberto, J. Harris, and R. Miranda: "On the Surjectivity of the Wahl Map". Duke Mathematical Journal, Vol. 57, No.3 (1988), 829-858.

[CKM] H. Clemens, J. Kollár, and S. Mori: *Higher Dimensional Complex Geometry.* Astérisque No. 166, Société Mathématique de France (1988).

[CD] F. Cossec and I. Dolgachev: *Enriques Surfaces I.* Progress in Mathematics, Vol. 76, Birkhauser Press (1989).

[D] A. Durfee: "Fifteen characterizations of rational double points and simple critical points". L'Enseignement math., T. XXV, fasc. 1-2 (1979), 131-163.

[E] F. Enriques: *Le Superficie Algebriche.* Nicola Zanichelli Editore, Bologna (1949).

[F] M. Freedman: "The topology of four-dimensional manifolds". J. Diff. Geom., Vol. 17 (1982), 357-453.

[FM] R. Friedman and J. Morgan: "Algebraic Surfaces and Seiberg-Witten Invariants", preprint (1995).

[GH] P. Griffiths and J. Harris: *Principles of Algebraic Geometry.* Wiley Interscience (1978).

[Ha] R. Hartshorne: *Algebraic Geometry,* GTM 52, Springer-Verlag (1977).

[Hi] F. Hirzebruch: "Arrangements of lines and algebraic surfaces". In: *Arithmetic and Geometry, Papers Dedicated to I. R. Shafarevich on the Occasion of His Sixtieth Birthday. Volume II, Geometry* M. Artin and J. Tate, Editors. Progress in Mathematics, Vol. 36, Birkhauser Press (1984)

[Ho1] E. Horikawa: "On Deformations of quintic surfaces". Invent. Math., vol. 31 (1975), 43-85.

[Ho2] E. Horikawa: "Algebraic surfaces of general type with small c_1^2, I". Annals of Math., vol. 104 (1976), 357-387.

[K] K. Kodaira: "On compact analytic surfaces II, III". Annals of Math. 77, (1963), 563-626; 78 (1963), 1-40.

[Mi] R. Miranda: *The Basic Theory of Elliptic Surfaces.* Dottorato di Ricerca in Matematica, Dipartimento di Mathematica dell'Universiá di Pisa, EDS Editrice Pisa, (1989).

[Mu] D. Mumford: "An algebraic surface with K ample, $K^2 = 9$, $p_g = q = 0$". American J. of Math., Vol. 101 (1979), 233-244.

[P] U. Persson: "On Chern Invariants of Surfaces of General Type". Compositio Math., Vol. 43 (1981), 3-58.

[Rd] M. Reid: "Surfaces with $p_g = 0$, $K^2 = 1$". J. Fac. Sci., Univ. Tokyo, Sect. IA, 25 (1978), 75-92.

[Rr] I. Reider: "Vector bundles of rank two and linear systems on algebraic surfaces". Ann. of Math., Vol. 127 (1988), 309-316.

[S1] I. R. Shafarevich: *Basic Algebraic Geometry.* Grundlehren der mathematischen Wissenschaften, vol. 213, Springer-Verlag, Berlin-Heidelberg-New York (1977).

[S2] I. R. Shafarevich et. al.: *Algebraic Surfaces.* Proc. of the Steklov Institute, Vol. 75 (1965).

[W] J. Wahl: "The Jacobian of a graded Gorenstein singularity". Duke Mathematical Journal, Vol 55 (1987), 843-871.

[Z] O. Zariski: *Algebraic Surfaces.* Ergebnisse der Mathematik, Vol. 3, No. 5, Springer-Verlag, Berlin (1935). second supplemented edition, with appendices by S. S. Abyankar, J. Lipman, and D. Mumford, Ergebnisse der Mathematik, Vol. 61, Springer-Verlag, Berlin-Heidelberg-New York (1971).

Computing Invariants of Hypersurface Singularities

Donal O'Shea

Department of Mathematics
Mount Holyoke College
South Hadley
Mass 01075, USA

e-mail: doshea@mhc.mtholyoke.edu

1 Introduction

This set of notes is intended for distribution to the participants of the Algebraic Geometry Summer School in Bilkent University, Ankara, Turkey. The notes provide a brief overview of some invariants associated to hypersurface singularities and indicate how one goes about computing them. They are idiosyncratic, dealing with matters that my students and I have found interesting. I have tried to supply references to more complete (and balanced) treatments throughout the text. But the road to hell is paved with good intentions. Caveat lector.

I would like to thank Prof. Sinan Sertoz for the wonderful environment during the school and Prof. Dr. Ihsan Dogramaci for his extraordinary patronage. This work was partially supported by National Science Foundation grant DMS 9404497.

These notes are divided into five parts as follows, roughly corresponding to the contents of each lecture.

1. Brief overview of singularities
2. Limits of tangent spaces
3. Computational methods in polynomial rings
4. Milnor numbers
5. Computational methods in local rings

Good general references on singularity theory include [2], [10], [19] and [21]. Good general references on computation in polynomial rings include [3], [8], and [4].

2 Singularities: a brief overview

We begin with some basic definitions that we shall use throughout. Let **C** be the field of complex numbers.

Definition 2.1 *A subset* $V \subset \mathbf{C}^n$ *is an* **algebraic set** *if* V *is the set of common zeroes of polynomials in* $\mathbf{C}[x_1, \ldots, x_n]$.

Definition 2.2 *An algebraic set* V *is called a* **variety** *or an* **irreducible** *algebraic set if it cannot be expressed as the union of two proper algebraic subsets.*

The following proposition characterizes the property of being a variety algebraically.

Proposition 2.3 *Let $V \subset \mathbf{C}^n$ be an algebraic set and let $\mathbf{I}(V)$ be the ideal of $\mathbf{C}[x_1, \ldots, x_n]$ consisting of all polynomials which vanish at every point of V. Then*

$$V \text{ a variety} \iff \mathbf{I}(V) \text{ prime} \iff \mathbf{C}[x_1, \ldots, x_n]/\mathbf{I}(V) \text{ an integral domain.}$$

This characterization ensures that the following definition makes sense.

Definition 2.4 *If V is a variety, then the* **dimension** *of V, denoted* dim(V), *is the transcendence degree over $\mathbf{C}$ of the quotient field of* $\mathbf{C}[x_1, \ldots, x_n]/\mathbf{I}(V)$.

We are now ready to define a singular point.

Definition 2.5 *Let $V \subset \mathbf{C}^n$ be a nonempty algebraic set and suppose that $f_1, \ldots, f_k \in \mathbf{C}[x_1, \ldots, x_n]$, $k < \infty$ are such that $\mathbf{I}(V)$ is generated by $f_1 \ldots f_k$. We write $\mathbf{I}(V) = \langle f_1, \ldots, f_k \rangle$. Let ρ be the largest rank the $k \times m$ matrix $(\partial f_i / \partial x_j)$ attains at any point $x \in V$. A point $x \in V$ is* **nonsingular** *or* **simple** *if $(\partial f_i / \partial x_j)$ attains its maximal rank ρ at x and* **singular** *if*

$$\text{rank} \left(\frac{\partial f_i}{\partial x_j} \right) < \rho.$$

We remark that the Hilbert basis theorem guarantees that *any* ideal in $\mathbf{C}[x_1, \ldots, x_n]$ can be generated by a *finite* set of polynomials. Moreover, it is easy to check that the definition above is $\Sigma(V)$ denote the set of singular points of V. It is always a proper subset of V.

Theorem 2.6 *(***Whitney***) If V is a nonempty algebraic set, the set $V - \Sigma(V)$ is a smooth non-empty, complex manifold. It is complex analytic of dimension $n - \rho$ over $\mathbf{C}$.*

Our **main question** will be: what does V look like in a neighborhood of a singular point? To even begin to answer this question, we need to decide when two varieties look alike in some neighborhood. The definitions are easiest in the case when the variety can be defined by a single polynomial. Accordingly, in what follows, unless explicitly stated otherwise, we specialize to (complex) **hypersurfaces**.

Definition 2.7 *A variety $V \subset \mathbf{C}^n$ is called a* **hypersurface** *if V is the set of zeroes of a single polynomial.*

By a standard theorem in algebraic geometry, a variety in $\mathbf{C}^n$ is a hypersurface if and only if $\dim_{\mathbf{C}} V = n-1$. (This fails spectacularly over $\mathbf{R}$, where every algebraic set can be given by a single equation.)

Notice that if V is a hypersurface, $V = \{f = 0\}$, then

$$p \in V \quad \text{singular} \Longleftrightarrow \frac{\partial f}{\partial x_1}(p) = \ldots + \frac{\partial f}{\partial x_n}(p) = 0.$$

For hypersurfaces, we formalize the phrase "a singular point p of V **looks like** a singular point p' of V'" in one of the following ways.

Definition 2.8 *If $V, V' \subset \mathbf{C}^n$ are hypersurfaces with singular points p, p', respectively, then we say that the singular point p of V is* **topologically equivalent** *to p' of V' if there exists a neighborhood $B_\epsilon = \{x \in \mathbf{C}^n : |x - p| < \epsilon$ in $\mathbf{C}^n$ and a homeomorphism $h : B_\epsilon \longrightarrow h(B_\epsilon$ such that $h(p) = p'$ and $h(V \cap B_\epsilon) = V' \cap h(B_\epsilon)$. We say that p of V is* **analytically equivalent** *to p' of V' if there exist B_ϵ and H as above with h an analytic diffeomorphism.*

Other equivalences are possible: for example, we could insist that h be Lipschitz, or C^r for some r. Or we could drop the requirement that h be defined on the ambient space.

A usual tool in visualizing a singularity p is its link.

Definition 2.9 *Let $p \in V \subset \mathbf{C}^n$ and for each $\epsilon > 0$ let $B_\epsilon = \{x \in \mathbf{C}^n : |x-p| < \epsilon\}$ and $S_\epsilon = \partial B_\epsilon = \{x \in \mathbf{C}^n : |x-p| = \epsilon\}$. Then for ϵ sufficiently small, $V \cap S_\epsilon$ is called the* **link** *of the singularity.*

In order for this definition to make sense, it must be the case that for ϵ sufficiently small $V \cap S_\epsilon$ does not depend on ϵ in any important way. This is indeed the case as the following result will show. First, we recall that if K is a set in $\mathbf{C}^n$, then the **cone** over K based at the point p, denoted $\mathrm{Cone}_p(K)$, is the set $tp + (1-t)k : k \in K, t \in \mathbf{R}$. By the cone over a pair of sets (G, K), denoted $\mathrm{Cone}_p(G, K)$, we mean the pair of cones $(\mathrm{Cone}_p(G), \mathrm{Cone}_p(K))$. Note that the cone over a sphere based at the center of the sphere is just the ball centered on the same point.

Theorem 2.10 *Let V be a hypersurface and $p \in V$. Then for all ϵ sufficiently small, and S_ϵ as in definition 9, $p \in V$ is topologically equivalent to $\mathrm{Cone}_p(V \cap S_\epsilon$. In other words, if B_ϵ is the ball of radius ϵ centered on p, then $(B_\epsilon, B_\epsilon \cap V)$ and $Cone_p(S_\epsilon, S_\epsilon \cap V)$ are homeomorphic as pairs.*

This result tells us both that the link of a singularity is well-defined up to homeomorphisms, and that the link determines the singularity up to topological equivalence. A proof can be found in [21] in the case of isolated singularities (see below for the definition) and in [7] in general.

We say that a singularity $p \in V$ is **isolated** if there exists B_ϵ centered at p such that the only singularity of $V \cap B_\epsilon$ is p. In the case that p is isolated, then $V - \{p\}$ is a manifold near p and it is easy to see that S_ϵ intersects V transversely for small ϵ. It follows that the link of an isolated singularity is a smooth manifold. Here are some examples.

- If $V = \{x^2 - y^3 = 0\}$, then S_ϵ is the 3-sphere and $V \cap S_\epsilon$ is homeomorphic to a circle. It is, however, embedded in S_ϵ as a torus knot of type $(2,3)$. That is, the link is homeomorphic to the type of knot one obtains in the three-sphere, by winding a curve twice around a torus in one direction and three times in the other. If one thinks of the torus as obtained by identifying opposite sides of a unit square, then a torus knot of type (3,2) is traced out by a straight line of slope 3/2. In particular, there is no homeomorphism of S_ϵ which carries the link onto an unknotted circle.

- If $V = \{x^k + y^l = 0\}$ with k, l relatively prime, then the link is a torus knot of type (k, l).

- If $V = \{x^k + y^{kl} = 0\}$ with k, l relatively prime, then the link is a torus link consisting of k unknotted circles, any two of which have linking number l.

- The above three items are easily proved by parameterizing V. In general, if V is a plane curve with one branch, then the link is an iterated torus knot (a torus knot on the boundary of a tubular neighborhood of a knot which is a torus knot on the boundary of a tubular neighborhood of ... of a torus knot). See [12]

- If $V = \{2xy - z^2 = 0\}$, then the link is homeomorphic to $\mathbf{RP}^3$ embedded in the 5-sphere S_ϵ. One sees this by parameterizing: $V = \{(s^2, t^2, \sqrt{2}st) | s, t \in \mathbf{C}\}$ where (s, t) and $(-s, -t)$ give the same point. On V,

$$|x|^2 + |y|^2 + |z|^2 = (|s|^2 + |t|^2)^2,$$

so that $V \cap S_\epsilon$ is homeomorphic to the unit sphere in (s, t) space modulo identification of antipodal points, and this is just real projective 3-space.

- If $V = \{x_1^3 + x_2 + \ldots + x_n^2 = 0\}$ and n is even, then the link is a topological sphere, which may fail to be diffeomorphic to the standard sphere. This fact was discovered by Hirzebruch: see [21].

One studies singularities by associating various invariants to them. These invariants can be geometric, algebraic or topological or (in the best cases) some combination of two of these. We are going to concentrate on two sorts of invariants: geometric invariants related to the tangent cone and algebraic (and topological) invariants related to "Milnor numbers". To compute the former, we will have to have some knowledge of computation in polynomial rings. To compute the latter, we will need to look at how one can do computations in local rings.

3 Tangent cones and limits of tangent spaces

Suppose that $V \subset \mathbf{C}^n$ is a variety (not necessarily a hypersurface). Then at any point p of V we can consider the tangent space $T(V, p)$, which is defined as follows.

Definition 3.1 *If $f \in \mathbf{C}[x_1, \ldots, x_n]$ is a polynomial, the* **linear part** *of f at p, denoted $d_p(f)$, is defined to be the polynomial*

$$d_p(f) = \frac{\partial f}{\partial x_1}(p)(x_1 - p_1) + \cdots + \frac{\partial f}{\partial x_n}(p)(x_n - p_n) .$$

Note that $d_p(f)$ has total degree ≤ 1. $V \subset \mathbf{C}^n$ and $p = (p_1, \ldots, p_n) \in V$ then the **tangent space** *to V at p, denoted $T(V, p)$ is the variety*

$$T(V, p) = \mathbf{V}(d_p(f) : f \in \mathbf{I}(V)) .$$

If p is nonsingular, then V is a manifold near p and the definition of tangent space above coincides with the usual definition of the tangent space to a manifold at a point. However, if p is a singular point, then the dimension of the tangent space is "too big".

Example 3.2 *Suppose that $V = \{x^2 + y^3 = 0\}$. If $p = (1, 1)$ then $T(V, p)$ is the line $2(x - 1) + 3(y - 1) = 0$ through (1,1) normal to (2,3). However, if $p = (0, 0)$ is the singular point, then $T(v, p) = \{\mathbf{C}^2\}$ has dimension 2.*

To get something more useful we approximate V near p not by linear terms in the defining equations (which might well be zero), but by the lowest degree nonvanishing homogeneous pieces of the defining equations. More precisely, suppose that $p = (p_1, \ldots, p_n) \in \mathbf{C}^n$. If $\alpha = (\alpha_1, \ldots, \alpha_n) \in \mathbf{Z}^n_{>0}$, let

$$(x-p)^\alpha = (x_1 - p_1)^{\alpha_1} \cdots (x_n - p_n)^{\alpha_n} ,$$

and note that $(x-p)^\alpha$ has total degree $|\alpha| = \alpha_1 + \cdots + \alpha_n$. Now, given any polynomial $f \in \mathbf{C}[x_1, \ldots, x_n]$ of total degree d, we can write f as a polynomial in $x_i - p_i$, so that f is a $\mathbf{C}$-linear combination of $(x-p)^\alpha$ for $|\alpha| \leq d$. If we group terms according to total degree, we can write

$$f = f_{p,0} + f_{p,1} + \cdots + f_{p,d} ,$$

where $f_{p,j}$ is a $\mathbf{C}$-linear combination of $(x-p)^\alpha$ for $|\alpha| = j$. Note that $f_{p,0} = f(p)$ and $f_{p,1} = d_p(f)$ (as defined above). One can use Taylor's formula to express $f_{p,j}$ in terms of the partial derivatives of f at p. Notice also that if $p = 0$, then we are writing f as the sum of its homogeneous components. In most situations, it will be convenient to translate p to the origin. We now define the tangent cone.

Definition 3.3 *If $f \in \mathbf{C}[x_1, \ldots, x_n]$ is a nonzero polynomial, then $f_{p,min}$ is defined to be $f_{p,j}$, where j is the smallest integer such that $f_{p,j} \neq 0$ in the displayed formula above. The* **tangent cone** *of V at p, denoted $C(V, p)$, is the variety*

$$C(V,p) = \mathbf{V}(f_{p,min} : f \in \mathbf{I}(V)) .$$

The tangent cone of at a point on a hypersurface $V \subset \mathbf{C}^n$ is especially easy to compute. It is an easy exercise to show that if $\mathbf{I}(V) = \langle f \rangle$, then $C(V,p)$ is defined by the the single equation $f_{p,min} = 0$. However, when $\mathbf{I}(V) = \langle f_1, \ldots, f_s \rangle$ has more generators, it need *not* follow that $C(V,p) = \mathbf{V}((f_1)_{p,min}, \ldots, (f_s)_{p,min})$. For example, suppose that V is defined by $xy = xz + z(y^2 - z^2) = 0$. One shows without difficulty that $\mathbf{I}(V) = \langle xy, xz + z(y^2 - z^2) \rangle$. To see that $C(V, 0) \neq \mathbf{V}(xy, xz)$, note that $f = yz(y^2 - z^2) = y(xz + z(y^2 -$

$z^2)) - z(xy) \in \mathbf{I}(V)$. Then $f_{0,min} = yz(y^2 - z^2)$ vanishes on $C(V, 0)$, but not on all of $\mathbf{V}(xy, xz)$.

The tangent cone $C(V, p)$ to a variety V at the point p is made up of lines through p. To describe which lines through p lie in $C(V, p)$ we recall first that a line L in $\mathbf{C}^n$ through p is called a **secant line** if it meets V in a point distinct from p. If we take secant lines determined by points of V getting closer and closer to p, then the "limit" of the secant lines lies on the tangent cone. To make this idea precise, suppose we have parametrized L as $p + tv$, where $v \in \mathbf{C}^n$ is a nonzero vector parallel to L and $t \in \mathbf{C}$. We say that a line $L \in \mathbf{C}^n$ through a point $p \in \mathbf{C}^n$ is a **limit of lines** $\{L_k\}_{k=1}^{\infty}$ through p if given a parameterization $p + tv$ of L, there exist parametrizations $p + tv_k$ of L_k such that $\lim_{k\to\infty} v_k = v$ in $\mathbf{C}^n$. A standard result in algebraic geometry then asserts that the tangent cone consists precisely of limits of secant lines (see, for example, [26].)

Theorem 3.4 *Let $V \subset \mathbf{C}^n$ be a variety. Then a line L through p in $\mathbf{C}^n$ lies in the tangent cone $C(V, p)$ if and only if there exists a sequence $\{q_k\}_{k=1}^{\infty}$ of points in $V - \{p\}$ converging to p with the property that the secant lines joining p and q_k converge to the given line L.*

Instead of introducing the tangent cone using its defining equations, we could have begun with the geometric characterization afforded by the result above. First note that the tangent cone can be viewed as the subspace of projective space whose points correspond to lines through a prescribed point. Limits of lines then translate into limits of points in projective space.

Now, let $V \subset \mathbf{C}^n$ be a variety and $p \in V$ a singular point. If ϕ maps $V - \{0\}$ to the set $\mathbf{P}^{n-1}$ of lines through p:

$$\phi : x \mapsto \overline{0x}, \quad x \in V - \{0\},$$

and if we consider the graph of ϕ,

$$\operatorname{gr} \phi \subset (V - \{0\}) \times \mathbf{P}^{n-1},$$

then the closure

$$\mathrm{cl}\,(\mathrm{gr}\,\phi) \subset V \times \mathbf{P}^{n-1}$$

projects onto V. Call this projection π. Then, one checks easily that

$$\pi^{-1}(p) = \{p\} \times [C(V,p)],$$

where $[C(V,p)]$ denotes the canonical image of $C(V,p)$ in $\mathbf{P}^{n-1}$.

The tangent cone provides an approximation to a surface near a singular point in the sense that, given any conical neighborhood of the tangent cone, a sufficiently small neighborhood of the surface lies entirely within the conical neighborhood (see [23] for a precise statement). It is also easy to see that the tangent cone transforms well under analytic diffeomorphism. Thus the tangent cone is an invariant of a singularity which is easy to compute. The difficulty with the tangent cone is that it is too coarse – it gives no indication of a number of features of a variety about which one would want to know, and entirely dissimilar varieties may have the same tangent cone. For example the surfaces $\{y^2+x^3-z^2x^2 = 0\}$ and $\{y-x^2-z^2 = 0\}$ have the same tangent cone, namely the plane $y = 0$.

A more subtle invariant is the space $K(V,p)$ of limiting tangent spaces to V at p. Since we shall restrict ourselves to hypersurfaces in what follows, we only provide a definition in the case that $V \subset \mathbf{C}^n$ is a hypersurface. Suppose that this is the case. Then, if $x \in V$ is a nonsingular point of V, the tangent space $T(V,x)$ to V at x is a hyperplane, which we label by its "conormal vector", $df(x)$, whose components are just the partials of f evaluated at x (its complex conjugate is normal to $T(V,x)$ with respect to the usual Hermitian metric on $\mathbf{C}^n$). Let Σ denote the singular set of V and let ψ map $V - \Sigma$ to the set $\mathbf{P}^{n-1}$ of planes in $\mathbf{C}^n$ through p by

$$\psi : x \mapsto [df(x)], \quad x \in V - \Sigma,$$

where $[df(x)]$ denotes the canonical image of the vector $df(x)$ in projective space (that is, the set of all nonzero scalar multiples of $df(x)$). Again, we consider the closure of the graph of ψ and let π be the map onto V induced by projection. We define $K(V,p)$ by demanding that its image $[K(V,p)]$ in $\mathbf{P}^{n-1}$ satisfy

$$\pi^{-1}(p) = \{p\} \times [K(V,p)].$$

We call an element of $K(V, p)$ a *limiting tangent space* to V at p.

An element of $K(V, p)$ consists of the set of vectors v in $\mathbf{C}^n$ based at the point p with the property that there exists a sequence of points $x_k \in V - \Sigma$, $x_k \rightarrow p$, for which the lines through x_k in the direction $df(x_k)$ converge to the line thru p in the direction v. Each such vector determines a hyperplane T in $\mathbf{C}^n$ through p (if $v = (v_1, \ldots, v_n)$ and $p = (p_1, \ldots, p_n)$, the equation of $T = \{u = (u_1, \ldots, u_n) \in \mathbf{C}^n : v_1(u_1 - p_1) + \ldots + v_n(u_n - p_n) = 0\}$, and each such hyperplane is the limit of some sequence of tangent hyperplanes $T(V, x_k)$ as x_k tends to p, the convergence of the tangent hyperplanes being understood as convergence in the "dual" projective space $\check{\mathbf{P}}^{n-1}$. In order to avoid trivialities, we shall suppose that $p \in \Sigma$ (otherwise $K(V, p)$ is the one point set $\{T(V, p)\}$).

As we shall see shortly, the invariant $K(V, p)$ contains a great deal of information about the surface. Since it is so easy to find the equation for $C(V, p)$ given the equation for V, one might ask if the same is true for $K(V, p)$: can we find its equation(s) given the equation for V? Sadly, things are not so straightforward as they are for tangent cones. In order to compute equations for $K(V, p)$, we will need to develop some computational techniques in polynomial rings. Before doing so, we develop a few other facts about the geometry of $K(V, p)$.

A moment's reflection will convince one that a limiting tangent space to the tangent cone of a variety "ought" to be a limiting tangent space to the variety itself: that is, $K(C(V, p)) \subset K(V, p)$. This inclusion was established by [27]. However, as Whitney points out, this inclusion can be, and typically is, strict. The question of characterizing the "extra" limiting tangent spaces is answered by a structure theorem due to Lê and Teissier [18] for arbitrary varieties.

Theorem 3.5 *There exists a finite (possibly empty) set of proper subcones*
$C_1, \ldots, C_r \subset C(V, p)$, *$C_i$ cones over p for all (called* **exceptional cones***) such that*

$$K(V, p) = K(C(V, p), p) \cup \cup_{i=1}^{r} K(C_i),$$

where $K(C_i)$ *denotes the set of all hyperplanes in* $\mathbf{C}^n$ *tangent to* C_i *along some line in* C_i *through* p.

Thus, one knows $K(V,p)$, given $C(V,p)$ and the exceptional cones. The data {tangent cone, exceptional cones} are called the *aureole* of V at p and signal important geometric features of the variety near p. For example, the z-axis is an exceptional cone for $\{y^2 + x^3 - z^2x^2 = 0\}$, and this signals the "vanishing fold". The fact that the z-axis is a line of singular points is incidental: one can easily concoct similar examples in which the surface has an isolated singularity at the origin.

Lê and Teissier also characterize the exceptional cones as the non-moving parts of tangent cones to polar varieties. We explain briefly in the case of surfaces. So suppose, temporarily, that $V \subset \mathbf{C}^3$ is a surface. Then the exceptional subcones of $C(V,p)$ will be lines through p. To find them, let $\pi : \mathbf{C}^3 \longrightarrow \mathbf{C}^2$ be a linear projection and define the *polar curve* P_π of V with respect to π to be the closure of the critical locus of the restriction of π to the smooth part of V:

$$P_\pi \equiv \mathrm{cl}\,(\mathrm{crit}\,(\pi|_{V-\Sigma})).$$

Generically, P_π is either empty or one-dimensional (over $\mathbf{C}$). Consider the tangent cone at 0 of P_π. If P_π is one-dimensional, the tangent cone will be a union of lines in $\mathbf{C}^3$ through p. In fact, it is easy to see that, as sets, $C(P_\pi, p) \subset C(V,p)$. As π varies, some lines in $C(P_\pi, p)$ will vary, while others will stay fixed.

Theorem 3.6 *The exceptional lines (of* V *at* p*) are precisely those lines in* $C(P_\pi, p) \subset C(V,p)$ *which do not vary as* π *varies.*

(The result, above, for surfaces was first proved by Lê and Henry for surfaces with isolated singularities and by Lé for arbitrary surfaces.) This result gives us another approach to computing $K(V,p)$, at least for surfaces. Here, again, however, we will need to develop some computational techniques for working with polynomials, an issue to which we now turn.

4 Computational Methods in Polynomial Rings

The modern algorithmic theory of computation in polynomial rings begins with the constructive solution, in 1965, of the ideal membership problem. This problem, first posed by Macaulay in [20] asks the following:

(Ideal Membership Problem). *Given an ideal* $I = \langle f_1, \ldots, f_r \rangle \subset \mathbf{C}[x_1, \ldots, x_n]$ *and a polynomial* $g \in \mathbf{C}[x_1, \ldots, x_n]$, *determine whether* $g \in I$.

The problem is easy if $n = 1$. For in this case every ideal can be generated by a single polynomial and to decide if a polynomial g belongs to the ideal, one merely divides g by the generating polynomial – if the remainder is zero, then g belongs to the ideal; otherwise it doesn't. The problem is likewise easy if $f_1, \ldots, f_r$ are linear and n is arbitrary. The problem in the general case is more difficult and Macaulay never did solve it. A fully algorithmic solution to the ideal membership problem only appeared in 1965, when Bruno Buchberger, a student of Groebner, created a working algorithm to test ideal membership [6]. (A year earlier in 1964, Hironaka had established a solution in principle as a part of his work on resolution of singularities [16].)

The solution of the ideal membership problem consisted of three pieces:

1. the introduction of a linear multiplicative order on all monomials

2. the extension of the division algorithm from polynomials of one variable to polynomials in an arbitrary number of variables, and

3. the notion of a Groebner basis,

each of which we describe in turn.

1. Multiplicative Orders on Monomials

If $\alpha = (\alpha_1, \ldots, \alpha_n) \in \mathbf{Z}^n_{>0}$, we let x^α denote the monomial $x_1^{\alpha_1} \cdots x_n^{\alpha_n}$. The first ingredient in the Hironaka-Buchberger solution was to linearly order all monomials in $\mathbf{C}[x_1, \ldots, x_n]$ in such a way that if $x^\alpha > x^\beta$, where $\alpha, \beta \in \mathbf{Z}^n_{>0}$, then $x^{\alpha+\gamma} > x^{\beta+\gamma}$ for all $\gamma \in \mathbf{Z}^n_{>0}$. Such an ordering is called a **multiplicative ordering**. Multiplicative orderings were already used to great effect by Macaulay.

The best known example of a multiplicative ordering is the **lexicographic** or **lex** ordering. Here we say that $x^\alpha >_{lex} x^\beta$ (and drop the subscript *lex* when the order is understood) if the first nonzero term of $\alpha - \beta$ is greater than 0. Thus, $x_1^2 x_2 x_3^5 x_4 > x_1^2 x_2 x_3^2 x_4^9$ because the first nonzero term (reading from the left) of $(2,1,5,1)-(2,1,2,9) = (0,0,3,-8)$ is 3 which is greater than 0. As another example, in the lex ordering we have

$$x^4 > x^3y^3 > x^2y^{10} > x^2 > xy^9 > x > y^{100} > y > 1.$$

A different multiplicative ordering is the **graded lexicographic** or **glex** ordering. This ordering sorts first by total degree and then uses the lexicographic order to sort monomials of the same degree. That is, we say that $x^\alpha >_{glex} x^\beta$ if $|\alpha| > |\beta|$ (here, $|\alpha| = \alpha_1 + \ldots + \alpha_n$ and $|\beta|$ is defined similarly) or if $|\alpha| = |\beta|$ and the first nonzero term of $\alpha - \beta$ is nonzero. Again we drop the subscript *glex* when the ordering is clear from the context. With respect to the graded lex ordering we have

$$y^{100} > x^2y^{10} > xy^9 > x^3y^3 > x^4 > x^2 > x > y > 1.$$

In general, a multiplicative ordering is said to be **graded** if it first sorts by degree.

There are many other examples of monomial orderings, some better than others for different sorts of computations. In doing computations in polynomial rings, one usually restricts oneself to orderings that are **well-orderings**: that is, which are such that every non-empty subset of monomials has a least element. One can show that this is equivalent to the condition that $1 < x_i$ for all variables (see [8], §2.4). So, the lex and graded lex orders are well-orderings.

If $f = \sum_\rho a_\rho x^\rho \in \mathbf{C}[x_1, \ldots, x_n]$ is a polynomial, we say that a monomial $x^\rho = x_1^{\rho_1} \cdots x_n^{\rho_n}$ or a term $a_\rho x^\rho$ **belongs to** f if the corresponding coefficient a_ρ is nonzero. Once we have established an order $>$ on monomials, we define the **leading term** $\text{LT}_>(f)$ of f to be $a_\rho x^\rho$ where x^ρ is the largest monomial (with respect to the order $>$) which belongs to f. Thus, for example,

$$\text{LT}_{lex}(x^4 + xy^9 + x + 15y^{100} + y + 1) = x^4,$$

while

$$\text{LT}_{glex}(x^4 + xy^9 + x + 15y^{100} + y + 1) = 5y^{100}.$$

2. The Division Algorithm for Polynomials

The second ingredient of the solution to the ideal membership problem is the division algorithm for polynomials. This is a procedure for dividing a polynomial in any number of variables by an ordered collection of polynomials. This procedure works once a monomial order has been chosen. It is best explained by an example, which I have lifted from [24].

Consider the following question.

$$\text{Does } x^4y - xy^3 + y^4 + x^2y \in \langle xy + 1, x - y\rangle \text{ ?}$$

With the case of one variable in our mind, the natural thing to do is to try to divide $x^4y - xy^3 + y^4 + x^2y$ by $xy + 1$ and $x - y$. Suppose we choose the graded lex order with $x > y$. Then the terms in the three polynomials are written in descending order and

$$\text{LT}(x^4y - xy^3 + y^4 + x^2y) = x^4y, \quad \text{LT}(xy+1) = xy, \text{ and } \text{LT}(x-y) = x.$$

The leading monomial xy of $xy + 1$ divides the leading monomial x^4y of $x^4y - xy^3 + y^4 + x^2y$ to give x^3. Multiplying x^3 by $xy + 1$ and subtracting gives $-xy^3 + y^4 - x^3 + x^2y$. The leading monomial xy of

$xy+1$ divides the leading term $-xy^3$ of $-xy^3+y^4-x^3+x^2y$ to give $-y^2$. Multiplying $-y^2$ by $xy+1$ and subtracting gives $y^4-x^3+x^2y+y^2$. Neither the leading monomial of $xy+1$ nor the leading monomial x of $x-y$ divide the leading monomial y^4 of $y^4-x^3+x^2y+y^2$. Thus, we put y^4 into the remainder and subtract it from $y^4-x^3+x^2y+y^2$ to get $-x^3+x^2y+y^2$. Now the leading monomial xy of $xy+1$ does not divide the leading term $-x^3$ of $-x^3+x^2y+y^2$, but the leading monomial x of $x-y$ does. Dividing the latter (that is, x) into $-x^3$ gives $-x^2$; multiplying $-x^2$ by $x-y$ and subtracting the result from $-x^3+x^2y+y^2$ gives y^2. Neither the leading monomial of $xy+1$ nor the leading monomial x of $x-y$ divide the leading monomial y^2 of y^2, so we put y^2 into the remainder leaving nothing left to divide. Thus, the process terminates and we have found that

$$x^4y-xy^3+y^4+x^2y=(x^3-y^2)(xy+1)-x^2(x-y)+y^4+y^2.$$

We leave it as an exercise to show that if we divide in a different order, putting $x-y$ before $xy+1$, then we get

$$\begin{aligned}&x^4y-xy^3+y^4+x^2y=\\&(x^3y+x^2y^2+xy^3+y^4-y^3+xy+y^2)(x-y)+0(xy+1)+y^5+y^3.\end{aligned}$$

Notice that the answers and, in particular, the remainders are sensitive to the order in which we divide! This is bad or, at least, highly undesirable. In particular, the results of our efforts might lead us to carelessly conclude that $x^4y-xy^3+y^4+x^2y$ does not belong to the ideal $\langle x-y, xy+1\rangle$, which is false because

$$x^4y-xy^3+y^4+x^2y=x^3(xy+1)+(x^2+y^3)(x-y).$$

In fact, the reader may enjoy finding an example of polynomials $g, f_1, f_2 \in \mathbf{C}[x,y]$ such that dividing g by f_1, f_2 gives nonzero remainder, but dividing g by f_2, f_1 gives zero remainder!

The last ingredient which allows the division algorithm to to be used to determine ideal membership is the notion of a Groebner basis, together with Buchberger's algorithm which allows one to compute such bases.

3. Groebner Bases

Fix a multiplicative ordering $>$ on the monomials in $\mathbf{C}[x_1, \dots, x_n]$. If I is an ideal in $\mathbf{C}[x_1, \dots, x_n]$, we let $\text{LT}(I)$ denote the ideal generated by all leading terms of elements of I:

$$\text{LT}(I) = \langle \text{LT}(f) : f \in I \rangle.$$

A **Groebner basis** of an ideal I (with respect to the ordering $>$) is a set of polynomials $f_1, \dots, f_r \in I$ such that

$$\text{LT}(I) = \langle \text{LT}(f_1), \dots, \text{LT}(f_r) \rangle.$$

We leave it as an easy exercise to show that if $f_1, \dots, f_r$ is a Groebner basis of I, then $f_1, \dots, f_r$ is actually a basis of I; that is, $I = \langle f_1, \dots, f_r \rangle$.

Consider our example above and let $I = \langle xy + 1, x - y \rangle$. Note that

$$\text{LT}(\langle x - y, xy + 1 \rangle) \neq \langle \text{LT}(x - y), \text{LT}(xy + 1) \rangle = \langle x, xy \rangle = \langle x \rangle$$

because

$$-y(x - y) + (xy + 1) = y^2 + 1 \in I,$$

but

$$y^2 = \text{LT}(y^2 + 1) \notin \langle x \rangle.$$

An easy computation shows that $\text{LT}(I) = \langle x, y^2 \rangle$ so that $xy + 1, x - y, y^2 + 1$ is a Groebner basis of I. Dividing $x^4y - xy^3 + y^4 + x^2y$ by $xy + 1, x - y, y^2 + 1$ gives

$$x^4y - xy^3 + y^4 + x^2y = (x^3 - y^2)(xy + 1) + x^2(x - y) + y^2(y^2 + 1)$$

clearly showing that the former is in I. (In fact, the polynomial $xy + 1$ is redundant and $x - y, y^2 + 1$ is actually a Groebner basis of I.)

Bruno Buchberger introduced the notion of a Groebner basis in 1965 (naming in honor of his thesis supervisor) and, equally importantly, he discovered an algorithm for computing them. The idea behind the algorithm is simplicity itself. Given a set of generators $\{f_1, \dots, f_r\}$ of an ideal, one successively considers the polynomials

$$S(f_i, f_j) = \text{LT}(f_i)f_j - \text{LT}(f_j)f_i.$$

If $\text{LT}\, S(f_i, f_j)$ is not divisible by some $\text{LT}\, f_k$ (that is, if $\text{LT}\, S(f_i, f_j) \notin \langle f_1, \ldots f_r \rangle$) then one adds $\text{LT}\, S(f_i, f_j)$ to the list of generators $\{f_1, \ldots, f_r\}$, otherwise one does nothing. The actual algorithm is a few lines of code and relies on a result which characterizes Groebner bases as those bases which give zero remainder when divided into all possible S-polynomials (eee [3], Chap 2.7). Simple arguments show that the algorithm terminates.

Groebner bases allow one to solve the ideal membership problem because the remainder obtained by dividing any $f \in \mathbf{C}[x_1, \ldots, x_n]$ by a Groebner basis is unique (that is, independent of the order in which one takes the elements of the basis) and $f \in I$ if and only if the remainder is 0. Algorithmically, one first uses Buchberger's algorithm to compute a Groebner basis, then divides the polynomial in question by this basis, and checks whether the remainder is zero.

Buchberger's algorithm and the division algorithm are implemented on many computer algebra systems. These two algorithms allow one to explicitly carry out operations on polynomial ideals. In particular, they allow us to carry out the operations we will need to compute limits of tangent spaces: namely, elimination and intersection.

Suppose that want to eliminate variables in a system of equations. In other words, suppose we are given an ideal $I \subset \mathbf{C}[x_1, \ldots, x_n]$, and we want to find the ideal $I \cap \mathbf{C}[x_{n-r+1}, \ldots, x_n]$ generated by the elements of I containing only the last r, say, variables. (This is the algebraic operation corresponding to computing the closure of the projection of the variety defined by I onto the coordinate subspace of $\mathbf{C}^n$ spanned by the last r coordinate vectors.) The following result shows that this can be done by ordering the variables lexicographically and computing a Groebner basis with respect to this ordering.

Theorem 4.1 *If $g_1, \ldots, g_s$ is a Groebner basis of $I \subset \mathbf{C}[x_1, \ldots, x_n]$ with respect to the lexicographic order in which*

$$x_1 > x_2 > \ldots > x_n,$$

then, for some $k \leq s$, the polynomials $g_{k+1}, \ldots, g_s$ will only involve the last r variables $x_{n-r+1}, \ldots, x_n$ (and any polynomial g_i with $i <$

k, will involve at least one of the variables $x_1, \ldots, x_{n-r}$ in a term with nonvanishing coefficient). The polynomials $g_{k+1}, \ldots, g_s$ are a standard basis for $I \cap \mathbf{C}[x_{n-r+1}, \ldots, x_n]$:

$$I \cap \mathbf{C}[x_{n-r+1}, \ldots, x_n] = \langle f_{n-r+1}, \ldots, f_n \rangle.$$

The proof consists in examining the division algorithm carefully (see [8], Chap. 2.8). To return to our example above, if $I = \langle xy+1, x-y \rangle$ and we want to compute $I \cap \mathbf{C}[y]$, then we compute a standard basis for I, finding that $I = \langle x - y, y^2 + 1 \rangle$ so that $I \cap \mathbf{C}[y] = \langle y^2 + 1 \rangle$.

Being able to eliminate variables immediately allows us to compute the equations of the image of an algebraic set. To be more precise, let $I \subset \mathbf{C}[x_1, \ldots, x_n]$ be an ideal and $V = \mathbf{V}(I)$ the corresponding algebraic set in $\mathbf{C}^n$ (here, and henceforth, we use the notation $\mathbf{V}(I)$ to denote the set of common zeroes of the polynomials in I). If $\Phi : \mathbf{C}^n \longrightarrow \mathbf{C}^m$ is a polynomial map in the sense that $\Phi = (\phi_1, \ldots, \phi_m)$ with each $\phi_i \in \mathbf{C}[x_1, \ldots, x_n]$, then the the image of $\Phi(V)$ of V under the map Φ need not be algebraic set. However, it is the case that the topological closure of $\Phi(V)$ is algebraic (and, in fact, the smallest algebraic set containing $\Phi(V)$ – see [23]) We can use theorem 4.1 to compute an ideal which defines the closure of $\phi(V)$. Namely, the following holds.

Proposition 4.2 *Let $I = \langle f_1, \ldots, f_r \rangle$ be an ideal in $\mathbf{C}[x_1, \ldots, x_n]$ and $\Phi = (\phi_1, \ldots, \phi_m)$ an algebraic map. Consider the ideal*

$$J \equiv \langle f_1, \ldots, f_r, y_1 - \phi_1, \ldots, y_m - \phi_m \rangle \cap \mathbf{C}[y_1, \ldots, y_m].$$

Then $\mathbf{V}(J)$ is the closure of $\Phi(V)$.

The proof is straightforward: see [8] Chap. 2.8.

This already allows us to compute the limits of tangent spaces at an isolated singularity because of the following lovely observation, due to Hénaut [15].

Proposition 4.3 *Let $V = \{x \in \mathbf{C}^n : f(x) =\}$, $f \in \mathbf{C}[x_1, \ldots, x_n]$ be a hypersurface with an isolated singularity at the origin. Let $\phi : \mathbf{C}^n, \mathbf{0} \longrightarrow \mathbf{C}^n, \mathbf{0}$ be the Jacobian map $\phi(x) = (\frac{\partial f}{\partial x_1}(x), \ldots, \frac{\partial f}{\partial x_n}(x))$. In a sufficiently small neighborhood U of the origin, $\phi(V \cap U)$ is a hypersurface in $\phi(U)$ (which we denote somewhat carelessly as $\phi(V)$). Then, as sets, the limit of tangent spaces (thought of as normals) $K(V, 0)$ to V at 0 is the tangent cone $C(\phi(V), 0)$ to $\phi(V)$ at 0.*

The proposition is almost a tautology when we represent tangent spaces by their "normals". We leave the easy proof to the reader (or see [24]). As an example of the proposition, consider the hypersurface $V = \{\frac{1}{2}x^2 + \frac{1}{2}y^2 + \frac{1}{3}z^3 = 0\}$. Here, $\phi(x, y, z) = (x, y, z^2)$ and we can parameterize V as $\{(s, t, (-\frac{3}{2}s^2 - \frac{3}{2}t^2)^{1/3}\}$. Thus, $\phi(V) = \{(s, t, (-\frac{3}{2}s^2 - \frac{3}{2}t^2)^{2/3}\}$. A local equation for $\phi(V)$ is the hypersurface $z^3 = (-\frac{3}{2}x^2 - \frac{3}{2}y^2)^2$ and the reduced tangent cone is $\{z = 0\}$. So $K(V, 0)$ consists of all normals lying in the xy-plane (or all planes containing the z-axis), which is certainly plausible from the (real) picture.

Combining the two previous propositions allows us to write out a procedure for computing the equation for $K(V, 0)$ when V has an isolated singularity.

Proposition 4.4 *Let $V = \{x \in \mathbf{C}^n : f(x) =\}$, $f \in \mathbf{C}[x_1, \ldots, x_n]$ be a hypersurface with an isolated singularity at the origin. Then $K(V, 0)$ is a hypersurface in $\mathbf{C}^n$ with a local equation given by the initial form of the generator of (the radical of) the ideal*

$$\langle f, y_1 - \frac{\partial f}{\partial x_1}, \ldots, y_n - \frac{\partial f}{\partial x_n} \rangle \cap \mathbf{C}[y_1, \ldots, y_n].$$

In practice, one does not bother computing radicals. We leave it as an easy computation to redo our example above to show directly that

$$\langle \tfrac{1}{2}x^2 + \tfrac{1}{2}y^2 + \tfrac{1}{3}z^3, u - x, v - y, w - z^2 \rangle \cap \mathbf{C}[u, v, w] = \langle 4w^3 + 9u^4 + 18u^2v^2 + 9v^4 \rangle.$$

Being able to eliminate variables also allows us to compute intersections of ideals. This, in turn, allows us to compute equations for $K(V,0)$ in the case that 0 is not an isolated singularity. To pursue this, suppose that I and J are two ideals in $\mathbf{C}[x_1,\ldots,x_n]$ and let $t\cdot I$ denote the ideal in $\mathbf{C}[t,x_1,\ldots,x_n]$ generated by all multiples of elements of I by t and $(1-t)\cdot J$ the ideal generated by multiplying all elements of J by $1-t$). Then, the following holds.

Proposition 4.5 $I\cap J = t\cdot I + (1-t)\cdot J \cap \mathbf{C}[x_1,\ldots,x_n]$.

The proof is easy (see, for example, [8], §4). Moreover, because we can compute intersections, the following proposition is algorithmic.

Proposition 4.6 *Let $V=\{x\in\mathbf{C}^n : f(x)=\}$, $f\in\mathbf{C}[x_1,\ldots,x_n]$ be a hypersurface with an isolated singularity at the origin and let $K(V,0)\subset\mathbf{C}^n$ be the space of limiting tangent spaces. Let A denote the ideal in $\mathbf{C}[x_1,\ldots,x_n,s,y_1,\ldots,y_n]$ given by setting*

$$A=\langle f, y_1 - s\frac{\partial f}{\partial x_1},\ldots, y_n - s\frac{\partial f}{\partial x_n}\rangle.$$

Then the ideal $\mathbf{I}(K(V,0))$ of $K(V,0)$ is the radical of the ideal in $\mathbf{C}[y_1,\ldots,y_n]$ given by eliminating s and setting $x_1,\ldots,x_n$ equal to zero. That is,

$$\mathbf{I}(K(V,0))=\sqrt{A\cap\mathbf{C}[x_1,\ldots,x_n,y_1,\ldots,y_n]/\langle x_1,\ldots,x_n\rangle}.$$

For a proof, see [24]. As an example, suppose again that $V=\{\frac{1}{2}x^2+\frac{1}{2}y^2+\frac{1}{3}z^3=0\}$. Then

$$A=\langle x^2+y^2+\tfrac{2}{3}z^3, u-tx, v-ty, w-tz^2\rangle.$$

Note that $zt^2(x^2+y^2+\frac{2}{3}z^3)$ (which belongs to A) is congruent to $zu^2+zv^2+\frac{2}{3}w^2$ modulo A and thus belongs to A. But the latter does not depend on t and belongs, therefore, to $A\cap\mathbf{C}[x,y,z,u,v,w]$. Thus, $w^2\in(A\cap\mathbf{C}[x,y,z,u,v,w])/\langle x,y,z\rangle$ (we leave it as an exercise to show that the latter ideal is, in fact, equal to $\langle w^2\rangle$), in agreement with the calculation we did using Hénaut's observation.

One can also compute the exceptional cones, using their characterization as non-moving parts of tangent cones to polar varieties. However, the computational techniques needed for this would bring us beyond the scope of these lectures. For more information about this approach, see [24].

5 Milnor Numbers

Suppose henceforth that $V = \{x \in \mathbf{C}^n : f(x) = 0\}$, $f \in \mathbf{C}[x_1, \ldots, x_n]$ is a hypersurface with an isolated singularity at the origin. We have already seen several geometric invariants of V, namely $C(V, 0)$ and $K(V, 0)$. These are invariant under analytic equivalence of a singularity. Arguably, the most important invariant, however, of an isolated singularity of a hypersurface is its Milnor number.

Definition 5.1 *Let V and f be as above, then the* **Milnor number** *of V at 0 denoted $\mu(V, 0)$ is the number*

$$\dim_{\mathbf{C}} \mathbf{C}[[x_1, \ldots, x_n]]/\langle \frac{\partial f}{\partial x_1}, \ldots, \frac{\partial f}{\partial x_n}\rangle \mathbf{C}[[x_1, \ldots, x_n]]$$

where $\mathbf{C}[[x_1, \ldots, x_n]]$ denotes the ring of formal power series in $x_1, \ldots, x_n$.

An argument that we sketch later shows that the Milnor number is finite precisely when 0 is an isolated singularity.

This is a purely algebraic definition. However, the Milnor number can be characterized in a number of different ways. For $u = (u_1, \ldots, u_n)$ sufficiently close to the origin and generic, the Milnor number of f at the origin is precisely the number of critical points of $f_u = f(x) + u_1 x_1 + \ldots + u_n x_n$ near 0 (if the Milnor number is μ, what happens is that the critical point of f at 0 breaks up into a number of critical points near 0 which have smaller Milnor number. For u generic, these decomposed critical points all are nondegenerate (i.e. have Milnor number one), and there μ of them).

The Milnor number can also be characterized topologically. For ϵ sufficiently small, the ball $B_\epsilon = \{x \in \mathbf{C}^n : |x| \leq \epsilon\}$ contains no singularities of f other than the origin and is such that for all $\delta << \epsilon$ sufficiently small, $f^{-1}(z)$, $z \in \mathbf{C}$, $|z| < \delta$ meets S_ϵ transversally. For $z \neq 0, |z| < \delta$, the level hypersurfaces $f^{-1}(z) \cap B_\epsilon$ are all diffeomorphic, and nonsingular. Any one of them is called the **Milnor fiber** of f. They are $(2n-2)$-dimensional real manifolds with boundary, and no homology except in the middle dimension $n-1$. The rank of the homology group $H_{n-1}(F, \mathbf{Z})$, $F = f^{-1}(z) \cap B_\epsilon$ turns out to be the Milnor number. More is true: if μ is the Milnor number, then the Milnor fiber has the same homotopy type as a bouquet of μ real $(n-1)$-dimensional spheres. For proofs and more see [21].

Best of all, if two hypersurface singularities are topologically equivalent, then they have the same Milnor numbers [17]. Thus, the Milnor number is an invariant that has both topological and algebraic significance – hence, its importance.

There are a number of invariants of hypersurfaces that also appear as dimensions of algebras. Best known is the **Tjurina number** $\tau(V,0)$ of a hypersurface, which is defined to be

$$\dim_{\mathbf{C}} \mathbf{C}[[x_1, \ldots, x_n]]/\langle f, \frac{\partial f}{\partial x_1}, \ldots, \frac{\partial f}{\partial x_n}\rangle.$$

The Tjurina number is equal to the dimension of the base space of a so-called miniversal deformation of the singularity (see [19] for details), and is an analytic invariant.

Once the Milnor number of a hypersurface is defined, one can define a sequence of Milnor numbers as follows. If 0 is an isolated singularity of the hypersurface $V \subset \mathbf{C}^n$ and if $H \subset \mathbf{C}^n$ is a generic hyperplane passing through the origin, then $V \cap H$ is a hypersurface in $H \equiv \mathbf{C}^{n-1}$ with an isolated singularity at the origin. For almost all hyperplanes H, the Milnor number $\mu(V \cap H, 0)$ is constant (and a minimum). Call this number $\mu^{n-1}(V,0)$, Iterating, we get a sequence

$$\mu^*(V,0) = \{\mu \equiv \mu^n, \mu^{n-1}, \ldots, \mu^1\}$$

of numbers called the **μ^*-invariant**. We remark that μ^1 is the Milnor number of the intersection of a generic line through the origin with V – an easy exercise shows that this is just one less than the multiplicity $m(V,0)$ of V at the origin. Here, the multiplicity is the usual multiplicity of a hypersurface; namely, the degree of the lowest degree term that belongs to f. Equivalently, it is the intersection multiplicity of a generic line through the origin with V.

Incidentally, the hyperplanes H for which $\mu(V \cap H, 0) = \mu^{n-1}$ are precisely those which do not correspond to elements in $K(V,0)$. Similarly, the lines L for which $\mu(V \cap L, 0) = \mu^1$ (that is, lines whose intersection number with V at 0 is the $m(V,0)$) are precisely those which do not belong to $C(V,0)$. Thus, being able to compute Milnor numbers would give us another way to determine membership in $K(V,0)$ and $C(V,0)$!

Teissier, who introduced the μ^*-invariants, showed that the μ^*-invariant is an analytic invariant (see [25] p.315). It is known that it is not a topological invariant (two singularities with are topologically equivalent can have different $\mu*$-invariants (see [5]). One of the most stubbornly resistant unsolved problems in singularity theory is whether multiplicity is invariant under topological equivalence of hypersurfaces (this question was asked by Zariski in 1971 (see [28]) and the assertion that the multiplicity is a topological invariant is often referred to as *Zariski's conjecture*, notwithstanding the fact that Zariski did not take any position on the issue, other than musing in public that it ought to be easy to settle.) It would be a major advance if one could determine whether all members of family of singularities in which the Milnor number is constant necessarily have the same multiplicity (except in dimension 3, it is known that all members of such families are topologically equivalent).

We would like to have a way to compute Milnor numbers and related invariants.

6 Computations in Local Rings

In order to compute Milnor numbers, it is useful to be able to make computations in local rings. Recall that a local ring is a ring with exactly one maximal ideal.

The ring $\mathbf{C}[[x_1, \ldots, x_n]]$ of formal power series

$$\mathbf{C}[[x_1, \ldots, x_n]] = \{\sum_{\alpha \geq 0} a_\alpha x^\alpha, \alpha \in \mathbf{Z}_+^n, a_\alpha \in \mathbf{C}\},$$

with the usual definitions of addition and multiplication, is a local ring with maximal ideal $\langle x_1, \ldots, x_n \rangle$. Any element not in $\langle x_1, \ldots, x_n \rangle$ is invertible as a formal series and is, therefore, a unit (in particular, it cannot belong to any proper ideal). The ring $\mathbf{C}\{x_1, \ldots, x_n\}$ consisting of power series in $x_1, \ldots, x_n$ which converge in some neighborhood of the origin is also a local ring. Another example of a local ring is the ring of complex-valued rational functions defined at the origin, denoted $\mathbf{C}[x_1, \ldots, x_n]_{\langle x_1, \ldots, x_n \rangle}$,

$$\mathbf{C}[x_1, \ldots, x_n]_{\langle x_1, \ldots, x_n \rangle} = \{\frac{f}{g} : f, g \in \mathbf{C}[x_1, \ldots, x_n], g(0) \neq 0\},$$

with the usual operations of addition and multiplication. We have natural inclusions

$$\mathbf{C}[x_1, \ldots, x_n] \subset \mathbf{C}[x_1, \ldots, x_n]_{\langle x_1, \ldots, x_n \rangle} \subset \mathbf{C}\{x_1, \ldots, x_n\} \subset \mathbf{C}[[x_1, \ldots, x_n]]$$

and ideals in $\mathbf{C}[x_1, \ldots, x_n]$ containing $< x >$ extend to ideals in these local rings.

In computing Milnor numbers, we are called to compute the dimension of the quotient ring by a zero-dimensional ideal. We remark, that it doesn't matter in which local ring we compute the dimension of the quotient.

Proposition 6.1 *Suppose that $f_1, \ldots, f_n \in \mathbf{C}[x_1, \ldots, x_n]$ have a common isolated zero which we take to be at the origin (isolated, means that there is a ball about the origin containing no common* *zeroes of $f_1, \ldots, f_n$ other than the origin). Then, the the quotient by the ideal generated by the f_i's is a finite dimensional algebra and*

$$\begin{aligned} & \dim_{\mathbf{C}} \mathbf{C}[x_1, \ldots, x_n]_{\langle x_1, \ldots, x_n \rangle} / \langle f_1, \ldots, f_n \rangle \mathbf{C}[x_1, \ldots, x_n]_{\langle x_1, \ldots, x_n \rangle} = \\ = \; & \dim_{\mathbf{C}} \mathbf{C}[[x_1, \ldots, x_n]] / \langle f_1, \ldots, f_n \rangle \mathbf{C}[[x_1, \ldots, x_n]] = \\ = \; & \dim_{\mathbf{C}} \mathbf{C}\{x_1, \ldots, x_n\} / \langle f_1, \ldots, f_n \rangle \mathbf{C}\{x_1, \ldots, x_n\}. \end{aligned}$$

To see why this is so, suppose that we want to compute the quotient by the ideal generated by $x^2 + x^3$ and y^2. Then $\dim \mathbf{C}[x, y]_{\langle x,y \rangle} / \langle x^2 + x^3, y^2 \rangle \mathbf{C}[x, y]_{\langle x,y \rangle} = 4$ because $\langle x^2 + x^3, y^2 \rangle \mathbf{C}[x, y]_{\langle x,y \rangle} = \langle x^2, y^2 \rangle \mathbf{C}[x, y]_{\langle x,y \rangle}$ since $1/(1 + x) \in \mathbf{C}[x, y]_{\langle x,y \rangle}$ and because the monomials $1, x, y, xy$ are a vector space basis of $\dim \mathbf{C}[x, y]_{\langle x,y \rangle} / \langle x^2, y^2 \rangle$. We can represent $1/(1 + x)$ as the formal power series $1 - x + x^2 - x^3 + x^4 - \ldots \in \mathbf{C}[[x, y]]$ and then

$$(x^2 + x^3)(1 - x + x^2 - x^3 + x^4 - \ldots) = x^2$$

in $\mathbf{C}[[x, y]]$. This shows that $\langle x^2 + x^3, y^2 \rangle \mathbf{C}[[x, y]] = \langle x^2, y^2 \rangle \mathbf{C}[[x, y]]$ from which it follows that

$$\dim_{\mathbf{C}} \mathbf{C}[[x, y]] / \langle x^2, y^2 \rangle \mathbf{C}[[x, y]] = 4$$

(as before, the monomials $1, x, y, xy$ are a vector space basis of $\mathbf{C}[[x, y]] / \langle x^2, y^2 \rangle$). However, the power series $1 - x + x^2 - x^3 + x^4 - \ldots$ is, in fact, convergent, and precisely the same reasoning shows that $\langle x^2 + x^3, y^2 \rangle \mathbf{C}\{x, y\} = \langle x^2, y^2 \rangle \mathbf{C}\{x, y\}$ and, therefore,

$$\dim_{\mathbf{C}} \mathbf{C}\{x, y]\} \langle x^2, y^2 \rangle \mathbf{C}\{x, y\} = 4.$$

One can generalize from this example to give an clumsy, but valid proof of the proposition assuming that one of the dimensions is finite. The dimension $\dim_{\mathbf{C}} \mathbf{C}\{x_1, \ldots, x_n\} / \langle f_1, \ldots, f_n \rangle \mathbf{C}\{x_1, \ldots, x_n\}$ is finite by the local analytic Nullstellensatz (the hypothesis that the origin is isolated and the local analytic Nullstellensatz say that $\sqrt{\langle f_1, \ldots, f_n \rangle} = \langle x_1, \ldots, x_n \rangle$ so that some power of the ideal $\langle f_1, \ldots, f_n \rangle$ must lie in the maximal ideal). The proposition is true, however, for quite general reasons. See [11].

In the case of the usual polynomial ring, the corresponding dimension bounds the number of common zeroes of the generators of the ideal. That is, if the polynomials $f_1, \ldots, f_m$ have a finite number of zeroes, then the ring $\mathbf{C}[x_1, \ldots, x_n]/\langle f_1, \ldots, f_m\rangle$ is finite dimensional and the number of common zeroes is less or equal to $\dim_{\mathbf{C}} \mathbf{C}[x_1, \ldots, x_n]/\langle f_1, \ldots, f_m\rangle$. More is true: if $m = n$ (so the number of polynomials is the same as the number of variables), then $\dim_{\mathbf{C}} \mathbf{C}[x_1, \ldots, x_n]/\langle f_1, \ldots, f_n\rangle$ is exactly the number of common zeroes of $f_1, \ldots, f_n$ counted with multiplicity. (See [13] §2.9, Proposition 6.)

Happily, we know how to compute the dimension in the case of polynomial rings. For any ordering and any ideal I such that $\dim_{\mathbf{C}} \mathbf{C}[x_1, \ldots, x_n]/I < \infty$, we have

$$\dim_{\mathbf{C}} \mathbf{C}[x_1, \ldots, x_n]/I = \dim_{\mathbf{C}} \mathbf{C}[x_1, \ldots, x_n]/\langle \text{LT} I\rangle,$$

and the latter is just the number of monomials

$$x^\alpha : x^\alpha \notin < \langle \text{LT} I\rangle.$$

(See, for example, [8], Chapter 5.3.)

So, for example, if

$$I = \langle x^2 + x^3, y^2\rangle \subset \mathbf{C}[x, y],$$

and we are using lexicographic order, then

$$\dim \mathbf{C}[x, y]/I = \dim \mathbf{C}[x, y]/\langle \text{LT} I\rangle = \dim \mathbf{C}[x, y]/\langle x^3, y^2\rangle = 6.$$

The rightmost equality follows because $1, x, x^2, y, xy, xy^2$ project to a vector space basis of $\mathbf{C}[x, y]/\langle x^3, y^2\rangle$. So there are at most 6 common zeroes of $x^2 + x^3$ and y^2, in A_k^2. In fact, we see there are two solutions $(-1, 0)$ and $(0, 0)$.

So, in the polynomial case, the heart of the matter is to compute $\langle \text{LT} I\rangle$ given generators $f_1, \ldots, f_s$ of I and this is done by computing a Groebner basis of I.

Our computation in the local case suggests that a similar strategy works for quotient of local rings. If $I = \langle x^2 + x^3, y^2\rangle$ and R were

$\mathbf{C}\{x,y\}_{\langle x,y\rangle}$ (or $\mathbf{C}[[x,y]]$ or $\mathbf{C}\{x,y\}$), we computed $\dim R/IR$ by replacing I by the monomial ideal $\tilde{I} = \langle x^2, y^2\rangle$. Note that $\tilde{I}$ consists of the *lowest degree terms* of I in contrast to the situation in which we compute $\dim \mathbf{C}[x,y]/I$ by replacing I by $\langle \text{LT} I\rangle = \langle x^2, y^3\rangle$, the highest degree terms in the generators.

It should come as no surprise to us that we need to examine lowest degree terms. When dealing with a power series (or a rational function), it is natural to define the initial form to be the lowest degree homogeneous piece.

There are two ways of proceeding: we could retain the term orderings we have been using with polynomials, but re-examine what we did in picking out terms which were maximal with respect to that ordering with a view to modifying the procedures to pick out terms minimal with respect to the ordering. There is a trick, often credited to Lazard, which allows one to do this easily. Or, we could admit orderings in which the maximal elements would have least degree, and try to carry over the theory of Groebner bases to that situation. We adopt the latter procedure, see[9] for more on the Lazard procedure.

So, we will we consider degree-anticompatible (or anti-graded) orderings:

$$|\alpha| < |\beta| \Longrightarrow x^\alpha > x^\beta.$$

We still insist that such orderings be total orderings and multiplicative.

Definition 6.2 *A* **degree-anticompatible ordering** *on* $\mathbf{C}[x_1,\ldots,x_n]$, $\mathbf{C}[x_1,\ldots,x_n]_{\langle x_1,\ldots,x_n\rangle}$, $\mathbf{C}\{x_1,\ldots,x_n\}$, *or* $\mathbf{C}[[x_1,\ldots,x_n]]$ *is a relation* $>$ *on* $\mathbf{Z}^n_{>0}$, *or equivalently, on the set of monomials* $x^\alpha, \alpha \in \mathbf{Z}^n_{>0}$, *satisfying:*

(i) $>$ *is a total ordering on* $\mathbf{Z}^n_{>0}$

(ii) $>$ *is a multiplicative ordering on* $\mathbf{Z}^n_{>0}$

(iii) $>$ *is degree-anticompatible on* $\mathbf{Z}^n_{>0}$.

Recall that the first property means that for any $\alpha, \beta \in \mathbf{Z}^n_{>0}$, exactly one of the following is true:

$$x^\alpha > x^\beta, \qquad , x^\alpha = x^\beta, \quad x^\alpha > x^\beta;$$

and the second means that

$$\text{for any } \gamma \in \mathbf{Z}^n_{>0}, \ \text{ if } \ x^\alpha > x^\beta, \text{ then } \ x^{\alpha+\gamma} > x^{\beta+\gamma}.$$

Notice that the third property implies that $1 > x_i$ for all $i, 1 \leq i \leq n$. Any ordering satisfying (i), (ii) and this latter property is called a **local ordering**. So a degree-anticompatible ordering is a local ordering (but not conversely).

Perhaps the simplest example is degree-anticompatible lexicographic order, abbreviated **alex**, which first sorts by degree, lower degree terms preceding higher degree terms, and then sorts monomials of the same degree lexicographically.

Definition 6.3 (Anti-graded Lex Order) *Let $\alpha, \beta \in \mathbf{Z}^n_{>0}$. We say $x^\alpha >_{alex} x^\beta$ if*

$$|\alpha| = \sum_{i=1}^{n} \alpha_i < |\beta| = \sum_{i=1}^{n} \beta_i, \quad \textit{or} \quad |\alpha| = |\beta| \quad \textit{and} \quad x^\alpha >_{lex} x^\beta.$$

Thus, for example, we have:

$$1 >_{alex} x >_{alex} y >_{alex} x^2 >_{alex} xy >_{alex} y^2 >_{alex} x^3 >_{alex} \cdots$$

Notice that a degree-anticompatible (and, more generally, any local) ordering is not a well-ordering and, hence, not a monomial ordering in the sense of most books on Groebner bases (which require that monomial orderings satisfy (i), (ii) and be well-orderings – see, for example, [8], Definition 2.2.1)). Since the property of being a well ordering is often used to assert that algorithms terminate, one needs to be especially careful in checking that procedures which use degree-anticompatible orderings terminate.

If $f = \sum_\alpha a_\alpha x^\alpha$ is a non-zero power series and $<$ a local order, we define the multidegree, the leading coefficient, the leading monomial, and the leading term of f exactly as for a well-ordering:

$$\text{multideg}(f) = \max_{<}\{\alpha \in \mathbf{Z}^n_{>0} : a_\alpha \neq 0\},$$

$$\text{LC}(f) = a_{\text{multideg}(f)}, \quad \text{LM}(f) = x^{\text{multideg}(f)}, \quad \text{and} \quad \text{LT}(f) = lc(f) \cdot \text{LM}(f).$$

If $f \in \mathbf{C}[x_1, \ldots, x_n]_{\langle x_1,\ldots,x_n\rangle}$, then we first re-write f (uniquely) in the form

$$f = \frac{g}{1+h} \quad \text{with} \quad g, h \in \mathbf{C}[x_1, \ldots, x_n], \quad h(0) = 0,$$

and set

$$\text{multideg}(f) = \text{multideg}(g), \quad \text{LC}(f) = \text{LC}(g),$$
$$\text{LM}(f) = \text{LM}(g), \quad \text{and} \quad \text{LT}(f) = \text{LT}(g).$$

Notice that the multidegree and leading term of $f \in \mathbf{C}[x_1, \ldots, x_n]_{\langle x_1,\ldots,x_n\rangle}$ agree with what one obtains upon viewing f as a power series.

Lemma 6.4 *Let f, g both be in $\mathbf{C}[x_1, \ldots, x_n]_{\langle x_1,\ldots,x_n\rangle}$, (or $\mathbf{C}\{x_1, \ldots, x_n\}$ or $\mathbf{C}[[x_1, \ldots, x_n]]$) and both be non-zero, Then*

i. multidegree*(fg)* = multidegree*(f)* + multidegree*(g)*.

ii. *If $f + g \neq 0$, then*

$$\text{multidegree}(f + g) \leq \max(\text{multidegree}(f), \text{multidegree}(g))$$

with equality if $\text{multidegree}(f) \neq \text{multidegree}(g)$.

Now, just as for well-orderings, given an ideal I in a local ring R, where $R = \mathbf{C}[x_1, \ldots, x_n]_{\langle x_1,\ldots,x_n\rangle}, \mathbf{C}[[x_1, \ldots, x_n]]$, or $\mathbf{C}\{x_1, \ldots, x_n\}$, and an ordering $<$, we define the *the set of leading terms of I*, denoted $\text{LT}(I)$, to be the set of leading terms of elements of I with respect to $<$, and define the ideal of leading terms of I, denoted

$\langle \text{LT} I\rangle$, to be the ideal generated by the set $\text{LT}(I)$. Just as for ideals in polynomial rings, it can happen that $I = \langle f_1, \ldots, f_s\rangle$, but $\text{LT}(I) \neq \langle \text{LT}(f_1), \ldots \text{LT}(f_s)\rangle$. In analogy to the notion of a Groebner basis, we define a **Groebner** or **standard basis** of an ideal I to be a set $f_1, \ldots, f_s \in I$ such that $\langle \text{LT} I\rangle = \langle \text{LT}(f_1), \ldots \text{LT}(f_s)\rangle$. It is easy to show that such a set necessarily generates the ideal I. In the literature, the phrase "standard basis" is more common than the expression "Groebner basis" when referring to local orderings, so that is what we use here.

It is also not too difficult to show that

$$\begin{aligned} & \dim_{\mathbf{C}} \mathbf{C}[x_1, \ldots, x_n]_{\langle x_1, \ldots, x_n\rangle} / I\mathbf{C}[x_1, \ldots, x_n]_{\langle x_1, \ldots, x_n\rangle} = \\ = \ & \dim_{\mathbf{C}} \mathbf{C}[x_1, \ldots, x_n]_{\langle x_1, \ldots, x_n\rangle} / \langle \text{LT} I\rangle \mathbf{C}[x_1, \ldots, x_n]_{\langle x_1, \ldots, x_n\rangle} \end{aligned}$$

where $\langle \text{LT} I\rangle$ is taken with respect to a graded anti-compatible order. The argument is similar to that in the polynomial case (see [9]).

Given generators of an ideal, how can we compute a standard basis for the ideal? Recall that the key elements in the polynomial case were the division algorithm and Buchberger's algorithm. Let us review each of them here and see what needs to be modified to allow us to use them with local orderings.

The key step in the division algorithm is the reduction of one polynomial f by another g. If $\text{LT}(f) = t\,\text{LT}(g)$, we define

$$\text{Red}\,(f, g) = f - \frac{\text{LC}(f)}{\text{LC}(g)} tg,$$

and say that we have *reduced* f *by* g. The polynomial $\text{Red}\,(f, g)$ is just what is left after the first step in dividing f by g – it is the first partial dividend. In the one variable case, the division algorithm in which one divides a polynomial f by another g is just the process of repeatedly reducing by g until either one gets 0 or a polynomial that cannot be reduced by g (because its leading term is not divisible $\text{LT}(g)$). In the case of several variables, the division algorithm, in which one divides a polynomial by a set of other polynomials, is just the process of repeatedly reducing the polynomial by members of the set, adding leading terms to the remainder, when no reductions are possible. This terminates in the case of polynomials because successive leading terms form a strictly decreasing sequence, and such sequences always terminate for well-orderings.

In the case of power series and local orders, one defines $\mathrm{Red}\,(f, g)$ exactly as above. However, a sequence of successive reductions need no longer terminate. For example, suppose $f = x$ and we decide to divide by $g = x - x^2$, so that we successively reduce by $x - x^2$. We get

$$\begin{aligned}
f_0 &\equiv f = x, \\
f_1 &= \mathrm{Red}\,(f_0, g) = x^2, \\
f_2 &= \mathrm{Red}\,(f_1, g) = x^3, \\
f_3 &= \mathrm{Red}\,(f_2, g) = x^4, \\
&\vdots \\
f_n &= \mathrm{Red}\,(f_{n-1}, g) = x^{n+1}, \\
&\vdots
\end{aligned}$$

which clearly does not terminate. The difficulty, of course, is that $x > x^2 > x^3 > \ldots$ is a strictly decreasing sequence of terms which does not terminate.

We can evade this difficulty, with a splendid idea of Mora's: when dividing f by a single polynomial or power series g say, we allow ourselves to reduce not just by g, but by the result of any previous reduction. More generally, when dividing a set of polynomials or power series, we allow ourselves to reduce by the original set together with the results of any previous reductions. So, in our example, where we are dividing $f = x$ by $g = x - x^2$, after the first reduction, we we allow ourselves to reduce with f (the result of the zeroth reduction) as well as g. The first reduction by g tells us that

$$f = 1 \cdot g + x^2.$$

Reducing the result x^2 with f instead of g gives $\mathrm{Red}\,(x^2, f) = 0$ so that we halt. Moreover, the fact that the reduction is zero gives

$x^2 = xf$ from which we obtain the relation

$$f = g + xf.$$

But this relation tells us that in $\mathbf{C}[x_1, \ldots, x_n]_{\langle x_1,\ldots,x_n\rangle}$

$$f = \frac{1}{1-x} g.$$

Looking at the above example, one might ask whether it would always suffice to first reduce by g, then subsequently reduce by f. Sadly, this is not the case: it is easy to construct examples where the sequence of reductions does not terminate. Suppose, for example, that we wish to divide $f \equiv x + x^2$ by $g \equiv x + x^3 + x^5$. Reducing f by g and then subsequently reducing the results by f gives the sequence

$$\begin{aligned}
f_0 &\equiv f = x + x^2, \\
f_1 &= \operatorname{Red}(f_0, g) = x^2 - x^3 - x^5, \\
f_2 &= \operatorname{Red}(f_1, f) = -2x^3 - x^5, \\
f_3 &= \operatorname{Red}(f_2, f) = 2x^4 - x^5, \\
f_4 &= \operatorname{Red}(f_3, f) = -3x^5, \\
f_5 &= \operatorname{Red}(f_4, f) = 3x^6, \\
f_6 &= \operatorname{Red}(f_5, f) = -3x^7, \\
&\vdots
\end{aligned}$$

which clearly does not terminate. We get something which terminates by reducing f_5 by f_4:

$$\begin{aligned}
f_5 &= \operatorname{Red}(f_4, f) = -3x^6, \\
\tilde{f}_6 &= \operatorname{Red}(f_5, f_4) = 0.
\end{aligned}$$

From this, we can easily write out an expression for f as a sum of a polynomial times g, plus a polynomial which vanishes at the origin times f plus a polynomial which vanishes at the origin times f_5:

$$\begin{aligned}
f &= (\qquad) \cdot g + (\qquad) \cdot f + (\qquad) \cdot f_5 \\
x + x^2 &= 1 \cdot (x + x^3 + x^5) + (x - 2x^2 + 2x^3 - 3x^4) \cdot (x + x^2) \\
&\quad - x \cdot (-3x^5)
\end{aligned}$$

Since, f_5 can be expressed as a sum of a polynomial times g plus a polynomial times f, we can actually express f as a sum of a polynomial times g plus a polynomial which vanishes at the origin times f by backsubstituting:

$$\begin{aligned}
x+x^2 &= 1\cdot(x+x^3+x^5)\\
&+ (x-2x^2+2x^3-3x^4)\cdot(x+x^2)\\
&- x\cdot(-3x^5)\\
-3x^5 &= -1\cdot(x+x^3+x^5)+(1-x+2x^2-2x^3)\cdot(x+x^2)\\
\Longrightarrow &\\
x+x^2 &= (1-x)(x+x^3+x^5)+(-x^2-x^4)(x+x^2)\\
\text{so} &\\
f &= (\text{polynomial})\cdot g+(\text{polynomial vanishing at }0)\cdot f+(\quad).
\end{aligned}$$

This, of course, is what we want because, upon transposing, we have

$$(\text{unit})\cdot f=(\text{polynomial})\cdot g.$$

In order to create an algorithm, we have to specify a way to choose a sequence of elements by which to reduce so that the reduction terminates, either by giving 0 or a polynomial whose leading coefficient is not divisible by the leading term of anything by which we can reduce.

Looking at the sequence $f_0, f_1, \ldots f_6$ in our division of $f = x + x^2$ by $g = x + x^3 x^5$ above gives us some clue how to proceed. Notice that neither the degree of the dividends nor the order necessarily decreases. However, the difference between the degrees of the highest degree term and lowest degree terms in the partial dividends does decrease. Is there someday to guarantee that this will happen? Before addressing this, let us fix notation.

Definition 6.5 *Let $g \in \mathbf{C}[x_1, \ldots, x_n], g \neq 0$, and write g as a finite sum of homogeneous, non-zero polynomials of different degrees: $g = \sum_{i=1}^{k} g_i$, $g_i \neq$ homogeneous,* $\deg(g_1) < \ldots < \deg(g_k)$. *Define the* **order** *of g, denoted* $\operatorname{ord}(g)$ *to be the degree of g_1 and the* **écart** *of*

g, denoted $E(g)$, to be the difference of the degree of g and the order of g:

$$E(g) = \deg(g) - \text{ord}\,(f).$$

By convention, we set $E(0) = -1$,

A simple argument which we leave as an exercise shows the following.

Lemma 6.6 *Let f and g be two nonzero polynomials such that* $\text{LT}(g)$ *divides* $\text{LT}(f)$*. Then*

$$E(\text{Red}\,(f, g)) \le \max\{E(f), E(g)\}.$$

If f, g are polynomials of a single variable, then the inequality is always strict. If, in the case of several variables, $E(g) \le E(f)$ and $E(\text{Red}\,(f, g)) = E(f)$, then $\text{ord}\,(\text{Red}\,(f, g)) = \text{ord}\,(f)$.

In the one variable case, this gives us a strategy that guarantees termination. Namely, at each stage, among all the polynomials by which we can reduce, we reduce by the polynomial whose écart is least. This will ensure that the écarts of the sequence of partial dividends strictly decreases to zero, at which point we have a monomial which can be used to reduce any subsequent monomial to 0. In the multivariable, case we use the same strategy, but the écart may not strictly decrease – however, we can again assert that the sequence of reductions will terminate after a finite number of steps, essentially because whenever the écart stays fixed, the order must, too, and there are only finitely many monomials of a given order.

Making these remarks precise, gives us an algorithm for successively reducing one polynomial by others using an anti-graded local order, a proof that the algorithm is correct, and a proof that the algorithm terminates. see [14], [22], or [1].

With this in place, we now have a way of computing a standard basis of an ideal with respect to an anti-graded order, and hence the Milnor number of a singularity. Given an isolated singularity f, we compute the ideal of leading terms (with respect to an anti-graded order) of the ideal generated by the first partial derivatives of f. The Milnor number is just the number of monomials not contained in this (monomial) ideal. The same strategy allows is to compute Tjurina numbers and the μ^*-invariant.

References

[1] M.E. Alonso, T. Mora, M. Raimondo, A computational model for algebraic power series, *J. Pure Appl. Alg.*, **77** (1992) 1-38.

[2] V.I. Arnold, S.M. Gusein-Zade, A.N. Varchenko, *Singularities of Differentiable Maps, Vols 1 and 2*, Monographs in Mathematics 82 and 83, Birkhäuser, Boston, 1985 and 1988.

[3] W.W. Adams, P. Loustanau, *An Introduction to Gröbner Bases*, American Mathematical Society, Providence, 1994.

[4] T. Becker, V. Weispfennig, *Gröbner Bases, A Computational Approach to Commutative Algebra*, Spriner-Verlag, New York, Berlin, 1993.

[5] J. Briançon, J.P. Speder, La trivialité topologique n'implique pas les conditions d Whitney, *C. R. Acad. Sci. Paris, Sér. A*, **280** (1975) 365-367.

[6] B. Buchberger, *Ein Algorithmus zum Auffinden der Basiselemente des Restklassenringes nach einem nulldimensionalem Polynomideal,* Thesis, Univ. Innsbruck, 1965.

[7] D. Burghelea, A. Verona, Local homological properties of analytic sets, *Manuscripta Math.* **7** (1972) 55-66.

[8] D. Cox, J. Little, D. O'Shea, *Ideals, Varieties, Algorithms*, Springer-Verlag, Berlin, New York, 1992.

[9] D. Cox, J. Little, D. O'Shea, *Using Algebraic Geometry*, in preparation.

[10] A. Dimca, *Singularities and Topology of Hypersurfaces*, Springer-Verlag, New York, Berlin, 1992.

[11] D. Eisenbud, *Commutative Algebra with a view towards Algebraic Geometry*, Springer-Verlag, Berlin, New York, 1995.

[12] D. Eisenbud, W. Neumann, *Three-Dimensional Link Theory and Invariants of Plane Curve Singularities*, Princeton Univ. Press, Princeton, 1992.

[13] W. Fulton, *Algebraic Curves*, W.A. Benjamin, Reading, 1969.

[14] H.-G. Gräbe, Algorithms in local algebra, preprint, Univ. Leipzig, 1993.

[15] A. Hénaut, Cycle exceptionnel de l'éclatement de Nash d'une hypersurface analytique complex à singularité isolée, *Bull. Soc. Math. de France*, **109** (1981) 475-481.

[16] H. Hironaka, Resolution of singularities of an algebraic variety over a field of characteristic zero. *Ann. Math.*, **79** (1964) 109-326.

[17] Lê Dũng Tráng, Topologie des singularités des hypersurfaces complexes, *Astérisque*, **7/8** (1973) 171-182.

[18] Lê Dũng Tráng, B. Teissier, Limites d'espaces tangents en géometrie analytique, *Comment. Math. Helv.*, **63** (1988), 540-578.

[19] E. Looijenga, *Isolated Singular Points on Complete Intersections*, LMS Lecture Notes 77, Cambridge University Press, Cambridge, 1984.

[20] F.S. Macaulay, *The Algebraic Theory of Modular Systems*, Cambridge Univ. Press, Cambridge, 1916.

[21] J. Milnor, *Singular Points of Complex Hypersurfaces*, Annals of Math. Studies 61, Princeton University Press, Princeton, 1968.

[22] T. Mora, An algorithm to compute the equations of tangent cones, in *Proc. EUROSAM-82*, Lect. Notes in Comp. Sci., **144**, Springer-Verlag, Berlin (1982) 24-31.

[23] D. Mumford, *Algebraic Geometry I: Complex Projective Varieties*, Springer-Verlag, Berlin, New York, 1976.

[24] D. O'Shea, Computing limits of tangent spaces: effective computation in singularity theory, in D.T. Lê, K. Saito, B. Teissier (eds) *Singularity Theory*, World Scientific, Singapore, New Jersey, London, Hong Kong, 1995.

[25] B. Teissier, Cycles évanescents, sections planes et conditions de Whitney, *Astérisque*, **7/8** (1973) 285-362.

[26] H. Whitney, *Complex Analytic Varieties*, Addison-Wesley, Reading,1972.

[27] H. Whitney, Tangents to an analytic variety, *Ann. of Math.* **81** (1965), 496-549.

[28] O. Zariski, Some open questions in the theory of singularities, *Bull. Amer. Math. Soc.*, **77** (1971) 481-491.

Some Milestones of Lemniscatomy

Norbert Schappacher

UFR de mathématique et d'informatique,
7 rue René Descartes,
67084 Strasbourg Cedex, France

e-mail: `schappa@math.u-strasbg.fr`

This little article is an attempt to kindle the reader's interest in some modern research in the domain of complex multiplication by sketching a few episodes from the history of the very first example of this theory, the lemniscatic elliptic integral and the corresponding elliptic functions. Even though treating such an example offers elementary proofs of some key properties, our exposition is not mathematically self-contained. All we hope is that it will motivate the reader to look up some of the further literature which we indicate. At the same time, we have tried to draw attention to the differences between the various periods of the history of lemniscatomy, and to a few historical details which may not be generally known.

1 Prelude in Antiquity.[1]

The story from which we are about to tell a few instructive episodes does not really begin, but is vaguely suggested to begin, in the dawn of times, or rather in the twilight of Greek mathematics, where the sources which are extant today do not quite permit us to know what was said and thought, and at what time.

The leading Neo-Platonist of his time, *Proclus* from Lykia (on today's maps, this is roughly the region between Fethiye and Antalya) lived from 410 to 485 AD, and went from Alexandria, where he got his education, to Athens. His numerous works, most of which are strongly philosophical or even religious in character, also include a Commentary on Book I of Euclid's Elements. These were basically notes for his own elementary mathematics courses in the Neo-Platonic Academy, but for us they constitute the richest extant source on the history of Greek geometry. At one point Proclus tells us about the work of a certain *Perseus*, about whom we do not know anything else today. Proclus quotes a distich of Perseus's,

Τρεις γραμμας επι πεντε τομαις ευρων • • – •
Περσευς των δ'ενεκεν δαιμονας ιλασατο.

"Finding three ... curves upon five sections, Perseus thanked the gods therefor." The precise interpretation of these lines and of some of Proclus's explanations, as well as the missing word(s) at the end of the first verse are open to discussion. But Proclus seems to suggest that Perseus had considered what reminds us today of pictures one draws to motivate Morse theory: Imagine a torus, not the way it would float on the water, but being immersed vertically into water, and look at the curves marked on the torus by the water level. Increasing the depth of the immersion, one finds sometimes an oval, sometimes a pair of circles, but if the water touches precisely the inner hole of the torus, the resulting curve looks somewhat like a figure-eight, or, as Proclus puts it, like a horse fetter: *εοικυα τη του ιππου πεδη*, locking, say, the ankles of the horse's

[1]See [Thomas (1941), pp. 360–365], [Heath (1921), pp. 203–206, 529–535], [Brieskorn & Knörrer (1986), §1.6].

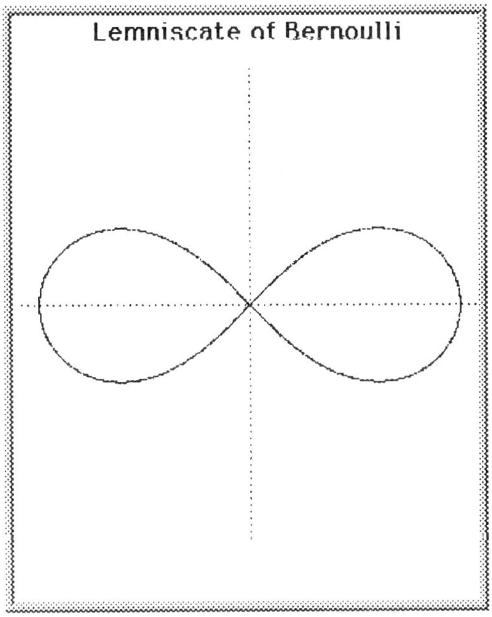

forelegs. In the case where the diameter of the hole in the torus equals that of the ring of the torus, this curve is precisely

2 The Lemniscate of the Bernoulli Brothers.[2]

The Swiss brothers Jakob and Johann Bernoulli hit upon the lemniscate almost exactly three hundred years ago from a completely different point of view, which brings us quite a bit closer to what we want to explain in this paper. They proposed the curve as the solution to an integral, more precisely to the differential equation

$$\frac{dr}{\sqrt{ar}} = \frac{2adz}{\sqrt{az(a^2 - z^2)}}, \tag{1}$$

where a is a (positive) parameter. This differential equation itself was the answer, found by the Bernoullis, to the so-called problem of

[2]See [Bos 1993] and references given there. For the computations at the end of the section see [Siegel 1969, §1.1].

paracentric isochrones which had been posed by Leibniz. The curve "integrates" (1) in the sense that the integral of the right-hand side of (1) measures the arc-length of the lemniscate.

We see that, at the end of the seventeenth century, it was considered desirable and satisfactory to describe an integral as giving the arc-length of a curve—which in turn could be given by an explicit equation or be mechanically constructed. This attachment to rectilinear integration is quite different from our modern standards. Today we usually insist on expressing an integral formally, for instance, expand it into a power series if no closed formula is available. Still, the increasing use of computer packages like MATHEMATICA or MAPLE which also produce beautiful pictures may bring back a bit of the 17th century's predilection for geometric solutions to analytic problems.

To actually get the lemniscate from integrating the right-hand side of (1), Johann Bernoulli renormalized slightly, and looked for functions $x(z), y(z)$ satisfying

$$\frac{a}{\sqrt{2}}\frac{2adz}{\sqrt{az(a^2-z^2)}} = \sqrt{(dx)^2+(dy)^2}. \tag{2}$$

It is not too hard to guess (and even easier to check) that

$$x(z) = \sqrt{az+z^2} \qquad y(z) = \sqrt{az-z^2}$$

do the job. So we see that x and y satisfy the following algebraic equation identically in z.

$$(x^2+y^2)^2 = 2a^2(x^2-y^2) \tag{3}$$

This is the standard equation of the lemniscate (with parameter a). The name, by the way, was chosen by Jakob Bernoulli, in an article published in 1694. The Greek word λημνισκος is related to *το* λῆνος, wool—cf. the French *laine*—, and originally means a woolen band. This is a softer metaphor than Perseus's horse-fetter.

The points $P(x, y)$ in the (x, y)–plane satisfying equation (3) are characterized geometrically by the fact that the product of the distances from P to the point $(-a, 0)$ and from P to the point $(+a, 0)$ is constant, equal to a^2.

We will not need to look at more general curves which also go under the name of lemniscates. In fact, for our purposes we will even normalize the parameter and assume that $2a^2 = 1$. Doing this and rewriting (3) in terms of $r = \sqrt{x^2 + y^2}$ gives

$$2x^2 = r^2 + r^4 \qquad \text{and} \qquad 2y^2 = r^2 - r^4.$$

This clearly yields a parametrization of the lemniscatic arc in terms of the variable r. Considering the leaf in the first quadrant, the integral giving the arc-length on the lemniscate then comes out to be

$$(4) \qquad s(r) = \int_0^r \frac{dr}{\sqrt{1 - r^4}} \qquad (0 \leq r \leq 1).$$

3 Elementary Arithmetic of Elliptic Integrals: the Count of Fagnano, Euler, and Legendre.[3]

We now enter an era of the history of our subject where new discoveries were made through the formal manipulation of integrals. This is especially visible in the works of the count G.C. di Fagnano (1682–1766). In fact, they are somewhat disconcerting for the modern reader because of the great number of formulas which do not easily betray the line of thought that led to them. But his various investigations about the lemniscate constitute the first milestone in the history of lemniscatomy, i.e., of the division of the lemnsicate.

[3]We do not know any comprehensive, detailed reference for this section. We are not convinced, for instance, by the speculations in [Ayoub 1984]. Of course, all historical accounts of the theory of elliptic functions—see for instance the encyclopedic [Brill, Noether 1892–93], and in particular the thorough study [Houzel 1978]—talk about Fagnano, Euler, Lagrange, Legendre, etc. Still, it is not superfluous to actually refer to the collected works [Fagnano 1750], second volume, esp. no.s xxxii, xxxiii, xxxiv. There is a nice discussion of Fagnano and Euler in the (somewhat rare) book [Enneper 1876, pp. 477ff]. More accessible is the account of Fagnano's doubling of the lemniscatic arc in [Siegel 1969, 1.1] and in the related short note [Siegel 1959].

This is not only so from our point of view, but Fagnano himself must have realized the importance of his discoveries. In fact, a lemniscate appears in the inscription on his tombstone in the Church of Santa Maria Magdalena in Senigallia: *Veritas Deo* ∞ *gloria....*

In 1718, Fagnano published the following seminal result (which is easily checked):

$$\int_0^z \frac{dz}{\sqrt{1-z^4}} = 2\int_0^u \frac{du}{\sqrt{1-u^4}}, \qquad \text{if } z = \frac{2u\sqrt{1-u^4}}{1+u^4}. \tag{5}$$

Before explaining its significance, let us discuss C.L. Siegel's comments about this result—see [Siegel 1959]. He tried to divine how Fagnano obtained (5). We mention his comments not because we find this kind of mathematical psychology across the centuries very convincing. But Siegel's observations prepare us for later stages of the history of our subject.

Siegel's first conjecture is that Fagnano may have found (5) by analogy with the arclength of the circle

$$\text{arc}\sin(r) = \int_0^r \frac{dr}{\sqrt{1-r^2}} \qquad (0 \le r \le 1). \tag{6.0}$$

Here, the substitutions

$$r = \frac{2s}{1+s^2}, \qquad s = \frac{2t}{1-t^2} \tag{6.1}$$

lead, respectively, to

$$\frac{dr}{\sqrt{1-r^2}} = 2\frac{ds}{1+s^2}, \qquad \frac{ds}{\sqrt{1+s^2}} = 2\frac{dt}{1-t^2}. \tag{6.2}$$

The essential difference between the circle and the lemniscate is that (6.2) reduces (6.0) to an elementary integral, whereas in (5) we find the same lemniscatic arclength on both sides.

Siegel also writes [Siegel 1959, p. 250] that (5) "in fact belongs to the theory of complex multiplication", because we may break up the relation between z and u using the Gaussian integer $\alpha = 1 + i$ and its complex conjugate $\overline{\alpha} = 1 - i$ (thus $\alpha\overline{\alpha} = 2$, $\alpha^4 = \overline{\alpha}^4 = -4$):

$$(7.1) \qquad z = \alpha\frac{t}{\sqrt{1-t^4}}, \qquad t = \overline{\alpha}\frac{u}{\sqrt{1-u^4}},$$

and these two algebraic relations yield, respectively, the differential equations:

$$(7.2) \qquad \frac{dz}{\sqrt{1-z^4}} = \alpha\frac{dt}{\sqrt{1-t^4}}, \qquad \frac{dt}{\sqrt{1-t^4}} = \overline{\alpha}\frac{du}{\sqrt{1-u^4}}.$$

Siegel claims [Siegel 1959, p. 251] that the possibility to decompose (5) into the simpler steps (7) was "obviously relevant to Fagnano's success". Since there is no mention of complex numbers or of formulas like (7) in Fagnano, this probably says more about Siegel's insights than about Fagnano's.

Instead of trying to slip further into Fagnano's mind let us now explain why his result was such a breakthrough. Fagnano's theorem (5) means that the differential equation

$$\frac{dz}{\sqrt{1-z^4}} = 2\frac{du}{\sqrt{1-u^4}}$$

can be *integrated algebraically*. Indeed, given u, there is an algebraic formula (5) for z such that the lemniscatic integral from 0 to z is twice that from 0 to u. We may also look at (5) the other way around: suppose we are given z, or more precisely (remembering the way we obtained (4) above), suppose we are given a point (x, y), say in the first quadrant, which lies on the lemniscate, so that $2x^2 = z^2 + z^4$ and $2y^2 = z^2 - z^4$. Then, if u satisfies the algebraic equation deduced from (5),

$$(8) \qquad z^2u^8 + 4u^6 + 2z^2u^4 - 4u^2 + z^2 = 0,$$

and if (x', y') lies on the lemniscate and in the first quadrant satisfying $2x'^2 = u^2 + u^4, 2y'^2 = u^2 - u^4$, then the lemniscatic arc between the origin and (x', y') has half the length of that between the origin and (x, y).

In this sense, Fagnano found the multiplication by 2 of the lemniscatic integral.[4] The method also leads to the trisection of the total lemniscatic arc in the first quadrant. To see this, note that the relation between u and z in (5) really should be written

$$z^2 = \frac{4u^2(1-u^4)}{(1+u^4)^2}.$$

This leaves two possible signs for the root, and the minus sign yields the equation

$$\frac{dz}{\sqrt{1-z^4}} = -2\frac{du}{\sqrt{1-u^4}}.$$

In this case, the points $(x,y),(x',y')$ on the lemniscate and in the first quadrant such that $2x^2 = z^2 + z^4$ and $2y^2 = z^2 - z^4$, resp. $2x'^2 = u^2 + u^4, 2y'^2 = u^2 - u^4$, are situated such that the arc between the origin and (x,y) has half the length of the arc from (x',y') to the point $(1,0)$. Thus, if $z = u$ in this case, $(x,y) = (x',y')$ marks one third of the arc from $(0,0)$ to $(1,0)$. The condition $z = u$ implies that $z^4 = 2\sqrt{3} - 3$.

As mentioned before, Fagnano's algebraic divisions by 2 and by 3 of the lemniscatic arc seem to be the first explicit examples of lemniscatomy in the history of mathematics. It is, however, amusing to note that some eighty years before Fagnano, P. de Fermat, working in an entirely different context, had developed a proof by infinite descent which, when translated into our language of elliptic curves, uses the bisection of the lemniscatic arc—see [Goldstein 1995, pp. 91–103], and the end of section 5 below.

According to Jacobi [Stäckel, Ahrens 1908, p. 23], the 23rd of December 1751 was the birthday of the theory of elliptic functions, because it was on this day that Euler received the collection [Fagnano 1750] from the Berlin Academy for internal review. Euler immediately started to generalize Fagnano's results and presented his first publication on elliptic integrals on January 17, 1752—see [Euler

[4]Siegel's formulas (7) do the same for the multiplication by $1+i$ and its complex conjugate. This possibility of arithmetic operations involving irrational integers, i.e., the presence of 'complex multiplication' is a special feature of the lemniscatic integral which we will understand more clearly further on.

1761]. This note was the beginning of a whole sequence of articles by Euler on this topic which, taken together, run to almost 400 well-filled pages. Of all this production, history has essentially retained only the general *addition theorem of elliptic integrals.*[5] In the special case of the lemniscatic integral, this fundamental result reads as follows—see [Siegel 1969, p. 9]:

$$(9) \qquad \int_0^u \frac{du}{\sqrt{1-u^4}} + \int_0^v \frac{dv}{\sqrt{1-v^4}} = \int_0^r \frac{dr}{\sqrt{1-r^4}},$$

$$\text{if } r = \frac{u\sqrt{1-v^4} + v\sqrt{1-u^4}}{1+u^2v^2}.$$

It expresses r algebraically in terms of u and v, if the arc from the double point of the lemniscate all the way to the point of modulus r is as long as those to u and to v added together.

In the spirit of this article, we refrain from discussing more general *elliptic integrals.* After Euler these were classified and systematized in particular by Legendre—see [Legendre 1825–28]. The unfortunate name for this class reflects the fact that the arclength of an ellipse is also given by an "elliptic integral". However, it is not an integral "of the first kind" like (4), i.e., it is not an integral of the form $\int \frac{dx}{\sqrt{P(x)}}$ for a polynomial P of degree 3 or 4, but rather "of the second kind", i.e., of the form $\int \frac{t^2\,dx}{\sqrt{P(x)}}$ for a polynomial P of degree 4. An important concrete question which does lead to an elliptic integral of the first kind is the mathematical theory of the pendulum.

The elliptic integrals constitute the first examples of integrals arising naturally which (exceptions apart) cannot be integrated in terms of elementary functions. By the way, as R. Remmert pointed out to me, it would be nice to see an explicit elementary *proof* of this fact, for instance for our lemniscatic integral (4).

---

[5]This selective injustice of history has infuriated some later writers who criticize that too much credit was given in particular to Legendre, to the detriment of Euler's contributions—see for example [Enneper 1876, pp. 490f], [Plana 1863]. These are emotions of the past. But in the otherwise excellent recent article [Belhoste 1996, p. 3], there is a surprising sentence creating the wrong impression that Euler treated only the lemniscatic integral.

Let us close this section with a few variations on the lemniscatic integral (4):

$$(10)\quad \int_1^\infty \frac{dx}{\sqrt{4x^3-4x}} = \int_0^1 \frac{dr}{\sqrt{1-r^4}} = \frac{1}{4}\int_0^1 \frac{dt}{t^{\frac{3}{4}}(1-t)^{\frac{1}{2}}} = \frac{\Gamma(\frac{1}{4})^2}{4\sqrt{2\pi}}.$$

These identities, which we leave as an exercise, follow from the substitutions: $x = \frac{1}{\sqrt{r}}$ and $r^4 = t$, and from standard properties of Euler's beta integral. The last expression occurs in [Legendre 1811, p. 209f]. It is the first special case of a general formula (due to Lerch, and known as the formula of Chowla and Selberg) which relates elliptic integrals with complex multiplication to special values of the gamma function—see [Schappacher 1988, p. 123–125], see also [Henniart 1987].

4 The Galois Theory of Lemniscatomy: from Gauss to Abel.

Carl Friedrich Gauss finished writing his momentous book *Disquisitiones Arithmeticæ* [Gauss 1801] when he was 21 years old. It transported arithmetic to a new theoretical level. It also contained the complete solution of a longstanding problem: in the final section VII [Gauss 1801, pp. 592–665], Gauss determined precisely the regular n-gons which can be constructed by ruler and a pair of compasses alone. The well-known answer is: those for which n is of the form $2^\nu p_1 \cdots p_r$ where $\nu \geq 1$ and the p_i are distinct odd prime numbers of the form $2^m + 1$.

This result is obtained from an algebraic analysis of what we call today the field extension $\mathbb{Q}(\zeta_p)/\mathbb{Q}$, for any odd prime p. Indeed, Gauss carries out the "Galois theory" (Galois was -13 years old when Gauss's book was completed) of this particular cyclic extension, exhibiting distinguished primitive generators of all intermediate fields and what we would call their behaviour under automorphisms.

In the field $\mathbb{C}$ of complex numbers, the n-th roots of unity can be given analytically as $\exp(2\pi ir/n)$ with $r = 0, 1, \ldots, n-1$. So the seventh section of the *Disquisitiones Arithmeticæ* develops the theory of the extensions generated by the division values of the function $x \mapsto e^{\pi ix}$. Now, in the introduction to this section, Gauss writes: "The principles of the theory we are about to set out extend to a much broader context... For they may be applied not only to the circular functions, but with equal success to many other transcendental functions, e.g., to those which are related to the integral $\int \frac{dx}{\sqrt{1-x^4}}$."[6] He also announces a forthcoming study of those transcendental functions (which was never published). Gauss probably had in mind (at least) the elliptic integrals of the first kind and their inverse functions—see for instance [Houzel 1978, p. 17].

Gauss's *Nachlass* (papers published only after his death) contains various sets of notes on the lemniscatic integral, its inverse function, and other related functions [Gauss (III), pp. 404ff]. But it was Niels Henrik Abel who first published a systematic theory along these lines—see [Abel 1827–28]. In this seminal article, Abel shifts attention from the elliptic integral to its inverse function. Formally, his theory deals with the function φ characterized by the equivalence

$$\text{(11)} \qquad \varphi(\alpha) = x \qquad \Longleftrightarrow \qquad \alpha = \int_0^x \frac{dx}{\sqrt{(1-c^2x^2)(1+e^2x^2)}}.$$

The special case of the lemniscatic integral corresponds to the choice of parameters $c = e = 1$. Abel calls the functions φ *elliptic functions*—much to Legendre's chagrin, who used this name for the corresponding integrals, and did not like to see his terminology chucked.

To be sure, formula (11) presents the problem of the dependence of this integral on choices of the root and of the path of integration. These questions were solved in general only after Riemann—see for instance [Siegel 1969, pp. 29–37]. Abel proceeds with care, in a rather piecemeal way. We specialize right away to the lemniscatic case $c = e = 1$ which interests us here. There Abel puts (see also

[6] [Gauss 1801, p. 593]: "Ceterum principia theoriæ, quam exponere aggredimur, multo latius patent, quam hic extenduntur. Namque non solum ad functiones circulares, sed pari successu ad multas alias functiones transcendentes applicari possunt, e. g. ad eas quæ ab integrali $\int \frac{dx}{\sqrt{1-x^4}}$ pendent."

(10) above):

$$(12) \qquad \omega = 2\int_0^1 \frac{dr}{\sqrt{1-r^4}}.$$

Then he defines φ first on the interval $[0, \frac{\omega}{2}]$ of positive real numbers. After this he passes to the imaginary interval $i \cdot [0, \frac{\omega}{2}]$ by the rule

$$(13) \qquad \varphi(\alpha i) = i\varphi(\alpha) \qquad (i = \sqrt{-1}),$$

and finally he uses parity, and double periodicity modulo 2ω and $2\omega i$, to define φ on all of $\mathbb{C}$. We will not follow Abel's developments in detail here, but rather present things from Weierstrass's point of view in the next section. This will break the chronological order of our exposition, but should make it more accessible to the reader.

The main result about lemniscatomy that Abel finds in [Abel 1827–28] is analogous to Gauss's cyclotomic result: the division of the entire lemniscate into n equal parts is possible by ruler and a pair of compasses alone if n is of the form $2^\nu p_1 \cdots p_r$ where $\nu \geq 1$ and the p_i are distinct odd prime numbers of the form $2^m + 1$. The main reason for this is that Abel can exhibit a cyclic "Galois action" on the $\varphi(\frac{r\omega}{n})$ $(r = 1, \ldots, n-1)$ if n is a prime number of the form $4\nu + 1$—see [Abel 1827–28, no.s 36, 37].

Those who like rare plants and pointless beauty may appreciate to see explicit geometric constructions of the division of the circle or the lemniscate into, say, $17 = 2^{2^2} + 1$ equal parts. (Gauss himself did not give this but computed algebraically, up to 10 decimal digits, the 19th and the 17th roots of unity in [Gauss 1801, no.s 353, 354].) It so happens that both constructions can be looked up in the 75th volume of *Crelle's Journal* from 1873—see [Schröter 1873] and [Kiepert 1873].

5 Lemniscatic Elliptic Functions from the Point of View of Weierstrass Theory and Elliptic Curves.

In order to have a modern analytic language for what follows, let us briefly recall here the Weierstrass theory of elliptic functions as far as we need it to treat the lemniscatic integral and its inverse function. In using this theory from now on, rather than Jacobi's formalism which dominated the greater part of the 19th century, we choose to skip an important part of the historical development. Even though Jacobi's system was rooted in a period where the definition of classes of functions by abstract properties was not yet established, his formalism could capture anything that can be couched in Weierstrass's approach, and Jacobi's theta series sometimes provide greater flexibility.[7]

We assume some basic familiarity with Weierstrass theory—see for instance [Siegel 1969, pp. 56–89], [Silverman 1986, chap. VI], [Freitag, Busam 1995, chap. V], or [Remmert 1991, pp. 71–74].

The theory starts with a lattice. In order to get out the lemniscatic integral at the end, we take $\Lambda = \mathbb{Z} + \mathbb{Z}i \subset \mathbb{C}$. For every $\nu \geq 3$, the series

$$G_\nu(\Lambda) = {\sum_{\lambda \in \Lambda}}' \frac{1}{\lambda^\nu} = {\sum_{m,n \in \mathbb{Z}}}' \frac{1}{(n+mi)^\nu}$$

converges absolutely. Here the prime signifies that undefined terms are deleted from the sum, i.e., we restrict to nonzero $\lambda = n + mi$. The G_ν are called Eisenstein series, but this is another story...

Our lattice admits *complex multiplication* in the sense that multiplication by i takes the lattice Λ into itself. Therefore we find: $G_\nu(\Lambda) = 0$ *unless* ν *is divisible by 4.* Also, the lattice Λ is invariant under complex conjugation, so that all the $G_{4k}(\Lambda)$ are real numbers.

[7]Let us note in passing that the Chelsea reprint of Jacobi's Collected Works has recently been made available again by the AMS. See also the 550 page almost day-to-day scientific biography of Jacobi, [Königsberger 1904].

Proposition. *For all $k = 1, 2, 3 \ldots$, one has $G_{4k}(\Lambda) > 0$.*

The proof follows from the theory of modular forms, more precisely, from the Fourier development of G_ν considered as functions on all complex lattices—see for instance [Serre 1970, VII.4.2].

In Weierstrass's setup one defines

$$g_2 = 60G_4, \qquad g_3 = 140G_6.$$

Thus, for our lattice Λ, we have $g_3(\Lambda) = 0$, and we can define the positive real number $\omega > 0$ by the relation

$$g_2(\Lambda) = 4\omega^4. \tag{14a}$$

In other words, defining $L = \omega \cdot \Lambda$ to be the lattice with basis $\omega, i\omega$, we have

$$g_2(L) = 4. \tag{14b}$$

Note that $g_3(L) = g_3(\Lambda) = 0$.

Recall that, in modern terminology, an *elliptic function* with respect to a lattice $\Gamma \subset \mathbb{C}$ is a meromorphic function f on $\mathbb{C}$ such that for all $z \in \mathbb{C}$ and $\gamma \in \Gamma$, one has $f(z+\gamma) = f(z)$. The first example of such an elliptic function, which is holomorphic except for double poles at the lattice points, is the *Weierstrass-P-function*:

$$\wp(z, \Gamma) = \frac{1}{z^2} + \sum_{\gamma \in \Gamma}{}' \frac{1}{(z-\gamma)^2} - \frac{1}{\gamma^2}.$$

It satisfies the fundamental differential equation

$$\wp'(z,\Gamma)^2 = 4\ \wp(z,\Gamma)^3 - g_2(\Gamma)\ \wp(z,\Gamma) - g_3(\Gamma).$$

For the particular lattice $L = \omega\Lambda$ defined above, it reads:

$$\wp'(z,L)^2 = 4\ \wp(z,L)^3 - 4\ \wp(z,L). \tag{15}$$

This allows us to establish the link between (10), (12), and (14a):

$$(16) \qquad \omega = 2\int_{-\frac{\omega}{2}}^{0} dz = 2\int_{1}^{\infty} \frac{dx}{\sqrt{4x^3 - 4x}} = 2\int_{0}^{1} \frac{dr}{\sqrt{1-r^4}}$$

$$= \frac{\Gamma(\frac{1}{4})^2}{2\sqrt{2\pi}} = 2.62205755\ldots.$$

In fact, all we have to check is the second equality. It follows from the substitution $x = \wp(z, L)$ once we know that $\wp(-\frac{\omega}{2}, L) = \wp(\frac{\omega}{2}, L) = 1$. To see this, recall that $\wp$ is an even function of z, so $\wp'$ is odd. $\wp'$ has a pole of order three at each lattice point. Its zeros (mod L) are therefore $\frac{\omega}{2}, \frac{i\omega}{2}, \frac{\omega+i\omega}{2}$. By (15), the corresponding values of $\wp(z, L)$ are $0, \pm 1$. But the above proposition and the well-known Taylor expansion [Serre 1970, VII.2.3]:

$$(17) \qquad \wp(z, L) = \frac{1}{z^2} + \sum_{k=1}^{\infty} (4k-1) G_{4k}(L) z^{4k-2}$$

show that $\wp(\frac{\omega}{2}, L) > 0$.

The function $\wp(z, L)$ is the inverse function of the lemniscatic integral in the context of Weierstrass theory, where one considers the integral $\int dx/y$ related to the Weierstrass equation (15): $y^2 = 4x^3 - 4x$. It must no be confused with the inverse function φ for the integral (4) studied by Abel and Eisenstein, and also already by Gauss in his unpublished papers [Gauss III, pp. 404ff] where it is written sin lemn. The relation with Weierstrass's formalism is as follows—see [Hurwitz 1899, §3]:

$$(18) \qquad \sin \mathrm{lemn}(z) = \varphi(z) = \sqrt{\frac{1}{\wp(z, L)}}.$$

The following theorem holds in general, for any lattice $L \subset \mathbb{C}$.

Addition theorem. *For $u, v \in \mathbb{C}$ such that $u - v \notin L$, one has*

$$\wp(u+v, L) = -\wp(u, L) - \wp(v, L) + \frac{1}{4}\Big(\frac{\wp'(u, L) - \wp'(v, L)}{\wp(u, L) - \wp(v, L)}\Big)^2.$$

And for $z \in \mathbb{C}$, one has

$$\wp(2z, L) = -2\wp(z, L) + \frac{1}{4}\Big(\frac{\wp''(z, L)}{\wp'(z, L)}\Big)^2.$$

The way we stated it here, this seems purely analytic. In fact, it can be deduced from the observation that the difference of both sides of the first equation is a holomorphic function of u, once v is fixed; it is then constant, and indeed zero. But if we combine it with the differential equation (15), we get the following consequence which we state only for the case of our lattice $L = \omega\Lambda$.

Corollary. *The map*

$$\Phi : \mathbb{C}/L \longrightarrow E(\mathbb{C}) = \{(x:y:z) \in \mathbb{P}_2(\mathbb{C}) \mid y^2z = 4x^3 - 4xz^2\} \subset \mathbb{P}_2(\mathbb{C})$$

which sends $z + L$ for $z \notin L$ to the point $(\wp(z, L) : \wp'(z, L) : 1)$, and $0+L$ to the point $\mathbf{0} = (0 : 1 : 0)$, is a bijection from the complex torus $\mathbb{C}/L$ onto the complex points of the elliptic curve $E : y^2 = 4x^3 - 4x$. Φ transforms the addition on $\mathbb{C}/L$ into the geometric group law on the cubic E (the chord and tangent process) with neutral element $\mathbf{0}$.

Bijectivity is an easy consequence of the basic properties of meromorphic functions on $\mathbb{C}/L$. The geometric group law on a cubic curve in the projective plane works as follows. Given two points P and Q on the curve, draw the line through both. (If $P = Q$, take the tangent to the curve at this point). Since the curve is cubic and we work in the projective plane, this line will meet the curve in 3 points (counting multiplicities): P, Q, R. Now take for the "sum" $P \oplus Q$ the third point of intersection with the curve of the line through R and $\mathbf{0}$. This last twist in the construction is necessary to make $\mathbf{0}$ the neutral element, and in fact turn the whole operation into an (abelian) group law—see [Silverman 1986, III.2], [Schappacher 1991]. But note that, in view of the position of our point $\mathbf{0}$ at infinity, lines passing through $\mathbf{0}$ are parallel to the y-axis. So this last step from R to $P \oplus Q$ leaves the x-coordinate unchanged, and only switches y to $-y$. To check, on the level of the x-coordinates, that Φ transforms the addition of complex numbers into this geometric group law, one has to recognize in the two formulas of the addition theorem above the equations for the line through the given points $\Phi(u), \Phi(v)$ (resp. the tangent at $\Phi(z)$). The fact that the y-coordinates also work out, is left to the reader.

The curve E to which the lemniscatic elliptic integral has led us, is an elliptic curve over $\mathbb{Q}$—i.e., a nonsingular cubic with a distinguished rational point, namely $\mathbf{0}$. It may also be given by the simpler equation $y^2 = x^3 - x$. It is a famous elliptic curve—although a good part of 'its' history consists of studies that only our modern viewpoint acknowledges as being related to this curve. We are alluding here to the tradition of *diophantine analysis 'over $\mathbb{Z}$'*. Contrary to Diophant himself, this tradition asked for integral solutions—or for the proof that there are none, or only 'trivial' ones—of polynomial equations with integer coefficients. It starts essentially in the Arab world in the 10th century. Its first eminent representative was al-Khæzin—see [Houzel 1995, p. 6f], [Rashed 1996, p. 74f].

Fermat (1601–1665) lifted this branch of mathematics into a new era by inventing the method of infinite descent. The only proof by infinite descent of which Fermat has left us a rather detailed verbal sketch (in the margin of his Bachet-edition of Diophant's Arithmetics) concerns (among other results) the following theorem.

A theorem proved by Fermat. *There is no right triangle with rational sides whose area is a rational square.*

Since the problem alluded to is invariant under scaling the triangle, it actually is, despite its formulation, a problem 'over $\mathbb{Z}$', not 'over $\mathbb{Q}$'. Thus the theorem claims that there are no positive (or, equivalently, nonzero) integers a, b, c such that $a^2 + b^2 = c^2$ and $ab = 2A^2$, for some integer A (whose square is the area of the triangle (a, b, c)). Making the substitutions

$$x = \frac{a+c}{b} \qquad\qquad y = 2A\frac{a+c}{b^2},$$

gives a rational point (x, y), $x, y \in \mathbb{Q}$, on our elliptic curve $y^2 = x^3 - x$. Since we assumed $abc \neq 0$, we never get the trivial solutions

$(x, y) = (0, 0), (\pm 1, 0)$ in this way, nor the point at infinity $\mathbf{0}$. Therefore Fermat's result shows that the elliptic curve E has exactly four rational points. In particular, the group of rational points $E(\mathbb{Q})$ is a finite torsion group. Translating Fermat's proof to the elliptic curve, one shows that a hypothetical non-trivial element of $E(\mathbb{Q})$ can be divided again and again by 2. This process strictly decreases the size (technically speaking, the "height") of the point involved. But the heights of rational points form a discrete set bounded below; a contradiction.[8]

Probably the first—and for a long time, the only—publication about the interrelation between elliptic integrals and diophantine analysis is due to Jacobi: [Jacobi 1835]. It was inspired by an unpublished manuscript of Euler and simply states the potential usefulness of the addition theorem for diophantine questions, without giving examples—see [Schappacher 1991, §3].

6 The Higher Arithmetic of the Lemniscatic Function: Gotthold Eisenstein (and Gauss again).

Only about 20 years separate Abel's article from Eisenstein's work on the arithmetic of the lemniscatic function. The world of number theory was still under the spell of Gauss's *Disquisitiones Arithmeticæ*. But this book contains a lot more than just cyclotomy. In fact, its chapter VII on cyclotomy is the least arithmetic, the most algebraic of all. One mildly arithmetic aspect of chapter VII concerns the irreducibility of the cyclotomic polynomial $X^{p-1} + \ldots + 1$. The analogue of this for lemniscatomy, i.e., the irreducibility of the equation for

[8]See [Goldstein 1995] for a thorough investigation of this whole circle of ideas. See also, for instance, [Weil 1984, chap. I, §XI]. Fermat's result concerns a special case of the problem of congruent numbers which, in the modern analysis, leads to an infinite family of elliptic curves all of which are isomorphic to E over $\mathbb{C}$ (in fact, over the algebraic numbers $\overline{Q}$). There is a whole introductory book on this subject: [Koblitz 1984]. See also [Schappacher 1989].

the division of the lemniscate into m equal parts for an odd prime number m in the ring $\mathbb{Z}[i]$, was *not* proved in Abel's article. This is one of the critical remarks with which Eisenstein opens his sequence of articles [Eisenstein 1850], the milestone treated in this section. Of course, one voices such criticism only if one can do better; Eisenstein shows the irreducibility of the lemniscatomic equation by applying what is well-known today as "Eisenstein's irreducibility criterion". It was developed for this purpose. [9]

Eisenstein's Irreducibility Criterion (for $\mathbb{Z}[i]$). *Let $m \in \mathbb{Z}[i]$ be a prime element, and let $f = \sum_{\nu=0}^{n} a_{n-\nu} X^n \in \mathbb{Z}[i][X]$. Assume that $a_0 = 1$, a_ν is divisible by m for all $\nu = 1, \ldots, n$, and $a_n = i^a m$ for some $a = 1, \ldots, 4$. Then f is irreducible.*

There are three types of prime numbers m in the euclidean ring $\mathbb{Z}[i]$. First there is $1 + i$ which divides the rational prime 2. Indeed, one has $(1 + i)(1 - i) = 2$. Now, $1 - i = -i(1 + i)$, so, up to units, there is only one prime number dividing 2 in $\mathbb{Z}[i]$. Second, given a rational prime number p of the form $4k + 1$, Gauss had shown it is a sum of two squares in a unique way: $p = a^2 + b^2 = (a + bi)(a - bi)$. Here, the two factors are not the same up to units, and we have two different primes of $\mathbb{Z}[i]$ dividing p. Eisenstein calls these primes $a + bi$ *zweigliedrig* (because they have a real and an imaginary term). Third, the rational prime numbers p of the form $4k + 3$ are also prime elements in $\mathbb{Z}[i]$. Eisenstein calls them *eingliedrig*. Now, let $m \in \mathbb{Z}[i]$ be any prime element not dividing 2, either *zweigliedrig* or *eingliedrig*. Gauss had already proposed a way to privilege one of the four elements $i^a \cdot m$, $a = 1, \ldots, 4$, representing the same prime. He had called m a *primary prime* ("*primäre Primzahl*") if $m \equiv 1 \pmod{(2 + 2i)}$. Given m, there is always a unique unit i^a satisfying $i^a m \equiv 1 \pmod{(2 + 2i)}$ because the unit group of residues

[9] The criterion is formulated, proved, and slightly generalized in [Eisenstein 1850, pp. 541–544]. Eisenstein also transfers "Gauss's lemma" from $\mathbb{Z}[X]$ to $\mathbb{Z}[i][X]$: the product of two monic polynomials with rational coefficients that are not all integers cannot have integer coefficients [Gauss 1801, no. 42]. Eisenstein had already stated and used his criterion in [Eisenstein 1846]. But at that time he managed to apply it only to the lemniscatomic equation for primes $m \in \mathbb{Z}[i]$ dividing a rational prime of the form $p = 4k + 1$. He settled the general case before August 1847—see [Eisenstein 1847].

$(\mathbb{Z}[i]/(2+2i))^* = (\mathbb{Z}[i]/(1+i)^3)^*$ consists precisely of the classes of the four units i^a.

In proving that the m-th lemniscatomic equation satisfies the irreducibility criterion Eisenstein discovered and emphasized [Eisenstein 1850, p. 549 and *passim*] the following result about the lemniscatic function (18):

Eisenstein's Theorem. *If m is a primary prime number in $\mathbb{Z}[i]$ and $\mathbb{N}m = m\overline{m} \in \mathbb{Z}$, then there exist polynomials $P, Q \in \mathbb{Z}[i][X]$ such that one has the following identity of complex functions:*

$$\varphi(mt) = \frac{\varphi(t)^{\mathbb{N}m} + m\ P(\varphi(t))}{1 + m\ Q(\varphi(t))}. \tag{19}$$

Eisenstein stresses how important it is that there are no undetermined units in this formula. Part of the reason for this remark seems to be related to his latent—and sometimes not so latent—rivalry with Jacobi. The latter had published a two page note in Crelle's journal [Jacobi 1846] dismissing the earlier partial version of Eisenstein's theorem in [Eisenstein 1846, p. 301] as an easy consequence of his general theory. But this human aspect should not distract us from Eisenstein's remarkable insight. In fact, his theorem says that we have the following (coefficientwise) congruence between rational functions of $\varphi(t)$:

$$\varphi(mt) \equiv \varphi(t)^{\mathbb{N}m} \pmod{m}. \tag{20}$$

Note that the right-hand side of (20) depends only on the prime ideal[10] $\mathfrak{p} = (m) = m \cdot \mathbb{Z}[i]$, not on the particular generator m, whereas the left-hand side does depend on the condition that m be primary; otherwise the congruence does not hold in general.

In other words, Eisenstein shows (and obviously appreciates) that Gauss's normalization of the prime numbers in $\mathbb{Z}[i]$ defines a map-

[10] Ideals were introduced only by Dedekind. On the other hand, Eisenstein alludes explicitly [Eisenstein 1850, p. 595] to Kummer's works [Kummer 1847] which introduced ideal numbers—see also below.

ping from ideals prime to $(1+i)$, to complex multipliers of the lemniscatic function,

$$(21)\quad \psi : \mathfrak{p} \quad \mapsto \quad m, \qquad\qquad \text{where } \mathfrak{p} = m\mathbb{Z}[i],\ m \equiv 1\ (2+2i),$$

which describes the reduction modulo $\mathfrak{p}$ of the action of the normalized generator m in terms of the q-th power automorphism $x \mapsto x^q$ mod $\mathfrak{p}$ for $q = \mathbb{N}\mathfrak{p}$. Following our trained reflexes, we call this map the *Frobenius automorphism* relative to the finite field $\mathbb{F} = \mathbb{Z}[i]/\mathfrak{p}$. It generates the Galois group of any finite extension of $\mathbb{F}$. (When Eisenstein's paper went to press, Georg Frobenius was less than one year old. On the other hand, the "Frobenius" automorphism is explicitly visible on roots of unity, and this was the model that Eisenstein tried to generalize.) To project even further into the future, the map ψ is a very explicit and simple example of an (algebraic) Hecke character—see [Hecke 1918], [Hecke 1920], [Neukirch 1992, Kap. VII]—, i.e., a *Größencharakter* of type A_0 in the sense of [Weil 1955c].

Before continuing these observations, let us try to stay closer to the text. It is not obvious what exactly intrigued Eisenstein in his formula (19). Compared to other papers of his, the sequence of articles [Eisenstein 1850] proceeds at a rather slow pace. At times, the 27 year old man—who had just about two more years to live at the time of this publication—simply alludes to striking applications of his result without actually carrying them out—see for instance [Eisenstein 1850, p. 559].

Let us put this work into the perspective of his earlier paper [Eisenstein 1846]. Recall that the supreme challenge in higher arithmetic after the *Disquisitiones Arithmeticæ* was to construct the theory of *higher reciprocity laws*. Now, Eisenstein, in [Eisenstein 1846], managed to deduce the biquadratic reciprocity law from the special case of the above theorem where m is *zweigliedrig*.

For two distinct primary prime numbers m, n such that n is *zweigliedrig*, the 4-th power residue symbol $\left[\frac{m}{n}\right]$ is defined to be the power of i such that $\left[\frac{m}{n}\right] \equiv n^{\frac{\mathbb{N}_{m-1}}{4}}$ (mod m). In [Eisenstein 1846, p.

306], this symbol is written in the form

$$\left[\frac{m}{n}\right] = \frac{\prod_r \varphi(\frac{2nr\omega}{m})}{\prod_r \varphi(\frac{r\omega}{m})},$$

with r ranging over a certain set of residues mod m. Then the reciprocity law

$$\left[\frac{n}{m}\right] = (-1)^{\frac{\mathbb{N}m-1}{4}\cdot\frac{\mathbb{N}n-1}{4}} \left[\frac{m}{n}\right]$$

follows easily from Eisenstein's theorem. (The case of two *eingliedrig* primes m, n, for which the theorem was not available in 1846, is elementary.)

In the same vein, the later series of articles culminates in a final section [Eisenstein 1850, pp. 613–619] proving the reciprocity law for 8-th power residues. Preceding parts [Eisenstein 1850, pp. 575–613] set the stage for this, preparing in particular for the arithmetic of the field of 8-th roots of unities.

Let us now come back to the general lemniscatic arithmetic in Eisenstein's work of 1850. In [Eisenstein 1850, p. 558], Eisenstein arrives at the following easy consequence of his theorem (19).

Corollary. *Let m and n be distinct primary prime numbers and put $q = \mathbb{N}n$. Let $k \in \mathbb{Z}[i]$, and let $F(k)$ be a polynomial expression with coefficients in $\mathbb{Z}[i]$, in the division values $\varphi(\frac{rk(1+i)\omega}{m})$, where r ranges modulo m. Then there exists a polynomial expression T in the same values such that*

$$F(k)^q = F(nk) + nT.$$

He recognized this corollary—see [Eisenstein 1850, p. 559]—as the essential tool to understand the arithmetic of the extensions generated by division values of φ in the same way as Kummer had developed the arithmetic of cyclotomic fields in [Kummer 1847]. This encourages us to sketch a few consequences of the corollary using some of the further development of algebraic number theory. Let us switch at the same time from $\varphi(z)$ to $\wp(z, L)$—recall the comparison (18)—, and also treat $\wp'(z, L)$ along with $\wp(z, L)$.

Call $K = \mathbb{Q}(i)$ our base field and recall the notation of the corollary of section 4 above. Given a primary prime m, we want to study the arithmetic of the field $K_m = K(E_m)$ generated over K by the coordinates of all m-division points $E_m = \Phi(\frac{1}{m}L/L)$. Let n be any primary prime of K which is unramified in K_m. Then n acts on the points of E_m via:

$$\Phi\Big(\frac{(a+bi)\omega}{m}\Big) \qquad \mapsto \qquad \Phi\Big(n\,\frac{(a+bi)\omega}{m}\Big).$$

Here, $a+bi$ ranges over $\mathbb{Z}[i]/(m)$. Eisenstein's corollary means that we can describe this action by reducing modulo the ideal $n \cdot \mathfrak{o}_{K_m}$ in the ring of integers $\mathfrak{o}_{K_m}$ of K_m, i.e. (as n is unramified), modulo any prime ideal $\mathfrak{P}$ of $\mathfrak{o}_{K_m}$ dividing $\mathfrak{p} = n\mathbb{Z}[i]$. In fact, the points of E_m reduce injectively modulo $\mathfrak{P}$ because m and n are relatively prime. We may do better if we use the fact that K_m/K is an abelian extension. For the field generated just by the x-coordinates of points in E_m, this follows already from Abel's work. To see it for our field K_m, one may adapt the proof in [Silverman, Tate 1992, VI.5] to our elliptic curve.

Since K_m/K is abelian, the Frobenius automorphism of the extension of finite fields $(\mathfrak{o}_{K_m}/\mathfrak{P})$ / $(\mathbb{Z}[i]/\mathfrak{p})$ lifts to a unique automorphism $\sigma_\mathfrak{p} \in \mathrm{Gal}(K_m/K)$, and, in the notation of (21), we find for all m-division points that

$$(22) \qquad \Phi\Big(\frac{(a+bi)\omega}{m}\Big)^{\sigma_\mathfrak{p}} = \Phi\Big(\psi(\mathfrak{p})\cdot\frac{(a+bi)\omega}{m}\Big),$$

where ψ was defined in (21). In modern parlance—much used recently in the context of Wiles's proof of Fermat's Last Theorem—, this just means that ψ gives the (abelian) mod m Galois representation of the elliptic curve E.

Let us note in passing that the extensions generated by division values of the elliptic curve $y^2 = x^3 - x$ generate in fact *all* abelian extensions of $\mathbb{Q}(i)$. This special case of what is known as "Kronecker's Jugentraum" was proved (with slight incorrectnesses) by Takagi in his thesis [Takagi 1903]; see also [Schappacher 1997].

Eisenstein remarks that the corollary extends to powers n^μ instead of n, with q replaced by q^μ, and in particular: "one has $F(k)^{q^\mu} =$

$F(k) + nT$ if $n^\mu \equiv 1 \pmod{m}$" [Eisenstein 1850, p. 558]. (Note the analogy with what happens for roots of unity.) Even without taking powers, we get the same conclusion if we take m dividing $n-1$. In other words, we see that all the $n-1$-division points on the elliptic curve $y^2 = x^3 - x$ are invariant under Frobenius, i.e., their reduction is defined already over the finite field $\mathbb{Z}[i]/\mathfrak{p}$. Since there are $\mathbb{N}(n-1)$ torsion points, and they remain distinct after reduction, the curve has at least that many rational points over this finite field—in fact, these are all.

This result about the number of points of E over $\mathbb{Z}[i]/\mathfrak{p}$ (counting also points at infinity!) was conjectured after numerical experiments by Gauss: see the last entry, dated 9 July 1814, of his mathematical diary [Gauss 1976], where he calls this conjecture a "most important observation, found by induction, which links the theory of biquadratic residues with the lemniscatic functions in the most elegant manner" (*Observatio per inductionem facta gravissima theoriam residuorum biquadraticorum cum functionibus lemniscatis elegantissime nectens*). In fact, Gauss worked with the model $x^2 + y^2 + x^2y^2 = 1$ of the curve E which corresponds to the equation satisfied by his two basic lemniscatic functions sin lemn and cos lemn, whereas we took $y^2 = x^3 - x$, or equivalently $y^2 = 4x^3 - 4x$, because it corresponds to the Weierstrass equation (15) relating the basic functions $\wp, \wp'$.

Eisenstein met Gauss once. It is not clear whether he learned about Gauss's conjecture. It was published only after Gauss's death (and Gauss died three years after Eisenstein). When Gauss's entry became known later in the 19th century, people did apparently not link it with Eisenstein's work. Dedekind for instance checked the conjecture numerically for all primes of norm less than 100. Fricke recognized the equation of the curve as the one linking Gauss's lemniscatic functions, but it was not until 1921 that the conjecture was proved, by Gustav Herglotz (the teacher of Emil Artin). His concise article [Herglotz 1921] uses the Weierstrass theory, albeit with notation that has not quite survived to the present day. He does allude to Eisenstein's irreducibility criterion in the original lemniscatic context, but otherwise does not seem to derive inspiration from

Eisenstein's paper—however, see also [Herglotz 1922, p. 455f]. A modern account, in the spirit of the Weil conjectures can be found in [Ireland, Rosen 1982, 11-5].

Today one defines the L-function of the elliptic curve E over $K = \mathbb{Q}(i)$ by the infinite product over prime ideals of $\mathbb{Z}[i]$ different from $(1+i)$

$$(23) \qquad L(E/K, s) = \prod_{\mathfrak{p}} \frac{1}{1 - a_{\mathfrak{p}} \mathbb{N}\mathfrak{p}^{-s} + \mathbb{N}\mathfrak{p}^{1-2s}} \; (s \in \mathbb{C}, \; \Re(s) > \frac{3}{2}),$$

where $a_{\mathfrak{p}} = \mathbb{N}\mathfrak{p} + 1 - N_{\mathfrak{p}}$ for $N_{\mathfrak{p}}$ the number of points of E over the finite field $\mathbb{Z}[i]/\mathfrak{p}$. Using Gauss's conjecture proved by Herglotz, we find for a *zweigliedrig* prime $\mathfrak{p} = n\mathbb{Z}[i]$, $n = a + bi$ primary: $a_{\mathfrak{p}} = a^2 + b^2 + 1 - (a-1)^2 - b^2 = 2a = \psi(\mathfrak{p}) + \overline{\psi(\mathfrak{p})}$. And for an *eingliedrig* prime $\mathfrak{p} = p\mathbb{Z}[i]$: $a_{\mathfrak{p}} = p^2 + 1 - (p-1)^2 = 2p = 2\psi(\mathfrak{p}) = \psi(\mathfrak{p}) + \overline{\psi(\mathfrak{p})}$. Since for any $\mathfrak{p}$, one clearly has $\psi(\mathfrak{p})\overline{\psi(\mathfrak{p})} = \mathbb{N}\mathfrak{p}$, this proves that

$$(24) \qquad L(E/K, s) = L(\psi, s) \cdot L(\overline{\psi}, s),$$

where we define the Hecke L-series of a Hecke character by the rule

$$(25) \qquad L(\psi, s) = \prod_{\mathfrak{p}} \frac{1}{1 - \psi(\mathfrak{p})^s \mathbb{N}\mathfrak{p}^{-s}}.$$

For a general elliptic curve with complex multiplication, the analogous result was obtained by Deuring approximately 100 years after Eisenstein's work—see [Deuring 1953+]. In the 1950's the time was ripe because the progress of abstract algebraic geometry and, more specifically, Weil's work on points of (abelian) varieties over finite fields, had provided the necessary theoretical basis for discussing arithmetic applications and in particular the reduction of algebraic curves modulo primes which we naïvely took for granted in the above discussion of our special case.

The beginning 1950s also saw the generalization of the theory of complex multiplication to higher dimensional abelian varieties by Taniyama and Shimura. Eisenstein's theorem was thus generalized in this higher dimensional theory to the crucial congruence relation of

[Shimura, Taniyama 1961, chap. III, §13]—see also [Giraud 1968] for the grothendieckian digest of this relation. But Eisenstein's work had not been entirely forgotten between 1850 and 1950. In fact, Leopold Kronecker had not only generalized Eisenstein's congruence to arbitrary transformations between elliptic curves [Kronecker 1886, p. 439], but also greatly emphasized the importance of this "fundamental congruence". Thus, Shimura and Taniyama [*loc. cit.*, p. 110] call their theorem a "generalization of Kronecker's congruence formula".

In the vast motivic generalization of the theory of Shimura and Taniyama, one may use Eisenstein's theorem as the one special check which implies the general identity of the "cocycles" describing on the one hand Langlands's Taniyama group and on the other, the motivic Galois group of a category of motives constructed from abelian varieties with complex multiplication—see [Schappacher 1994, §**0**, and §4.4.4].

7 Special L-values and p-adic L-functions: Adolf Hurwitz.

Let us come back to the analytic theory of section 5 above. Differentiating (17) twice in z, we get

$$(26) \qquad \wp''(z,L) = \frac{6}{z^4} + \sum_{k=1}^{\infty}(4k-1)(4k-2)(4k-3)G_{4k}(L)z^{4k-4}.$$

On the other hand, differentiating (15) once and dividing by $\wp'$ yields

$$(27) \qquad \wp''(z,L) = 6\wp^2(z,L) - 2.$$

Developing both sides of this equation using (17), resp. (26), we obtain recursive formulas for the coefficients $G_{4k}(L)$. For simplicity, let us drop the fixed argument L from the notation. First, comparing the constant terms in (27), we deduce $30G_4 = 2$ which is equivalent to (14b). In z^4, we find $G_8 = \frac{6}{7}G_4^2$. Comparison at z^8 yields $G_{12} = \frac{7}{22}G_4G_8$. Continuing in this way, we get the recursion valid for all

$k \geq 2$:

$$\Big((4k-1)(4k-2)(4k-3)-12(4k-1)\Big)G_{4k} = 6\sum_{l+m=k}(4l-1)(4m-1)G_{4l}G_{4m},$$

where the sum is over positive integers l, m that add up to k. Therefore the $G_{4k}(L)$ are all rational numbers. Equivalently, passing from L to Λ, we obtain the following result of Hurwitz—see [Hurwitz 1897, p. 339], [Hurwitz 1899, p. 342]:

Proposition. *For every $k \geq 1$ there exists a rational number e_k such that one has*

$$\sum_{a,b\in\mathbb{Z}}{}' \frac{1}{(a+bi)^{4k}} = e_k \cdot \omega^{4k}. \tag{28}$$

Hurwitz observes the striking analogy of this statement with the well-known rationality of $\zeta(2k)/\pi^{2k}$ for the Riemann zeta function $\zeta(s)$: note that $\pi = 2\int_0^1 dx/\sqrt{1-x^2}$ and rewrite the special ζ-value as

$$\sum_{n\in\mathbb{Z}}{}' \frac{1}{n^{2k}} = \frac{(2\pi)^{2k}}{(2k)!}B_{2k}$$

with the $2k$-th Bernoulli number B_{2k}—see for instance [Neukirch 1992, VII.1.10].

In today's perspective, there is no single natural generalization of the Riemann zeta function whose special values would give the expressions studied by Hurwitz. Rather, the modern approach adduces a different Hecke L-series for every value of k, namely that attached to the $4k$-th power of the Hecke character ψ:

$$L(\psi^{4k}, 4k) = \prod_{\mathfrak{p}} \frac{1}{1-\psi(\mathfrak{p})^{4k}\mathbb{N}\mathfrak{p}^{-4k}} = \frac{1}{4}\sum_{a,b\in\mathbb{Z}}{}' \frac{1}{(a+bi)^{4k}}.$$

See for instance [Harder, Schappacher 1984, §1].

In his perspective, Hurwitz tried to carry over as many properties as possible from the Bernoulli numbers to the e_k (after suitably normalizing the latter). He manages in particular for the congruences proved in [v. Staudt 1845]. These congruences are seen today—following [Kubota, Leopoldt 1964]—as giving the possibility to p-adically interpolate the rational parts of special values of $\zeta(s)$ (and of similar Dirichlet series), thus constructing *p-adic zeta functions*. Likewise, Hurwitz's result is at the origin of the construction of p-adic L-functions for elliptic curves with complex multiplication—see [de Shalit 1987]. These p-adic considerations were crucial for the first major breakthrough by Coates and Wiles towards the Conjecture of Birch and Swinnerton-Dyer—see [Coates, Wiles 1977]. The Coates-Wiles theorem was in a sense completed (following, as it were, a new idea of Thaine in the theory of cyclotomic fields) by Karl Rubin in 1987. There, our elliptic curve $E : y^2 = x^3 - x$ derived from the lemniscatic integral is among the very first for which the conjecture of Birch and Swinnerton-Dyer could be verified completely: [Rubin 1987, p. 528]. In this conjecture, the finiteness of the group of rational points $E(\mathbb{Q})$ corresponds to the fact that $L(\psi, 1) \neq 0$ ($L(\psi, s)$ can be analytically continued to the whole complex plane; this was proved by Hecke), and the quotient $L(\psi, 1)/\omega$ is interpreted arithmetically in terms of homogeneous spaces of E.

Acknowledgements. With this paper, I want to thank all my Turkish colleagues and friends for their excellent hospitality and company over the years, from 1987 through 1995. Special thanks are due to Sinan Sertöz for the Ankara meeting in 1995 and these proceedings. On the other hand, I am grateful to Jean-Pierre Friedelmeyer who organized an *Atelier de lecture, histoire des mathématiques* at Strasbourg in 1996/97 where some of the texts treated here were read. While writing this article, I profited from comments by Harold Edwards, Jean-Pierre Friedelmeyer, John McCleary, Jean-Yves Merindol.

References

N.H. Abel (1827–28), Recherches sur les fonctions elliptiques, *Journal f.d. reine & angew. Math.* **2**, 101–181, **3**, 160–190 [= Œuvres complètes (Sylow, Lie, ed.s), vol. I, pp. 263–388]

R. Ayoub (1984), The lemniscate and Fagnano's contributions to elliptic integrals, *Archive Hist. Exact. Sciences* **29**, no. 2, 131–149

B. Belhoste (1996), Autour d'un mémoire inédit:la contribution d'Hermite au développement de la théorie des fonctions elliptiques; Revue d'histoires des mathématiques **2**, 1–66

H. Bos (1993), Lectures in the History of Mathematics, Hist. of Math. series, vol. 7, American Mathematical Society, *in particular:* The lemniscate of Bernoulli, pp. 101–111, *and:* The concept of construction and the representation of curves in seventeenth-century mathematics, pp. 23–36

E. Brieskorn & H. Knörrer (1986), Plane Algebraic Curves, Basel etc. (Birkhäuser)

A. Brill & M. Noether (1892–93), Die Entwicklung der Theorie der algebraischen Functionen in älterer und neuerer Zeit, Bericht erstattet der Deutschen Mathematiker-Vereinigung, *Jahresbericht DMV* **3**, 107–566

J. Coates & A. Wiles (1977), On the conjecture of Birch and Swinnerton-Dyer, *Inventiones math.* **39**, 223–251

M. Deuring (1953+), Die Zetafunktion einer algebraischen Kurve vom Geschlechte Eins, *Nachr. Akademie Wiss. Gött., Math.-Phys. Klasse* 1953, 95–104; Zweite Mitteilung: 1955, 13–42; Dritte Mitteilung: 1956, 37–76; Vierte Mitteilung: 1957, 55–80

G. Eisenstein (1846), Beiträge zur Theorie der elliptischen Funktionen: I. Ableitung des biquadratischen Fundamentaltheorems aus der Theorie der Lemniscatenfunctionen, nebst Bemerkungen zu den Multiplications- und Transformationsformeln, *Journal f.d. reine & angew. Math.* **30** 1846, 185–210 [= Mathematische Werke, vol. I, pp. 299–324]

G. Eisenstein (1847), Letter to C.F. Gauss dated 18 August 1847, *in:* Mathematische Werke, vol. II, pp. 845–855

G. Eisenstein (1850), Über die Irreductibilität und einige andere Eigenschaften der Gleichung, von welcher die Theilung der ganzen Lemniscate abhängt, *Journal f.d. reine & angew. Math.* **39**, 160–179; *and the sequel:* Über einige allgemeine Eigenschaften der Gleichung, von welcher die Theilung der ganzen Lemniscate abhangt, nebst Anwendungen derselben auf die Zahlentheorie, *Journal f.d. reine & angew. Math.* **39**, 224–287 [= Mathematische Werke, vol. II, pp. 536–619]

A. Enneper (1876), Elliptische Functionen, Theorie und Geschichte, Halle (Nebert)

L. Euler (1761), Observationes de comparatione arcuum curvarum irrectificabilium. Novi comm. acad. sci. Petrop. **6**, 58–84 [= Mathematische Werke (Leipzig, Berlin 1912) **20**, 80–107]

G.C. di Fagnano (1750), Produzioni matematiche, vol. II, Pesaro [= Opere matematiche 1–2, Milano-Roma-Napoli 1911]

E. Freitag & R. Busam (1995), Funktionentheorie (2nd edition), *Springer Lehrbuch*, Berlin, Heidelberg

C.F. Gauss (1801), Disquisitiones Arithmeticæ, Leipzig (G. Fleischer)

C.F. Gauss (III), Werke, Göttingen (1870–1927), volume III

C.F. Gauss (1976), Mathematisches Tagebuch 1796–1814, Hrsg./Übers.: Biermann, Schuhmann, Wußing. Leipzig (Geest & Portig)

J. Giraud (1968), Remarque sur une formule de Shimura-Taniyama, *Inventiones Math.* **5**, 231–236

C. Goldstein (1995), Un théorème de Fermat et ses lecteurs, Saint Denis (Presses Universitaires de Vincennes)

G. Harder & N. Schappacher, Special Values of Hecke L-functions and Abelian Integrals; *in:* Arbeitstagung Bonn 1984, *Lecture Notes in Mathematics* **1111** (Springer) 1985, 17–49

T. Heath (1921), A History of Greek Mathematics, Oxford, vol. II

G. Henniart (1987), Cyclotomie et valeurs de la fonction Γ (d'après G. Anderson), Sém. Bourbaki no. 688; Astérisque **161–162**, 1988, pp. 53–72

G. Herglotz (1921), Zur letzten Eintragung im Gaußschen Tagebuch; *in: Berichte über die Verhandlungen der Sächsischen Akademie der Wissenschaften Math.-phys. Klasse* **73**, 271–276 [= Gesammelte Schriften (Schwerdtfeger, ed.), Göttingen 1979, 415–420]

G. Herglotz (1922), Über die Entwicklungskoeffizienten der Weierstrass'schen $\wp$-Funktion; *in: Berichte über die Verhandlungen der Sächsischen Akademie der Wissenschaften Math.-phys. Klasse* **74**, 269–289 [= Gesammelte Schriften (Schwerdtfeger, ed.), Göttingen 1979, 436–456]

C. Houzel (1978), Fonctions elliptiques et intégrales abéliennes, *in:* J. Dieudonné (ed.), Abrégé d'histoire des mathématiques 1700–1900, vol. II, Paris (Hermann), pp. 1–113

C. Houzel (1995), Théorème de Fermat, à travers l'histoire de l'analyse diophantienne, brochure (17 pages), Société Mathématique de France, Paris

A. Hurwitz (1897), Über die Entwicklungskoeffizienten der lemniskatischen Funktionen, *Nachr. Ges. Wiss. Göttingen*, 273–276 [= Mathematische Werke, vol. II, 338–341]

A. Hurwitz (1899), Über die Entwicklungskoeffizienten der lemniskatischen Funktionen, *Mathematische Annalen* **51**, 196–226 [= Mathematische Werke, vol. II, 342–373]

K. Ireland & M. Rosen (1982), A classical introduction to modern number theory, Graduate Texts in Mathematics **84**, New York, Berlin, etc. (Springer)

C.G.J. Jacobi (1835), De usu theoriæ integralium ellipticorum et integralium abelianorum in analysi diophantea, *Journal f.d. reine & angew. Math.* **13**, 353–355 [= Gesammelte Werke (Weierstrass, Hrsg.), vol. II, 53–55]

C.G.J. Jacobi (1846), Über einige die elliptischen Functionen betreffenden Formeln, *Journal f.d. reine & angew. Math.* **30**, 269–270 [= (with one correction) Gesammelte Werke (Weierstrass, Hrsg.), vol. I, 371–372]

L. Kiepert (1873), Siebzehntheilung des Lemniscatenumfangs durch alleinige Anwendung von Lineal und Cirkel, *Journal f.d. reine & angew. Math.* **75**, 255–263; *and:* Zusatz zu der Abhandlung über Siebzehntheilung der Lemniscate S. 255 dieses Bandes, *Journal f.d. reine & angew. Math.* **75**, 348

N. Koblitz (1984), Introduction to Elliptic Curves and Modular Forms, Graduate Texts in Mathematics **97**, New York, Berlin, etc. (Springer)

L. Königsberger (1904), Carl Gustav Jacob Jacobi, Festschrift zur hundertsten Wiederkehr seines Geburtstages, Leipzig (Teubner)

L. Kronecker (1886), Zur Theorie der elliptischen Functionen, XI, Sitzungsber. Akademie Berlin (27.5. & 29.7.1886), 701–780 = Werke IV, p. 389–471.

T. Kubota & H.W. Leopoldt (1964), Eine p-adische Theorie der Zetawerte, I. Einführung der p-adischen Dirichletschen L-Funktionen, *Journal f.d. reine & angew. Math.* **214/215**, 328–339

E.E. Kummer (1847), Zur Theorie der complexen Zahlen, *Journal f.d. reine & angew. Math.* **35**, 319–326; *and:* Über die Zerlegung der aus Wurzeln der Einheit gebildeten complexen Zahlen in ihre Primfactoren, *Journal f.d. reine & angew. Math.* **35**, 327–367 [= Collected Papers (A. Weil, ed.), vol. I (1975), Berlin, Heidelberg, New York (Springer), pp. 203–251]

A.M. Legendre (1811), Exercices de calcul intégral, Paris

A.M. Legendre (1825–28), Traité des fonctions elliptiques, 3 vol., Paris

T. Masahito (1995), Three aspects of Complex Multiplication; *in:* The intersection of history and mathematics (S. Chikara, S. Mitsuo, J.W. Dauben, ed.s), Science Networks, Historical Studies, vol. 15, Birkhäuser Verlag Basel, Boston, Wien, pp. 91–108.

J. Neukirch (1992), Algebraische Zahlentheorie, Berlin, Heidelberg, etc. (Springer)

Plana (1863), Mémoire sur la théorie des transcendantes elliptiques, *Memorie della reale Academia di Torino*, Serie seconda, **XX**, p. 292

R. Rashed (1996), Modernité classique et science arabe, *in:* L'Europe mathématique / Mathematical Europe (Goldstein, Gray, Ritter, ed.s), Paris (MSH), 67–81

R. Remmert (1991), Funktionentheorie 2, Berlin, Heidelberg etc. (Springer)

K. Rubin (1987), Tate-Shafarevich groups and L-functions of elliptic curves with complex multiplication, *Inventiones math.* **89**, 527–560

J.-P. Serre (1970), Cours d'arithmétique, Paris (PUF) [in English: Graduate Texts in Mathematics **7**, New York, Berlin, etc. (Springer)]

N. Schappacher (1988), Periods of Hecke Characters, Springer Lecture Notes Math. **1301**

N. Schappacher (1989), Neuere Forschungsergebnisse in der Arithmetik elliptischer Kurven, *Didaktik der Mathematik* **17**, 149–158

N. Schappacher (1991), Développement de la loi de groupe sur une cubique; Séminaire de Théorie des Nombres Paris 1988/89, *Progress in Mathematics* **91** (Birkhäuser), 159–184

N. Schappacher (1994), CM motives and the Taniyama group; *in:* Motives (Jannsen, Kleiman, Serre, eds.), Proceedings of Symposia in Pure Mathematics **55**, part I, AMS 1994, 485–508

N. Schappacher (1997), On the History of Hilbert's Twelfth Problem, I: Paris 1900 – Zürich 1932, a Comedy of Errors, *to appear*

H. Schröter (1873), Zur *v. Staudt*schen Construction des regulären Siebzehnecks, *Journal f.d. reine & angew. Math.* **75**, 13–24

E. de Shalit (1987), Iwasawa Theory of Elliptic Curves with Complex Multiplication, *Persepctives in Mathematics* **3**, Boston, Orlando, etc. (Academic Press)

G. Shimura & Y. Taniyama (1961), Complex multiplication of abelian varieties and its application to number theory, Publ. Math. Soc. Japan **6**

C.L. Siegel (1959), Zur Vorgeschichte des Eulerschen Additionstheorems; *in:* Sammelband Leonard Euler, Akademie-Verlag Berlin [= Gesammelte Abhandlungen, vol. III, 249–251]

C.L. Siegel (1969), Topics in Complex Function Theory, New York, London, etc. (Wiley)

J.H. Silverman (1986), The Arithmetic of Elliptic Curves, Graduate Texts in Mathematics **106**, New York, Berlin, etc. (Springer)

J.H. Silverman & J. Tate (1992), Rational Points, Undergraduate Texts in Mathematics, New York, Berlin, etc. (Springer)

P. Stäckel & W. Ahrens (ed.s) (1908), Der Briefwechsel zwischen C.G.J. Jacobi und P.H. Fuss über die Herausgabe der Werke Leonhard Eulers, Leipzig

C.G.C. v. Staudt (1845), De numeris Bernoullianis, and: De numeris Bernoullianis commentatio altera, Erlangen (Junge)

T. Takagi (1903), Ueber die im Bereiche der rationalen complexen Zahlen Abel'schen Körper, Journal of the College of Science, Tokyo Imperial University **19**, article 5, 42 pages [= Collected Papers (ed.s S. Iyanaga, K. Iwasawa, K. Kodaira, K. Yosida), Tokyo - Berlin - Heidelberg - New York - London - Paris - Hong Kong - Barcelona (Springer) 1990, no. 6, pp. 13–39]

I. Thomas (1941), Selections Illustrating the History of Greek Mathematics, vol. II, Loeb Classical Library, Cambridge (Mass.) and London

A. Weil (****), Œuvres scientifiques — Collected Papers, volume II (1951–1964), New York, Heidelberg, etc. (Springer) 1979; the individual articles are quoted according to the abbreviations of this edition.

A. Weil (1984), Number Theory, An Approach through History, from Hammurapi to Legendre, Boston etc. (Birkhäuser)

An Overview of the Search for Minimal Models of Algebraic Threefolds

Sinan Sertöz

Bilkent University
Department of Mathematics
06533 Ankara, Turkey

e-mail: `sertoz@fen.bilkent.edu.tr`

1 Introduction

The classification theory of algebraic threefolds had an exciting decade in the eighties and culminated in the major contributor Shigefumi Mori winning a Fields medal with [18] and [21]. The origins of the major ideas go back to the works of Hironaka [5] in the sixties. The road to the final accomplishment of the classification program is paved with the most prominent names of algebraic geometry. A list of the contributors suffices to awe a newcomer and the same applies to the beauty and delicacy of the theory.

My aim in this manuscript is to expose the grand classification scheme of threefolds with as little machinery as possible. In particular I make no attempt to motivate the subject by testing new ideas on surfaces, which the reader should do systematically. Also I omit a tremendous amount of information on several aspects of the theory in order to reach the destination; the description of the Minimal Model Program. In particular I made no mention of the log version of the program but I hope that after this paper the reader will be well equipped to read these from other sources.

Miles Reid wrote [25] "...you learn to get around your own village before you set about memorising the entire motorway network". Expounding on this allegory this exposition can be considered as an address, a description of a route from your own village to the capital where the MMP resides. Many beautiful scenic spots are only briefly observed and numerous chances for exciting excursions along the highway are carefully avoided in order to arrive on schedule. Having once surveyed the main route of the highway the reader, the venturer, can go back and take the first exit of his or her choice. That at least is what I intend to do ...

The plan of this survey is as follows: In the following Section 2 I give a long and tedious proof of the existence of rational curves contracted by a proper birational map following [13]. This is perhaps the most elementary part of the theory and I want a beginning graduate student to build up confidence by observing that a fundamental observation can be proved totally within his or her existing vocabulary. This is the only place where a formal proof is given. In Section 3 the reader will find a collection of new concepts which play a part in the theory. The list of definitions is short enough to make the manuscript readable but long enough to equip the reader with tools to appreciate what is to come. In Section 4 the reader is brought into contact with the necessity of admitting singularities to the program. Section 5 gives a brief description of these singularities *après* Reid. Again we exercise great will power in avoiding a deeper understanding of the singularity machinery of higher dimensional geometry. Section 6 describes the existence and finiteness of the flip map which is a way of avoiding singularities worse than those we are ready to handle. And finally in Section 7 a full statement of the classification theorem is given with its implications on the structure of the cone of curves.

Any article written on this topic carries a long list of references because the achievements in higher dimensional geometry result from a truly international collaboration. After each section I give some references to lead the graduate student into the literature. This has the effect of mostly choosing those articles which are either expository in nature or include abundant motivation and explanation to ease the student's way into the topic. It is inevitable that I have failed to mention numerous researchers whose work merits mention by any standards. This I hope will be forgiven given my explicit aim of paving a graduate student's way into the subject.

The final version of this manuscript owes its merit and accuracy to the careful reading and correcting of Miles Reid. All the inaccuracies owe their survival to my unenviable dexterity in concealing them. I am also grateful to Reid for coming to Ankara and lecturing on these topics. In preparing this note, among other sources, I occasionally made use of his Bilkent lecture notes but I hereby acquit him of any responsibility thereof ...

2 Contracting Curves

In this section a variety means a smooth projective variety over $\mathbb{C}$ of dimension two or more. We agree to say that a variety Y is 'simpler' than another variety X when there is a regular birational map $f\colon X \longrightarrow Y$ between them. We interpret this by imagining that more than one point of X is mapped to a certain point of Y so f^{-1} cannot be defined set theoretically, so X has 'more' in it than Y. In this section we will justify this notion by showing that X contains at least one rational curve which is mapped to a point in Y. Moreover it will be observed that the canonical divisor K_X of X has negative intersection with this rational curve and we will later investigate the sufficiency of this necessary condition.

We first compare the intersection numbers $[C] \cdot K_X$ and $[f(C)] \cdot K_Y$ for an irreducible curve C in X, where K_X and K_Y are the canonical divisors of X and Y respectively and we use $[\,]$ to denote equivalence class. Let s be a meromorphic section of the canonical sheaf ω_Y of Y. Then the zero set (s) of s defines a divisor which is linearly equivalent to K_Y. Pulling this section back by f defines a section of $f^*\omega_Y$. This then defines a section of ω_X under the natural map

$$f^*\omega_Y \longrightarrow \omega_X.$$

To describe this map locally let $(U, x = (x_1, \ldots, x_n))$ be a local coordinate chart on X and $(V, y = (y_1, \ldots, y_n))$ a coordinate chart on Y with $f(U) \subset V$. Then in these coordinates $s(y) = h(y)\mathrm{d}y_1 \wedge \cdots \wedge \mathrm{d}y_n$ for some meromorphic function h on V. $f^*s = h(f(x))\mathrm{d}y_1 \wedge \cdots \wedge \mathrm{d}y_n$ and the induced section on $\omega_X|U$ is $h(f(x)) \det \left|\dfrac{\partial f_i}{\partial x_j}\right| \mathrm{d}x_1 \wedge \cdots \wedge \mathrm{d}x_n$, where we take $f(x) = (f_1(x), \ldots, f_n(x))$. Since this is a section of $\omega_X|U$, its zero set defines a divisor equivalent to K_X and we have

$$\begin{aligned} K_X|U &= (h(f(x)) \det \left|\frac{\partial f_i}{\partial x_j}\right|) \\ &= (h(f(x))) + (\det \left|\frac{\partial f_i}{\partial x_j}\right|). \end{aligned}$$

Here we note that $(h(f(x))$ is the pullback of the divisor $K_Y|V$. On the other hand the above determinant, which is the Jacobian of the map f, vanishes where f is not a local isomorphism. Call this set E. Setting $E = \bigcup E_i$, where the E_i are irreducible codimension one components, we have

$$K_X|U = f^*K_Y|V + \sum a_i E_i \cap U.$$

Here the a_i denote the order of vanishing of the Jacobian and hence are nonnegative. Pasting this local data together we have

$$K_X = f^*K_Y + \sum a_i E_i.$$

Intersecting both sides of this equality with the irreducible curve C we have

$$[C] \cdot K_X = [f(C)] \cdot K_Y + \sum a_i \, [C] \cdot E_i.$$

Assume that $C \not\subset E$. Then C is not one of the E_i so $[C] \cdot E_i$ is nonnegative. Moreover $f(C)$ is a nontrivial curve in Y so that $[f(C)]$ is not identically equal to zero. We then have

$$\begin{aligned}[C] \cdot K_X - [f(C)] \cdot K_Y &= \sum a_i \, [C] \cdot E_i \\ &= \begin{cases} 0 & \text{if } C \cap E = \emptyset \\ > 0 & \text{if } C \cap E \neq \emptyset \end{cases}\end{aligned}$$

We record this for future reference:

Lemma 1 *Let $f: X \longrightarrow Y$ be a regular birational surjective map between smooth projective varieties of dimension $n \geq 2$ and C an irreducible curve in X not mapped to a point by f. Then*

$$\begin{aligned}[C] \cdot K_X &= [f(C)] \cdot K_Y \quad \text{if } C \cap E = \emptyset, \\ [C] \cdot K_X &> [f(C)] \cdot K_Y \quad \text{if } C \cap E \neq \emptyset,\end{aligned}$$

where E is the set consisting of points where f is not a local isomorphism.

Next we analyze the geometry around a point $y_0 \in Y$ where f^{-1} is not defined. First assume that Y is a smooth projective variety of dimension $n \geq 3$. Assume further that Y lies in the projective space $\mathbb{P}^N$. Consider the space G of all hyperplanes in $\mathbb{P}^N$ passing through y_0. G is isomorphic to a copy of $\mathbb{P}^{N-1}$. Let G_1 be the subset of G consisting of hyperplanes not containing $B - \{y_0\}$, where B is the subset of Y where f^{-1} is not defined. If $B = \{y_0\}$ then G_1 is all of G, otherwise it is a proper open subset of G. Further let G_2 be the subset of G consisting of those hyperplanes which intersect Y in a smooth locus. This involves the nonvanishing of certain Jacobians. Since Y is smooth these Jacobians do not vanish identically. Hence G_2 is an open subset of G. Then $G_1 \cap G_2$ is nonempty since any two nonempty open sets intersect in Zariski topology. Hence there is a hyperplane H in $\mathbb{P}^N$ which intersects Y through y_0 in a smooth locus. Note that codim $B \geq 2$ since f^{-1} is a rational map from a projective variety. Consequently codim $B \cap H \geq 2$ since H does not contain B

unless $B = \{y_0\}$. By induction we can construct a smooth surface S in Y such that $S \cap B = \{y_0\}$. Moreover by a similar argument there exist hypersurfaces H_0 and H_1 such that $C_0 = Y \cap H_0$ is a smooth curve in S passing through y_0 and $C_1 = Y \cap H_1$ is another smooth curve in S not passing through y_0. Since they are both hyperplane sections C_0 and C_1 represent the same divisor class in S. If $\dim Y = 2$, then we only construct C_0 and C_1 and set $S = Y$.

We now restrict f^{-1} to S to obtain the rational map

$$f^{-1}: S \dashrightarrow X.$$

Observe that f^{-1} is an isomorphism on $S - \{y_0\}$ into X. Any rational map on a surface can be resolved by a finite sequence of blowups. Let

$$\sigma: S' \longrightarrow S$$

be a composite of these blowups such that the composite

$$f' = f^{-1} \circ \sigma: S' \longrightarrow X$$

is a regular map. Denote the exceptional divisor $\sigma^{-1}\{y_0\}$ of σ by $D = \bigcup D_i$, where as usual the D_i are rational curves.

Our plan at this point involves three stages: **(i)** Pull the curves C_0 and C_1 back to S' by σ^{-1}. **(ii)** Send their preimages to X by f'. **(iii)** Intersect these by canonical classes using Lemma 1 with the setup $f: X \longrightarrow Y$. This plan is carried out below.

(i): Since $y_0 \in C_0$, the pullback of C_0 under σ^{-1} meets the exceptional divisor D but $y_0 \notin C_1$ so its pullback is disjoint from D. As equivalence classes of divisors we can write

$$\begin{aligned}\sigma^*[C_0] &= [C_0'] + \sum a_i[D_i], \quad a_i \geq 0, \\ \sigma^*[C_1] &= [C_1'],\end{aligned}$$

where C_j' is the monoidal transform of C_j under σ, i.e. the closure in S' of $\sigma^{-1}(C_j - \{y_0\})$, $j = 0, 1$. Note at this point that since $[C_0] = [C_1]$ in S we have

$$[C_0'] + \sum a_i[D_i] = [C_1'].$$

(ii): Sending both sides of the above equation by f' we have

$$[f'(C_0')] + \sum a_i[f'(D_i)] \quad = \quad [f'(C_1')]. \qquad (1)$$

Note that $f \circ f'(C_0') = C_0$ so $f'(C_0') \not\subset E$, but since $y_0 \in C_0$ the intersection $f'(C_0') \cap E$ is not empty. On the other hand $f'(C_1') \cap E = \emptyset$.

(iii): We can now apply Lemma 1 to the above class of divisors in X. Intersecting both sides of equation (1) by K_X we have

$$[f'(C_0')] \cdot K_X + \sum a_i[f'(D_i)] \cdot K_X \quad = \quad [f'(C_1')] \cdot K_X. \qquad (2)$$

The key point is to relate this equality with its image under the map $f: X \longrightarrow Y$ using Lemma 1:

$$\begin{aligned}
[f'(C_0')] \cdot K_X &> [f \circ f'(C_0')] \cdot K_Y \quad \text{(Lemma 1)} \\
&= [f \circ f^{-1} \circ \sigma(C_0')] \cdot K_Y \\
&= [C_0] \cdot K_Y \\
&= [C_1] \cdot K_Y \\
&= [f \circ f^{-1} \circ \sigma(C_1')] \cdot K_Y \\
&= [f \circ f'(C_1')] \cdot K_Y \\
&= [f'(C_1')] \cdot K_X \quad \text{(Lemma 1)}.
\end{aligned}$$

Together with equation (2) this implies that

$$\sum a_i[f'(D_i)] \cdot K_X \quad < \quad 0.$$

This means that at least one of the $f'(D_i)$, say $f'(D_{i_0})$, is not a point and hence is a rational curve, since the D_i are all rational curves, and

$$f'(D_{i_0}) \cdot K_X < 0.$$

Note also that $f(f'(D_{i_0})) = \sigma(D_{i_0}) = y_0$, so $f'(D_{i_0}) \subset E$.

We have thus proved the following theorem:

Theorem 2 *Let X and Y be smooth projective varieties and*

$$f: X \longrightarrow Y$$

a regular birational map. Let E be the subset of X where f is not a local isomorphism. Then there exists a rational curve in X whose intersection with the canonical divisor of X is strictly negative.

The converse of this theorem would start a classification program. Suppose that X contains a curve C which has negative intersection with the canonical divisor. If we knew that we could find a regular birational map $f: X \longrightarrow Y$ onto some variety Y such that the curve C is collapsed to a point by f, then we could take Y as a 'simpler' representative of the birational equivalence class of X. We also need to know when there are no more curves to collapse. To formalize the results we need to distinguish certain concepts as definition, as we will do in the following section.

References: [3], [13].

3 Curves and Divisors

We are interested in smooth projective threefolds but the following definitions make sense when X is only a normal projective variety. This will be useful later when we discover that we need to work with some singular threefolds too.

Definition 3

$\mathrm{Div}(X)$:= *Free Abelian group of Cartier divisors on X.*
$Z^1(X)$:= *Free Abelian group of Weil divisors on X.*
$Z_1(X)$:= *Free Abelian group of one cycles on X.*
$\mathrm{Pic}(X)$:= *Group of line bundles on X, also known as the Picard group.*

When L is a line bundle and C is an irreducible curve in X we can define an intersection number $D \cdot C$, or $C \cdot D$, as the degree of the pullback bundle $f^*(L)$ on $\tilde{C}$ where $f: \tilde{C} \longrightarrow C$ is a normalization of C. We then extend this to a product on $\mathrm{Pic}(X) \times Z_1(X)$ by linearity.

There is a natural morphism from $\mathrm{Div}(X)$ into $\mathrm{Pic}(X)$, sending D to its associated line bundle $\mathcal{O}(D)$. When X is projective this map is surjective. This allows us to define a product on $\mathrm{Div}(X) \times Z_1(X)$ by letting $D \cdot C$ be $\mathcal{O}(D) \cdot C$ for a Cartier divisor D.

Later we will be dealing with varieties X for which the canonical divisor K_X does not exist but mK_X is Cartier for some positive integer m. For an irreducible curve C in X we will then define $K_X \cdot C$ as $\frac{1}{m}(mK_X \cdot C)$. This makes it necessary to extend our definitions to divisors with rational coefficients, which we do in the definitions of the following objects.

We say that two Cartier divisors D_1 and D_2 are *numerically equivalent*, and we write $D_1 \equiv D_2$, if $D_1 \cdot C = D_2 \cdot C$ for every curve C in X. We similarly define numerical equivalence on $Z_1(X)$. Taking quotients by numerical equivalence and tensoring with the rationals we obtain the following objects:

Definition 4
$N^1(X) := (\mathrm{Div}(X)/ \equiv) \otimes \mathbb{Q}$
$N_1(X) := (Z_1(X)/ \equiv) \otimes \mathbb{Q}$

When X is projective $N^1(X)$ is isomorphic with the image of $\mathrm{Pic}(X) \otimes \mathbb{Q}$ in $H^2(X, \mathbb{Z}) \otimes \mathbb{Q}$ where the image is taken as the first Chern class of a line bundle. This image is of finite rank. We denote this rank by $\rho(X)$ and call it the *Picard number* of X. Since $N^1(X)$ and $N_1(X)$ are dual $\mathbb{Q}$-vector spaces, they are both of finite dimension $\rho(X)$.

Since $N_1(X)$ is a finite dimensional vector space we can define a topology on it by choosing a norm, and the topology is independent of the choice of the norm. We now come to the fundamental concept of the subject.

Definition 5

$\text{NE}(X) :=$ *cone generated in* $N_1(X)$ *by irreducible curves in* X.

$\overline{\text{NE}}(X) :=$ *the closure of* $\text{NE}(X)$ *in* $N_1(X)$. *This is known as the closed cone of curves in* X.

A subset F of $\overline{\text{NE}}(X)$ is called *extremal* if for every u and v in $\overline{\text{NE}}(X)$, $u+v$ cannot be in F unless both u and v are in F. A one dimensional extremal subset is called an *extremal ray.*

For any divisor class D in $N^1(X)$ we say that D is *nef* if $D \cdot C \geq 0$ for every C in $\overline{\text{NE}}(X)$. In particular we are interested when K_X is nef.

We have mentioned Weil divisors but did nothing with them so far. Every Cartier divisor defines a Weil divisor but the converse is not necessarily true. If for every Weil divisor D there is a positive integer m such that mD becomes Cartier then we say that X is $\mathbb{Q}$-factorial. This concept will play a vital role later when we start working on singular threefolds.

After this barrage of definitions we are finally ready to do some work!

References: [2], [3], [13].

4 Contraction Map

Going back to Theorem 2 we see that if $f: X \longrightarrow Y$ contracts a curve to a point then **(i)** $K_X \cdot C < 0$ and **(ii)** f contracts every curve which is numerically equivalent to C. We are asking for a converse: If $K_X \cdot C < 0$ for some curve C in X, then can we construct a map $f: X \longrightarrow Y$ into some smooth projective variety Y such that f contracts to a point every curve which is numerically equivalent to C?

If D is a Cartier divisor such that $\mathcal{O}(mD)$ is generated by global sections for some $m > 0$, then these sections define a map $\Psi_{mD}: X \longrightarrow \mathbb{P}^N$ into some projective space. Moreover we have $\Psi^*_{mD}(\mathcal{O}(1)) = \mathcal{O}(mD)$. This is expressed by saying that mD is *free* or *base point free* or that D is *eventually free.* If moreover Ψ_{mD} is a closed immersion, then we say that mD is *very ample*, or that D is *ample.*

Assume that D is a nef divisor on X and that mD is base point free. Define a subset of $\overline{\mathrm{NE}}(X)$ as

$$D^{\perp} = \{C \in \overline{\mathrm{NE}}(X) \mid D \cdot C = 0\,\}.$$

We then have for any $C \in D^{\perp}$

$$\begin{aligned} 0 &= D \cdot C \\ &= \frac{1}{m} mD \cdot C \\ &= \frac{1}{m} \Psi^*_{mD}(H) \cdot C \\ &= \frac{1}{m} H \cdot (\Psi_{mD})_*(C) \end{aligned}$$

where H is a hyperplane section in $\Psi_{mD}(X)$. But any nondegenerate curve has positive intersection with a hyperplane section, the intersection being the degree of the curve in general. Since we obtained zero in this case the curve C and all other curves numerically equivalent to it must have been contracted to a point by Ψ_{mD}.

Assume that $K_X \cdot C < 0$. Using the duality of the spaces $N^1(X)$ and $N_1(X)$ it is easy to find a nef divisor D such that $D^{\perp} = \{C\}$. The question is whether mD will be base point free for some $m > 0$. The affirmative answer was provided by Mori for smooth projective threefolds.

Theorem 6 (Mori) *Let X be a smooth projective threefold and C a curve on X such that C spans an extremal ray of $\overline{\mathrm{NE}}(X)$ and $K_X \cdot C < 0$. Then there is a map $f: X \longrightarrow Y$ onto some projective variety Y such that f contracts C and all other curves numerically equivalent to it. (f is the m-canonical map of some nef divisor D*

with $D^{\perp} = \{C\}$*.) We then have the following cases:*
(i): $\dim Y < \dim X$. *Then* X *is a fibred space.*
(ii): $\dim Y = \dim X$.
(a): Y *is smooth and* $\rho(Y) < \rho(X)$.
(b): Y *has an isolated hypersurface or quotient singularity.*

The classification program successfully terminates in the case of **(i)** and continues for finitely many more steps before stopping in the case of **(ii-a)**. But we have to stop unsuccessfully in the case of **(ii-b)** since we have reached to a class of varieties outside our accepted category. This is the point where we consent to admit some mild singularities into the theory.

References: [6], [11], [18].

5 Canonical Singularities

The singularities obtained by contracting an extremal ray in a smooth threefold are fully described by Reid. Since these singularities are obtained as the image of an m-canonical mapping associated to an effective divisor Reid called them "canonical singularities". As their surface counterparts are called DuVal singularities, I imagine that the future generations will with equal comfort and justice call them "Reid singularities".

The first obstacle for the program in the singular case is the definition of a suitable K_X. In fact we only need to know if some 'representative' of K_X has negative intersections with some curves. For that matter if K_X fails to exist but mK_X exists for some positive integer m, then we can still define $K_X \cdot C$ as $\frac{1}{m} mK_C \cdot C$. This way we can transport all the terminology about extremal rays into the singular case.

On the other hand the singular varieties obtained by contracting extremal rays are in the same birational class of varieties as the one we started with so any birational invariant defined on the singular space must be the same as the ones before contraction. These expectations are all met in the definition of canonical singularities:

Definition 7 (Reid) *Let X be a normal projective variety. We say that X has* canonical singularities *if*
(i): *rK_X exists as a Cartier divisor for some $r > 0$,*
(ii): *for any resolution of singularities $f: Y \longrightarrow X$ with $\{E_i\}$ the family of exceptional divisors, we have*

$$rK_Y = f^*(rK_X) + \sum a_i E_i,$$

where each $a_i \geq 0$. (This is equivalent to saying that the plurigenera of X and Y agree.)

If the requirement that each $a_i \geq 0$ in **(ii)** is replaced by $a_i > 0$, then the singularity is called *terminal.*

Terminal threefold singularities are isolated. If X has only canonical singularities, then there is a partial resolution $f: Y \longrightarrow X$ such that $K_Y = f^* K_X$ and Y has only terminal singularities.

Suppose that rK_X is Cartier for some positive integer r. We say that K_X is nef if $rK_X \cdot C \geq 0$ for every $C \in \overline{\mathrm{NE}}(X)$. In general we write $K_X \cdot C$ for $\frac{1}{r} rK_X \cdot C$.

Now we can go back to the considerations of the previous section and ask the same question: Given a nef divisor D on a projective normal threefold with at most canonical singularities, is it true that for some integer $m > 0$, mD is base point free?

By Kawamata's theorem mD is free for some $m > 0$ if $aD - K_X$ is nef for some $a > 0$ and $\dim \Psi_{aD-K_X}(X) = \dim X$ (i.e. $aD - K_X$ is *big*). Since $K_X \cdot C < 0$ the divisor $aD - K_X$ is going to be positive on $\overline{\mathrm{NE}}(X)$ for sufficiently large a. By Kleiman's criteria $aD - K_X$ is then ample. This implies in particular that $aD - K_X$ is big. Thus the map Ψ_{mD} is going to contract the curve C and all those numerically equivalent to it. Denote the contraction map by $f: X \longrightarrow Y$.

If rK_Y exists for some $r > 0$ we then need to check if Y has canonical singularities so that we know we are staying inside the category of varieties we started with. To ensure this we need to impose a condition on the canonical varieties: we want to work with $\mathbb{Q}$-factorial varieties, i.e. those in which every Weil divisor becomes Cartier after multiplying by a suitable positive integer.

We then have the following result:

Theorem 8 (Kawamata) *Let X be a $\mathbb{Q}$-factorial threefold with canonical singularities such that $K_X \cdot C < 0$ for some curve C. Then we have a map $f\colon X \longrightarrow Y$ onto some projective variety Y, $f = \Psi_{mD}$, such that one of the following holds:*
(i): $\dim Y < \dim X$. *Then X is a fibered space.*
(ii): $\dim Y = \dim X$ *and one of the following holds:*
(a): *Y is a $\mathbb{Q}$-factorial threefold with canonical singularities and f contracts a curve iff it is numerically equivalent to C, and we have $\rho(Y) = \rho(X) - 1$. In this case f is called a* divisorial contraction.
(b): *rK_Y does not exist for any $r > 0$. In this case f is called a* small contraction.

If we start with smooth X, then case (ii-b) does not materialize. Case (ii-b) occurs when X is singular and when f contracts a codimension 2 subvariety in X, i.e. when the collection of curves numerically equivalent to C fill out a 1-dimensional subvariety of X. In this case it is easy to check that rK_Y does not exist for any r: Assume that rK_Y exists for some sufficiently large and divisible $r > 0$, (divisibility is mentioned since we also want rK_X to exist as a Cartier divisor). Since the exceptional set E of f has codimension 2, $f^*(rK_Y)$ and rK_X are two extensions of rK_{X-E}. Since X is normal such an extension is unique so $f^*(rK_Y) = rK_X$. But since C is collapsed to a point by f we have $rK_Y \cdot f_*(C) = 0$ whereas $rK_Y \cdot f_*(C) = f^*(rK_Y) \cdot C = rK_X \cdot C < 0$, a contradiction. So rK_Y cannot exist for any $r > 0$.

We once again end up with a variety which is not in our accepted category. We need more magic!

References: [19], [22], [23], [25], [26].

6 Flip

In Section 4 we started with a smooth variety and contracting a negative extremal ray we obtained a singularity. We extended our category of acceptable varieties in Section 5 to include these singularities. However starting with these new set of varieties and contracting negative extremal rays we obtain varieties which are even worse as far as singularities are concerned. It is clear that we cannot continue to accept every stranger. We must find an alternate way of proceeding with the program whenever we meet a variety for which no multiple of the canonical divisor is Cartier. A major strategy for everyday life is to avoid disaster if we are not equipped to handle it. And that is what flip is all about.

Let $f: X \longrightarrow Y$ be a small contraction. Assume that there is an ample Cartier divisor H on Y such that for some sufficiently large integer $m > 0$ the ring

$$R(mf^*H + K_X) := \bigoplus_{n \geq 0} H^0(X, \mathcal{O}_X(n(mf^*H + K_X)))$$

is finitely generated. Then define a new projective variety X^+ as

$$X^+ := \mathbf{Proj}\ R(mf^*H + K_X).$$

We then have the following facts:
(i): X^+ is $\mathbb{Q}$-factorial and has canonical singularities.
(ii): There is a morphism $f^+: X^+ \longrightarrow Y$ contracting an extremal ray R and for any curve C in this ray we have $K_Y \cdot C > 0$.
(iii): There is a birational map $\phi: X \dashrightarrow X^+$ such that $f = f^+ \circ \phi$. In fact ϕ is an isomorphism outside the curves contracted by f and f^+.
(iv): $\rho(X) = \rho(X^+)$.

This set up is known as a *flip*. More often the map $\phi: X \dashrightarrow X^+$ is called a flip. The crucial point is that we have a

Theorem 9 (Mori) *If X is a threefold with at most canonical singularities and $f: X \longrightarrow Y$ is a small contraction which contracts the negative extremal ray R, then a flip $\phi: X \dashrightarrow X^+$ exists.*

Let X be a threefold with canonical singularities and $\pi: \tilde{X} \longrightarrow X$ a resolution. We have $K_{\tilde{X}} = \pi^* K_X + \sum a_i E_i$, where $\{E_i\}$ are exceptional divisors and $a_i \geq 0$.

Definition 10 (Shokurov) *The* difficulty $d(X)$ *of* X *is the number of* a_i *which are less than one, i.e.*

$$d(X) = \#\{i \mid a_i < 1\}.$$

This is independent of the resolution chosen and is an invariant of X. Note that if $f: X \longrightarrow Y$ is a regular birational map, such as a divisorial contraction, then $d(X) = d(Y)$ since a resolution of X is also a resolution for Y. However the crucial point is what happens to difficulty when we have a flip.

Theorem 11 (Shokurov) *If* $\phi: X \dashrightarrow X^+$ *is a flip, then* $d(X^+) < d(X)$.

This says that a threefold with canonical singularities cannot accept infinitely many flips.

We now have an algorithm to apply. Start with a normal projective $\mathbb{Q}$-factorial threefold which has at most canonical singularities. If K_X is nef, we stop. We accept X as a minimal model in the birational equivalence class of X. If however K_X is not nef, then there is a negative extremal ray which we contract to obtain Y. If $\dim Y < 3$ or if K_Y is nef, we stop. In the former case we can still accept X as a minimal model of its class since we can describe it as a fibered space. If Y has canonical singularities and K_Y is not nef, we continue the program with Y observing that $\rho(Y) < \rho(X)$ and $d(Y) = d(X)$. If no multiple of K_Y is Cartier, then we construct a flip X^+ of X and continue with X^+ observing that $\rho(X^+) = \rho(X)$ and $d(X^+) < d(X)$. With these observations we are assured that the algorithm will terminate after finitely many steps and when it terminates we will have at the last step a variety Z which is either of dimension less than 3 or which is a normal projective $\mathbb{Q}$-factorial threefold with at most canonical singularities and with K_Z nef.

When we start with a particular threefold X, then whether the above algorithm stops because we end up with a variety of lower dimension or because we obtain a threefold whose canonical divisor is nef depends only on the birational class X. However if we are going to end up with a threefold at the end of the algorithm, then it is not necessarily unique. The order of extremal rays chosen to contract or flip at each step makes a difference at the end. Suppose that starting with a fixed X but contracting and flipping extremal rays in different orders we obtain the threefolds Z_1 and Z_2 whose canonical divisors are nef. Then there exists a birational map $\psi\colon Z_1 \dashrightarrow Z_2$ which is an isomorphism in codimension one. ψ has the effect of cutting off some 1-cycles from Z_1 and gluing them back again with a different embedding. Observe that this is very much like a flip map except that the canonical divisors here have zero intersection with each of the curves on which ψ and its inverse are not defined. Such a map is called a *flop*.

References: [1], [7], [12], [14], [16], [17], [19], [21], [24], [27], [28], [31], [32].

7 The Minimal Model Program

In this section we summarize the classification algorithm, otherwise known as the Minimal Model Program, MMP for short.

Theorem 12 (Minimal Model Program) *Let X be a normal projective $\mathbb{Q}$-factorial threefold with at most canonical singularities. Assume that K_X is not nef. Let R be a negative extremal ray. Then we can construct a contraction map $f\colon X \longrightarrow Y$ onto a projective variety Y such that a curve C is mapped to a point in Y iff C belong to the ray R. Moreover only one of the following cases holds:*
(i): *f is a Fano contraction, i.e.* $\dim Y < 3$.
(ii): *f is a divisorial contraction, i.e. Y is a normal projective $\mathbb{Q}$-factorial threefold with at most canonical singularities. In this case there is a codimension one subvariety E in X such that $f(E)$ is a point in Y and $f\colon X \setminus E \longrightarrow Y \setminus f(E)$ is an isomorphism. As for the numerical invariants we have $\rho(Y) < \rho(X)$ and $d(Y) = d(X)$.*

(iii): *f is a small contraction, i.e. there is a codimension two subvariety E in X such that $f(E)$ is a point in Y and $f: X\backslash E \longrightarrow Y\backslash f(E)$ is an isomorphism. In this case no multiple of the canonical divisor of Y is Cartier. We then construct a flip $\phi: X \dashrightarrow X^+$, flipping the ray R. Then X^+ is a normal projective $\mathbb{Q}$-factorial threefold with at most canonical singularities. In this case we have $\rho(Y) = \rho(X^+)$ and $d(X^+) < d(X)$.*

If case **(i)** is encountered, then we stop. If case **(ii)** or **(iii)** is encountered, we continue the program with X replaced by Y or X^+ respectively. The observation on the numerical invariants assures us that the program will terminate after finitely many steps.

Thus if we start with an acceptable X with K_X not nef, then X is birationally isomorphic to a threefold Y such that either Y admits a Fano contraction or else K_Y is nef.

With this result we can give a full description of the closed cone of curves in X. First we have a

Definition 13 *If X is a normal $\mathbb{Q}$-factorial projective variety and K_X its canonical divisor, define*

$$\overline{\mathrm{NE}}_{K_X}(X) := \{R \in \overline{\mathrm{NE}}(X) | K_X \cdot R \geq 0\}.$$

We can now describe the structure of $\overline{\mathrm{NE}}(X)$.

Theorem 14 (Cone Theorem) *Let X be as in Theorem 12. Then there are a finite number of extremal rays R_i such that*

$$\overline{\mathrm{NE}}(X) = \overline{\mathrm{NE}}_{K_X}(X) + \sum \mathbb{Q}_+ R_i,$$

where $\mathbb{Q}_+$ denotes the positive rational numbers. (If we had chosen to define N^1 and N_1 as $\mathbb{R}$-vector bundles from the beginning (see Definition 4), then we would replace $\mathbb{Q}_+$ with $\mathbb{R}_+$.)

The choice of singularities to admit seems to vary with taste but the category of terminal singularities is the smallest category within which the MMP works.

It is time now to check if the reader is comfortable with these new ideas:

Exercise 15 *Classify smooth projective surfaces within the framework of MMP.*

References: [2], [4], [8], [9], [10], [11], [13], [15], [20], [25], [29], [30], [33].

References

[1] Atiyah, M., *On analytic surfaces with double points,* Proc. Royal Soc. A. **247** (1958), 237–244.

[2] Clemens, H., Kollár, J., Mori, S., *Higher dimensional complex geometry,* Astérisque **166** (1988).

[3] Hartshorne, R., *Algebraic geometry,* Springer-Verlag, Berlin, 1977.

[4] Esnault, H., *Classification des variétés de dimension 3 et plus,* Séminaire Bourbaki, Springer Lecture Notes in Math. **901**, (1981), 111–131.

[5] Hironaka, H., *On the theory of birational blowing-up,* Thesis, Harvard Univ., 1960.

[6] Kawamata, Y., *Elementary contractions of algebraic 3-folds,* Ann. of Math. **119** (1984), 95–110.

[7] ———, *On the finiteness of generators of the pluricanonical ring for a threefold of general type,* Amer. J. Math. **106** (1984), 1503–1512.

[8] ———, *Canonical and minimal models of algebraic varieties,* Proc. of Int. Congress of Math., Kyoto, Japan, 1990, 699–707.

[9] ———, *On the minimal model problem,* Springer Lecture Notes in Math. **1194** (1986), 41–55.

[10] ———, *The cone of curves of algebraic varieties,* Ann. of Math. **119** (1984), 603–633

[11] ———, Matsuda, K., Matsuki, K., *Introduction to the minimal model problem,* Proc. Symp. Algebraic Geometry, Sendai, Adv. in Pure Math. **10** (1987), 283–360.

[12] Kleiman, S., *Toward a numerical theory of ampleness,* Ann. of Math. **84** (1966), 293–344.

[13] Kollár, J., *The structure of algebraic threefolds: an introduction to Mori's program,* Bull. of Amer. Math. Soc. **17** (1987), 211–273.

[14] ———, *Flops,* Nagoya Math. J. **113** (1989), 15–36.

[15] ———, *Minimal models of algebraic threefolds: Mori's theorem,* Astérisque **177–178** (1990), 303–326.

[16] ———, *Flips, flops, minimal models, etc.,* Surveys in Diff. Geom. **1** (1991), 113–199.

[17] ———, et al., *Flips and abundance for algebraic threefolds,* Astérisque **211** (1992).

[18] Mori, S., *Threefolds whose canonical bundles are not numerically effective,* Ann. of Math. **116** (1982), 133–176.

[19] ———, *On 3-dimensional terminal singularities,* Nagoya Math. J. **98** (1985), 43-66.

[20] ———, *Classification of higher dimensional varieties,* Proc. Symp. in Pure Math. **46** (1987), 269–331.

[21] ———, *Flip theorem and the existence of minimal models for 3-folds,* J. Amer. Math. Soc. **1** (1988), 117–253.

[22] Reid, M., *Canonical 3-folds,* Journées de géométrie algébrique d'Angers, (1980), 273–310.

[23] ———, *Minimal models of canonical 3-folds,* Algebraic varieties and Analytic Varieties, Adv. Stud. Pure Math. bf1 (1983), 131–180.

[24] ———, *Decomposition of toric morphisms,* Progress in Math. **36** (1983), 395–418.

[25] ———, *Tendencious survey of 3-folds,* Proc. Symp. Pure Math. **46** (1987), 333–344.

[26] ———, *Young person's guide to canonical singularities,* Proc. Symp. Pure Math. **46** (1987), 345–414.

[27] ———, *What is a flip?* Warwick preprint (1992).

[28] Shokurov, V. V., *The nonvanishing theorem,* Izv. Akad. Nauk. SSSR Ser. Mat. **49** (1985), 635–651; Math. USSR Izv. **26** (1986), 591–604.

[29] ———, *Numerical geometry of algebraic varieties,* Proc. of Int. Congress of Math., Berkeley (1986), 672–681.

[30] Ueno, K., *Classification theory of algebraic compact complex spaces,* Springer Lecture Notes in Math. **439** (1975).

[31] Viehweg, E., *Vanishing theorems,* J. reine Angew Math. **335** (1982), 1–8.

[32] Wilson, P. M. H., *On the canonical ring of algebraic varieties,* Compositio Math. **43** (1981), 365–385.

[33] ———, *Toward birational classification of algebraic varieties,* Bull. London Math. Soc. **19** (1987), 1–48.

Character Sums, Algebraic Curves and Goppa Codes

Serguei A. Stepanov

Department of Mathematics
Bilkent University
06533 Bilkent,Ankara, Turkey
&
Department of Algebra
Steklov Mathematical Institute
Vavilov St. 42
Moscow GSP-I, 117966 Russia

e-mail: `stepanov@fen.bilkent.edu.tr`

1 Introduction

The purpose of this paper is to demonstrate a close interconnection between Number Theory, Algebraic Geometry and Coding Theory and construct considerably long geometric Goppa codes on fibre products with very good parameters.

Let F_p be a prime finite field of characteristic p, which we identify as a set with $\{1, 2, \ldots, p\}$, and F_q be an extension of F_p of degree $\nu \geq 1$, so that $q = p^\nu$. Let $s \geq 2$ be a positive divisor of $q - 1$ and

χ (resp. ψ) be a non-trivial multiplicative character of exponent s (resp. a non-trivial additive character) of the field F_q. For any two polynomials $f, g \in F_q[x]$ of degree $l \geq 1$, $m \geq 1$ respectively, consider the following character sums

$$S_q(f) = \sum_{x \in F_q} \chi(f(x))$$

and

$$T_q(g) = \sum_{x \in F_q} \psi(g(x)).$$

If $f = f_1^{s_1} \cdots f_r^{s_r}$ is the factorization of f into distinct irreducible polynomials $f_1, \ldots f_r \in F_q[x]$ with $\deg(f_1 \cdots f_r) = \mu$ and $(s, s_1, \ldots, s_r) = 1$, the well-known Weil's result [45] (see also Stepanov [33]) provides the upper bound

$$|S_p(f)| \leq (\mu - 1)q^{1/2} . \tag{1}$$

Similarly, if $(m, p) = 1$, we have

$$|T_q(g)| \leq (m - 1)q^{1/2}. \tag{2}$$

The upper bounds (1), (2) are not trivial only for $\mu, m \leq q^{1/2} + 1$, and we know several cases (see Stark [32], Korobov [18], Mit'kin [23], Serre [30] and Stöhr-Voloch [41]) when these bounds can be essentially sharpened using the author's elementary method (see Stepanov [33]). Unfortunately, until now the algebraic structure of the polynomials f and g providing such improvement of the Weil bound is not clear. If $\mu, m > q^{1/2} + 1$, we do not know practically any non-trivial upper bounds for absolute values of the sums $S_q(f)$ and $T_q(g)$, except in the case of polynomials f and g of very special form. Thus we have the following problem.

Problem 1. *Determine the class of polynomials $f, g \in F_q[x]$ of degree l, m respectively, $1 \leq l, m < q$, for which the upper bounds (1), (2) can be sharpened and absolute values of the sums $S_q(f)$, $T_q(g)$ can be estimated non-trivially for $\mu, m > q^{1/2} + 1$.*

Note that the bounds (1), (2) are sequences of the Riemann hypothesis for zeta-functions $\zeta(X,s)$ and $\zeta(Y,s)$ of the smooth projective curves defined over F_q by equations

$$y^s = f(x)$$

and

$$y^p - y = g(x).$$

According to this hypothesis all the zeros of the zeta-functions $\zeta(X,s)$ and $\zeta(Y,s)$ lie on the complex line $Re(s) = 1/2$. Under such interpretation we see that Problem 1 is related to the very deep question concerning the distribution of the zeros of $\zeta(X,s)$ and $\zeta(Y,s)$ on the critical line $Re(s) = 1/2$.

The problem becomes much more difficult if we consider the corresponding incomplete sums. Let $\omega_1, \ldots, \omega_\nu$ be a basis of F_q over F_p. Every element $x \in F_q$ can be uniquely written in the form

$$x = x_1\omega_1 + \cdots + x_\nu\omega_\nu$$

with $x_1, \ldots, x_\nu \in F_p$, and for any positive integer $N \leq p$ we can define the box $B_N \in F_q \simeq F_p^\nu$ of volume $V = N^\nu$ as

$$B_N = \{x = (x_1, \ldots, x_\nu) \in F_q \mid 1 \leq x_i \leq N, 1 \leq i \leq \nu\}.$$

Consider now the following incomplete character sums

$$S_N(f) = \sum_{x \in B_N} \chi(f(x))$$

and

$$T_N(g) = \sum_{x \in B_N} \psi(g(x)).$$

If $f(x) = x + a$ and χ is a non-trivial multiplicative character of the field F_q with $q = p^\nu$ elements, it follows from the well-known results of Burgess [4] and Davenport, Lewis [6] that for any $\varepsilon > 0$ there exists $\delta > 0$ such that for $p \geq p_0(\varepsilon)$ and $N > p^{(\nu/2(\nu+1))+\varepsilon}$ the estimate

$$|S_N(x)| < (Np^{-\delta})^\nu$$

holds. On the other hand, as was proved by Elliot [8] in the case when $q = p$ and χ is the non-trivial quadratic character of F_p, for any positive $\varepsilon < 1$ and any positive integer $N \leq c(\varepsilon) \log p$, the equality

$$|S_N(x)| = |\sum_{x=1}^{N} \chi(x)| = N$$

holds for at least $y^{1-\varepsilon}$ primes $p \leq y$.

In the general case when $q = p$ and $f, g \in F_p[x]$ are polynomials of degree $l > 1$, $m > 1$ respectively, we are able only to obtain upper bounds of the form (Burgess [5])

$$|S_N(f)| \leq c(l) p^{1/2} \log p \tag{3}$$

and (Korobov [17])

$$|T_N(g)| \leq c_0 N^{1-\frac{\gamma}{m^2}}, \tag{4}$$

where $N = p^{1/r}$, $1 \leq r \leq m$, $\min([r], [m - r + 1]) > \varepsilon m$ for any positive $\varepsilon < 1/2$, and the constants c_0 and γ depend only of ε. Thus we have the following problem, which is very important for many questions of analytic number theory.

Problem 2. *Extend estimates (3), (4) to the case of an arbitrary finite field F_q, and determine the class of polynomials $f, g \in F_q[x]$ of degree $l \geq 1$, $m \geq 1$ respectively, for which absolute values of the sums $S_N(f)$ and $T_N(g)$ can be estimated non-trivially for all $N > \log^{1+\varepsilon} q$.*

In view of the difficulties connected with the solution of Problems 1 and 2 there arises the problem on finding of lower (i.e. existence) bounds for absolute values of the sums $S_q(f)$ and $T_q(g)$. This problem is especially interesting in connection with its extraordinary significance for coding theory. Later on we shall consider two aspects of the problem, namely, coding theoretic and number theoretic ones. The first aspect concerns the obtaining of lower bounds for character sums by the use of concepts and results of coding theory, whilst the second one provides applications of the lower bounds for character sums, obtaining by the use of number theoretic methods, to the problem of construction of linear codes over F_q with extremely good parameters. Now we explain a close interconnection between the problem concerning lower bounds for character sums and the problem concerning construction of good linear codes. To do this we shall recall shortly the basic idea of the Goppa construction [14] of linear $[n,k,d]_q$-codes associated with a smooth projective curve over an algebraic closure $\overline{F}_q$ of the field F_q.

Let X be a smooth projective curve of genus $g = g(X)$ defined over a finite field F_q. Let $\{x_1, \ldots, x_n\}$ be a set of F_q-rational points of X and set

$$D_0 = x_1 + \cdots + x_n.$$

Let D be a F_q-rational divisor on X. We assume that D has support disjoint from D_0, i.e. the points x_i occur with multiplicity zero in D. Denote by $F_q(X)$ the field of functions on X rational over F_q and consider the following vector space over F_q:

$$L(D) = \{f \in F_q(X)^* \mid (f) + D \geq 0\} \cup \{0\}.$$

The linear $[n,k,d]_q$-code $C = C(D_0, D)$ associated with the pair (D_0, D) is the image of the linear evaluation map

$$\mathrm{Ev} : L(D) \to F_q^n, \qquad f \mapsto (f(x_1), \ldots, f(x_n)).$$

Such a q-ary linear code is called a geometric Goppa code.

Let us estimate the parameter of the code $C = C(D_0, D)$. The kernel of the map Ev is $L(D - D_0)$, hence $C \simeq L(D)/L(D - D_0)$ and

$$k = \dim C = \dim L(D) - \dim L(D - D_0) = l(D) - l(D - D_0).$$

In particular, if $\deg D < n$ then the map Ev is an embedding, and the Riemann-Roch theorem implies

$$k \geq \deg D - g + 1; \tag{5}$$

moreover, if $2g - 2 < \deg D < n$ then

$$k = \deg D - g + 1.$$

Now, if the weight of $\mathrm{Ev}(f)$ is d then f vanishes at $n-d$ points, say, $x_{i_1}, \ldots, x_{i_{n-d}}$, so $(f) + D - x_{i_1} - \cdots - x_{i_{n-d}} \geq 0$. By taking degrees we obtain

$$d \geq n - \deg D. \tag{6}$$

It follows from (5) and (6) that the relative parameters $R = k/n$ and $\delta = d/n$ of a geometric Goppa code $C = C(D_0, D)$ satisfy

$$R \geq 1 - \delta - \frac{g-1}{n}. \tag{7}$$

Thus to construct a long geometric Goppa $[n, k, d]_q$-code with rather good parameters we need to find a smooth projective curve X of genus g with a lot of F_q-rational points which ensures the condition that the quantity $(g-1)/n$ is small enough. Recall that for any linear $[n, k, d]_q$-codes we have the Singleton upper bound

$$R \leq 1 - \delta + \frac{1}{n}$$

and for $n \to \infty$ the asymptotic Gilbert-Varshamov lower bound

$$R \geq 1 - H_q(\delta),$$

where $H_q(\delta)$ is the q-ary entropy function defined as

$$H_q(\delta) = \delta \log_q(q-1) - \delta \log_q \delta - (1-\delta)\log_q(1-\delta).$$

The points (δ, R) for all linear $[n, k, d]_q$-codes form the set of code points $V_q^{lin} \subseteq [0,1]^2$. Let U_q^{lin} denote the subset of limit points of V_q^{lin}. In other terms, $(\delta, r) \in U_q^{lin}$ if and only if there exists an infinite sequence of different linear codes C_i with distinct relative parameters $\delta_i = \delta(C_i)$ and $R_i = R(C_i)$ such that

$$\lim_{i\to\infty} (\delta_i, R_i) = (\delta, R)\ .$$

If $\delta > 0$ and $R > 0$ such a family of codes C_i is called asymptotically good. The structure of the set U_q^{lin} can be described as follows (Aaltonen [1], Manin [22]): *there exist a continuous function* $\alpha_q^{lin}(\delta)$ *such that*

$$U_q^{lin} = \{ \quad (\delta, R) \mid 0 \leq R \leq \alpha_q^{lin}(\delta) \quad \}\ ;$$

moreover, $\alpha_q(0) = 1$, $\alpha_q(\delta) = 0$ *for* $(q-1)/q \le \delta \le 1$, *and* $\alpha_q^{lin}(\delta)$ *decreases on the interval* $[0, (q-1)/q]$.

In order to produce a family of asymptotically good geometric Goppa codes for which $R+\delta$ comes above the Gilbert-Varshamov bound one needs an infinite family of smooth projective curves with a lot of F_q-rational points compared to the genus. Examples of such families are provided by classical modular curves $X_0(N)$ and $X(N)$ (Ihara [16], Tsfasman-Vladut-Zink [44], or by Drinfeld modular curves (Tsfasman-Vladut [43, Chapters 4.1 and 4.2]). So, if $q = p^\nu$ is an even power of a prime p, there exists an infinite sequence of geometric Goppa codes C_i which gives the lower bound

$$\alpha_q^{lin}(\delta) \ge 1 - \delta - (\sqrt{q} - 1)^{-1} .$$

The line $R = 1 - \delta - (\sqrt{q}-1)^{-1}$ intersects the curve $R = 1 - H_q(\delta)$ for $q \ge 49$. Much easier proof of this result based on construction of a sequence of Artin-Schreier coverings of the projective line $\mathbb{P}^1(\bar{F}_q)$ was recently proposed by Garcia and Stichtenoth [9].

In terms of algebraic geometry the problem on construction of asymptotically good codes can be reformulated as follows. Let $N_q(g)$ denote the maximal number of F_q-rational points on a smooth projective curve X defined over F_q of genus $g = g(X)$, and

$$A(q) = \lim sup_{g\to\infty} \frac{N_q(g)}{g} .$$

It follows from the Hasse-Weil bound [45] that

$$A(q) \le 2\sqrt{q} .$$

The Serre bound [30]

$$|N_q - (q+1)| \le g[2\sqrt{q}]$$

yields

$$A(q) \le [2\sqrt{q}] .$$

Much stronger upper bound

$$A(q) \le \sqrt{q} - 1$$

was obtained by Drinfeld and Vladut [7] . This is the best possible upper bound, and construction of asymptotically good geometric codes is reduced to construction of a family of smooth projective curves over F_q, for which $A(q)$ is close to the Drinfeld-Vladut bound. So, if $q = p^\nu$ is an even power of a prime p , the result of Ihara [16]. Tsfasman-Vladut-Zink [44] (see also Garcia-Stichtenoth [9]) implies

$$A(q) \geq \sqrt{q} - 1 .$$

If q is an odd power of p, the result of Serre [30] provides existence of an absolute constant $c > 0$ such that

$$A(q) \geq c \log q .$$

In some cases the Serre bound was improved by Perret [28] and Zink [47]. In particular, the Zink result yields

$$A(q^3) \geq \frac{2(q^2 - 1)}{q + 2} .$$

In this paper we construct a family of sufficiently long geometric Goppa codes coming from smooth projective curves X_s given over F_q by equations

$$z_i^2 = f_i(u), \qquad 1 \leq i \leq s .$$

Each such curve is actually a fibre product of hyperelliptic curves. For some polynomials $f_i \in F_q[u]$ of a special form the curves X_s have a lot of F_q-rational points and provide a family of linear $[n, k, d]_q$-codes with fairly good parameters and very fast construction and decoding algorithms. For small values of s, the parameters of the codes are comparable with the parameters of codes on Artin-Schreier coverings introduced by Garcia and Stichtenoth [9]. Unfortunately, the parameter s in our construction is bounded by $q^{1/2}$, and as a result the genus $g(X_s)$ is bounded by

$$(q - 3)2^{\sqrt{q}-2} + 1 .$$

However, since the above bound is large enough for $q \geq q_0$, the curves X_s provide sufficiently long geometric Goppa codes.

A similar construction of non-singular projective curves with a lot of F_q-rational points based on the use of fibre products of some special Artin-Schreier curves was independently considered by van der Geer and van der Vlugt [10].

2 Coding Theoretic Aspects

Let $p > 2$ be a prime number, $N \leq p$ a positive integer and χ the unique non-trivial quadratic character of the prime finite field F_p, i.e. the Legendre symbol. Using simple coding-theoretic arguments we obtain the following result (Stepanov [33, p. 85] , [34]).

Theorem 1. *If m is an integer satisfying*

$$\frac{(N+1)\log 2}{\log p} + 1 < m \leq p$$

then there exists a square-free polynomial $f \in F_p[u]$ of degree $l = 2m$ for which

$$\sum_{u=1}^{N} \chi(f(u)) = N.$$

Proof. Consider the set $\{f_1(u), \ldots, f_n(u)\}$ of all distinct irreducible monic polynomials in $F_p[u]$ of degree $m > 1$ and the corresponding sequence $C = \{x_1, \ldots, x_n\}$ of vectors

$$x_i = (\chi(f_i(1)), \ldots, \chi(f_i(N))), \qquad 1 \leq i \leq n$$

with components $\chi(f_i(k)) = \pm 1$. We have

$$n = \frac{1}{m} \sum_{d|m} \mu(d) p^{m/d},$$

where $\mu(l)$ is the Möbius function, and hence $n \geq p^m/2m$. On the other hand, the supposition of the theorem implies $p^m/2m > 2^N$

and we arrive at the inequality $n > 2^N$. Since there is at most 2^N of distinct vectors x_i of length N it follows that at least two of the n vectors x_i must be equal. Let $x_i = x_{i'}$ for $i \neq i'$. The product $f = f_i(u)f_{i'}(u)$ is a square-free polynomial in $F_p[u]$ of degree $l = 2m$, and all the components of $(\chi(f(1)), \ldots, \chi(f(N)))$ are equal to 1. This.proves the theorem.

Note that the sequence $C = \{x_1, \ldots, x_n\}$ can be considered as a code of cardinality n in the Hamming space $\{-1, 1\}^N$. Now, the equality $x_i = x_j$ for $i \neq j$ follows from the fact that cardinality of the code C cannot exceed cardinality 2^N of the whole space. In this way the result was slightly improved by Levenshtein [21] by the use of the presently best known universal upper bounds for codes. It follows from Theorem 1 that the upper bound (1) cannot be improved essentially, so in general we cannot obtain any inequality of the form

$$|S_p(f)| \leq c \cdot (\mu p)^{1/2} .$$

The following theorem is more arithmetic in nature and essentially extends the above mentioned result of Elliot.

Theorem 2. *Let $p > 3$ be a prime number and $m \leq p/4$ be a positive integer. If*

$$1 \leq N < \frac{\log(p/2m)}{\log 2} \leq p/2$$

then there exists a square-free polynomial $f = (u + \alpha_1) \cdots (u + \alpha_{2m})$, $\alpha_i \in F_p$, of degree $l = 2m$ for which

$$\sum_{u=1}^{N} \chi(f(u)) = N.$$

Proof. Consider polynomials $u, u + 1, \ldots, u + p - N - 1$ in $F_p[u]$ of degree 1 and the corresponding vectors

$$x_i = (\chi(i), \ldots, \chi(N + i - 1)), \qquad 1 \leq i \leq p - N$$

with components $\chi(k+i) = \pm 1$. By the assumption of the theorem,

$$1 \leq N < \frac{\log p/2m}{\log 2} \leq p/2,$$

hence $p - N > m(2^N + 1)$. Now let us consider m subsequences $C_s = \{x_{(s-1)r+1}, \ldots, x_{sr}\}, 1 \leq s \leq m$, of the sequence $C = \{x_1, \ldots, x_{p-N}\}$ consisting of $r = 2^N + 1$ elements. Each subsequence C_s contains at least two equal vectors, say $x_{(s-1)r+i_s} = x_{(s-1)r+i'_s}$ for $i_s \neq i'_s, 1 \leq i_s, i'_s \leq r$. The product

$$f(u) = \prod_{s=1}^{m} (u + (s-)r + i_s)(u + (s-1)r + i'_s) = (u + \alpha_1) \cdots (u + \alpha_{2m})$$

is a square-free polynomial in $F_p[u]$ of degree $l = 2m$, and all the components of $(\chi(f(1)), \ldots, \chi(f(N)))$ are equal to 1. This completes the proof.

Similarly, if ψ is a non-trivial character of the field F_p, we have (Stepanov [33, p. 86]) the following result.

Theorem 3. *Let $p > 2$ be a prime number, $N \leq p$ a positive integer and $\varepsilon < 1/4\pi$ a positive number. For every integer*

$$m \geq \frac{N \log(1 + \varepsilon^{-1})}{\log p},$$

there exists a non-zero polynomial $g \in F_p[u]$ of degree at most m for which

$$|\sum_{u=1}^{N} \psi(f(u))| \geq (1 - 2\pi\varepsilon)N.$$

The results of Theorems 1 and 3 (in a slightly stronger form) were recently extended by Özbudak [26] to the case of an arbitrary finite field F_q and arbitrary non-trivial characters χ and ψ. More subtle technique of the coding theory, based on using some properties of Reed-Muller codes, arrives at the following result for complete exponential sums (Bassalygo, Zinov'ev, Litsyn [3]).

Theorem 4. *Let F_p be a prime finite field of characteristic $p > 2$, let F_q be an extension of F_p of odd degree $\nu > 1$ and ψ be a non-trivial additive character of F_q. Then there exists a polynomial $g \in F_q[u]$ of degree at most $p^{(\nu-1)/2} + 1$ such that tr $g(u)$ is not a constant and*

$$|T_q(g)| \geq q^{\nu-1/2}.$$

Below we prove this result (Theorem 8) in a stronger form through elementary arithmetical means.

3 Number Theoretic Aspects

Let p be a prime number, ν a positive integer and F_q a finite field with $q = p^\nu$ elements. The field F_q is a Galois extension of the prime finite field F_p of degree ν with a cyclic Galois group of order ν. The action of a generator θ of this group on an element $x \in F_q$ is given by the rule $\theta(x) = x^p$. The map

$$\text{norm } (x) = x \cdot \theta(x) \cdots \theta^{\nu-1}(x) = x \cdot x^p \cdots x^{p^{\nu-1}}$$

of F_q onto F_p is called a norm of the element x. Let χ be a multiplicative character of the field F_p and x be an element of F_q. Set

$$\chi'(x) = \chi(\text{norm } (x))$$

and call χ' a multiplicative character of the field F_q induced by the character χ.

Similarly, the map

$$\text{tr } (x) = x + \theta(x) + \cdots + \theta^{\nu-1}(x) = x + x^p + \cdots + x^{p^{\nu-1}}$$

is called a trace of the element $x \in F_q$. Let ψ be a non-trivial additive character of the field F_p. Set

$$\psi'(x) = \psi(\text{tr } (x))$$

and call ψ' the additive character of the field F_q induced by the character ψ.

Now let f be a square-free polynomial in the ring $F_q[u]$ of degree l and let g be a polynomial in $F_q[u]$ of degree m relatively prime to q. Let χ be the non-trivial quadratic character and ψ a non-trivial additive character of F_p. Consider the character sums

$$S_q(f) = \sum_{u \in F_q} \chi'(f(u)) = \sum_{u \in F_q} \chi(\text{norm } (f(u)),$$

and

$$T_q(g) = \sum_{u \in F_q} \psi'(g(u)) = \sum_{u \in F_q} \psi(\text{tr } (f(u)).$$

The following result of the author [35], [37] shows that the Weil bound cannot be sharpened essentially in any extension F_q of the field F_p.

Theorem 5. *Let F_q be a finite field of characteristic $p > 2$ consisting of $q = p^\nu$ elements and χ_ν be the character of F_q induced by the non-trivial quadratic character χ of the field F_p. If $\nu > 1$ then there exists a square-free polynomial $f \in F_p[u]$ of the form*

$$f(u) = \begin{cases} u + u^{p^{\nu/2}} & \text{if } \nu \equiv 0 \pmod 2 \\ (u + u^{p^{(\nu-1)/2}})(u + u^{p^{(\nu+1)/2}}) & \text{if } \nu \equiv 1 \pmod 2 \end{cases}$$

such that

$$S_q(f) = \begin{cases} (q^{1/2} - 1)q^{1/2} & \text{if } \nu \equiv 0 \pmod 2 \\ q - 1 & \text{if } \nu \equiv 1 \pmod 2 \end{cases}.$$

Proof. Let $\nu > 1$ be an even number. As $u^{p^\nu} = u$ for every $u \in F_q$, we have

$$\begin{aligned}\text{norm } f(u) &= \prod_{i=1}^{\nu}(u+u^{p^{\nu/2}})^{p^{i-1}} = \prod_{i=1}^{\nu}(u^{p^{i-1}}+u^{p^{\nu/2+i-1}})\\ &= \prod_{i=1}^{\nu/2}(u^{p^{i-1}}+u^{p^{\nu/2+i-1}})\prod_{j=1}^{\nu/2}(u^{p^{\nu+j-1}}+u^{p^{j-1}})\\ &= \prod_{i=1}^{\nu/2}(u^{p^{i-1}}+u^{p^{\nu/2+i-1}})^2.\end{aligned}$$

Therefore,

$$\sum_{u\in F_q}\chi'(f(u)) = \sum_{u\in F_q}\chi(\text{norm}(f(u))) = q-N,$$

where N is the number of roots of the polynomial $f(u)=u+u^{p^{\nu/2}}$ in the field F_q. We have $f(u)=u(1+u^{p^{\nu/2}-1})$; hence taking into account the equality

$$(p^{\nu/2}-1, p^{\nu}-1) = p^{\nu/2}-1,$$

we obtain by the Euler criterion that $N = 1+(p^{\nu/2}-1) = q^{1/2}$. Thus,

$$S_q(f) = \sum_{u\in F_q}\chi'(f(u)) = (q^{1/2}-1)q^{1/2},$$

which proves the theorem for ν an even positive integer.

Let now $\nu > 1$ be an odd number. In this case for any $u\in F_q$ we have

$$\begin{aligned}\text{norm }(f(u)) &= \prod_{i=1}^{\nu}(u^{p^{i-1}}+u^{p^{(\nu-1)/2+i-1}})(u^{p^{i-1}}+u^{p^{(\nu+1)/2+i-1}})\\ &= \prod_{i-1}^{(\nu-1)/2}(u^{p^{i-1}}+u^{p^{(\nu-1)/2+i-1}})\prod_{i=(\nu+1)/2}^{\nu}(u^{p^{i-1}}+u^{p^{(\nu+1)/2+i-1}})\\ &\times\prod_{i=1}^{(\nu-1)/2}(u^{p^{i-1}}+u^{p^{(\nu+1)/2+i-1}})\prod_{i=(\nu+1)/2}^{\nu}(u^{p^{i-1}}+u^{p^{(\nu+1)/2+i-1}})\end{aligned}$$

$$= \prod_{i=1}^{(\nu+1)/2} (u^{p^{i-1}} + u^{p^{(\nu-1)/2+i-1}}) \prod_{j=1}^{(\nu-1)/2} (u^{p^{(\nu+1)/2+j-1}} + u^{p^{j-1}})$$

$$\times \prod_{i=1}^{(\nu-1)/2} (u^{p^{i-1}} + u^{p^{(\nu+1)/2+i-1}}) \prod_{j=1}^{(\nu+1)/2} (u^{p^{(\nu-1)/2+j-1}} + u^{p^{j-1}})$$

$$= \prod_{i=1}^{(\nu+1)/2} (u(p^{i-1} + u^{p^{(\nu-1)/2+i-1}})^2 \prod_{i-1}^{(\nu-1)/2} (u^{p^{i-1}} + u^{p^{(\nu+1)/2+i-1}})^2$$

and hence

$$\sum_{u\in F_q} \chi'(f(u)) = \sum_{u\in F_q} \chi(\text{norm } (f(u))) = q - N',$$

where N' is the number of roots in F_q of the polynomial $f(u) - (u + u^{p^{(\nu-1)/2}})(u + u^{p^{(\nu+1)/2}})$. Clearly $N' = 1$ and therefore

$$\sum_{u\in F_q} \chi'(f(u)) = q - 1.$$

This completes the proof.

Note that the polynomial $f(u) = u + u^{p^{\nu/2}}$ has degree $l = p^{\nu/2}$; hence the Weil bound (1) is attained in any extension F_q of the field F_p of an even degree $\nu > 1$. The result of Theorem 5 can be extended as follows (Gluhov [11], [12], Gluhov, Özbudak [13]).

Theorem 6. *Let $\nu > 1$ be an integer, F_q the finite field of characteristic $p > 2$ consisting of $q = p^\nu$ elements, χ' the character of F_q induced by a non-trivial multiplicative character χ of the field F_p of exponent $s \geq 2$ and a, b positive integers satisfying $a + b = s$. Then*

(i) if $\nu \equiv 1 \pmod 2$, there exists a polynomial $f_1 \in F_p[u]$ of the form

$$f_1(u) = (u + u^{p^{(\nu-1)/2}})^a (u + u^{p^{(\nu+1)/2}})^b$$

such that

$$S_q(f_1) = \sum_{u\in F_q} \chi'(f_1(u)) = q - 1;$$

(*ii*) *if* $\nu \equiv 0 \pmod 2$, *there exists a polynomial* $f_2 \in F_p[u]$ *of the form*

$$f_2(u) = (u + u^{p^{\nu/2}-1})^a (u + u^{p^{\nu/2}+1})^b$$

such that

$$S_q(f_2) = \begin{cases} p^\nu - 1 & \textit{if} \quad 4 \nmid \nu \\ p^\nu - p & \textit{if} \quad 4|\nu \end{cases};$$

(*iii*) *if* $\nu \equiv 0 \pmod 2$ *and* $s \equiv 0 \pmod 2$, *there exists a polynomial* $f_3 \in F_p[u]$ *of the form*

$$f_3(u) = (u + u^{p^{\nu/2}})^{s/2}$$

such that

$$S_q(f_3) = (p^{\nu/2} - 1)p^{\nu/2}.$$

Theorem 7. *Let* $\nu > 1$ *be an integer,* F_q *the finite field of characteristic* $p = 2$ *consisting of* $q = p^\nu$ *elements,* χ' *the character of* F_q *induced by a non-trivial multiplicative character of the field* F_p *of exponent* $s \geq 3$ *and* a, b *positive integers satisfying* $a + b = s$. *Then*

(*i*) *if* $\nu \equiv 1 \pmod 2$, *there exists a polynomial* $f_1 \in F_p[u]$ *of the form*

$$f_1(u) = (u + u^{p^{(\nu-1)/2}})^a (u + u^{p^{(\nu+1)/2}})^b$$

such that

$$S_q(f_1) = \sum_{u \in F_q} \chi'(f_1(u)) = p^\nu - p;$$

(*ii*) *if* $\nu \equiv 0 \pmod 2$, *there exists a polynomial* $f_2 \in F_p[u]$ *of the form*

$$f_2(u) = (u + u^{p^{\nu/2}-1})^a (u + u^{p^{\nu/2}+1})^b$$

such that

$$S_q(f_2) = \begin{cases} p^\nu - p^2 & \text{if} \quad 4 \nmid \nu \\ p^\nu - p & \text{if} \quad 4|\nu \end{cases}.$$

Similar results can be obtained for the exponential sums $T_q(g)$ (Stepanov [35]).

Theorem 8. *Let $\nu > 1$ be an integer, F_q the finite field of characteristic $p > 2$ consisting of $q = p^\nu$ elements and ψ' the character of F_q induced by a non-trivial additive character ψ of the field F_p. There exists a polynomial $g \in F_p[u]$ of the form*

$$g(u) = \begin{cases} u^2 + 2 \sum_{i=1}^{(\nu-2)/2} u^{p^i+1} + u^{p^{\nu/2}+1} & \text{if} \quad \nu \equiv 0 \pmod 2 \\ u^2 + 2 \sum_{i=1}^{(\nu-1)/2} u^{p^i+1} & \text{if} \quad \nu \equiv 1 \pmod 2 \end{cases}$$

such that

$$|T_q(g)| = |\sum_{u \in F_q} \psi'(g(u))| = p^{\nu-1/2}.$$

Proof. Let $\omega_1, \ldots, \omega_\nu$ be a basis of the field F_q over F_p. Then every element $z \in F_q$ can be uniquely written as a linear combination

$$z = z_1\omega_1 + \cdots + z_\nu\omega_\nu$$

with coefficients z_i from F_p. Applying to both sides of the last equality the Frobenius automorphisms $\theta^j(z) = z^{p^j}$ for $j = 0, 1, \ldots, \nu - 1$ we arrive at the system of linear equations with respect to $z_1, \ldots, z_\nu$:

$$\begin{aligned} z &= z_1\omega_1 + \cdots + z_\nu\omega_\nu \\ \theta(z) &= z_1\theta(\omega_1) + \cdots + z_\nu\theta(\omega_\nu) \\ &\cdots \\ \theta^{\nu-1}(z) &= z_1\theta^{\nu-1}(\omega_1) + \cdots + z_\nu\theta^{\nu-1}(\omega_\nu). \end{aligned}$$

The determinant of the system

$$\Delta = \det(\theta^{j-1}(\omega_i))_{1 \le i,j \le \nu}$$

differs from zero, and we find

$$z_j = \Delta^{-1}(\Delta_{1j}z + \Delta_{2j}\theta(z) + \cdots + \Delta_{\nu j}\theta^{\nu-1}(z)), \qquad 1 \le j \le \nu,$$

where

$$\Delta_{ij} = (-1)^{i+j} \det(\theta^{l-1}(\omega_k))_{1 \le k,l \le \nu, k \ne i, l \ne j}.$$

It is clear that

$$\theta(\Delta) = (-1)^{\nu-1}\Delta,$$

and

$$\theta(\Delta_{i-1,j}) = (-1)^{\nu-2}\Delta_{ij}, \qquad 2 \le i \le \nu, \quad 1 \le j \le \nu.$$

Therefore

$$\Delta_{ij}/\Delta = \theta^{i-1}(\Delta_{1j}/\Delta), \qquad 1 \le i, j \le \nu$$

and hence

$$\begin{aligned} z_j &= (\alpha_j z) + \theta(\alpha_j z) + \cdots + \theta^{\nu-1}(\alpha_j z) \\ &= (\alpha_j z) + (\alpha_j z)^p + \cdots + (\alpha_j z)^{p^{\nu-1}} \\ &= \mathrm{tr}\ (\alpha_j z), \end{aligned}$$

where

$$\alpha_j = (-1)^{j+1}\Delta_{1j}/\Delta, \qquad 1 \le j \le \nu.$$

For $p > 2$ we have

$$z_j^2 = \mathrm{tr}\ (\alpha_j z)^2 + 2 \sum_{1 \le i < k \le \nu} (\alpha_j z)^{p^{i-1}} (\alpha_j z)^{p^{k-1}},$$

and setting $\alpha_j z = u$ for some $j = 1, 2, \ldots, \nu$ we obtain

$$z_j^2 = \mathrm{tr}\ u^2 + 2 \sum_{1 \le i < k \le \nu} u^{p^{i-1}} u^{p^{k-1}}.$$

Let $\nu > 1$ be an odd number. Represent the sum

$$\sum_{1 \le i < k \le \nu} u^{p^{i-1}} u^{p^{k-1}}$$

in the form

$$\begin{aligned} \sum_{1 \le i < k \le \nu} u^{p^{i-1}} u^{p^{k-1}} &= \sum_{i=1}^{\nu-1} \sum_{l=1}^{\nu-i} (u \cdot u^{p^l})^{p^{i-1}} \\ &= \sum_{i=1}^{(\nu-1)/2} \sum_{l=1}^{(\nu-1)/2} (u \cdot u^{p^l})^{p^{i-1}} \end{aligned}$$

$$+ \sum_{i=(\nu+1)/2}^{\nu-1} \sum_{l=1}^{\nu-i} (u \cdot u^{p^l})^{p^{i-1}} + \sum_{i=1}^{(\nu-1)/2} \sum_{l=(\nu+1)/2}^{\nu-i} (u \cdot u^{p^l})^{p^{i-1}}.$$

Since $u^{p^\nu} = u$ for every $u \in F_q$, we have

$$\sum_{i=1}^{(\nu-1)/2} \sum_{l=(\nu+1)/2}^{\nu-i} (u \cdot u^{p^l})^{p^{i-1}} = \sum_{i-1}^{(\nu-1)/2} \sum_{l=(\nu+1)/2}^{\nu-i} u^{p^{l-1}} u^{p^{l+i-1}}$$

$$= \sum_{k=(\nu+3)/2}^{\nu} \sum_{l=(\nu+1)/2}^{k-1} u^{p^{\nu-l+k-1}} u^{p^{k-1}} = \sum_{k=(\nu+3)/2}^{\nu} \sum_{j=\nu-k+1}^{(\nu-1)/2} (u \cdot u^{p^j})^{p^{k-1}}$$

$$= \sum_{i=(\nu+3)/2}^{\nu} \sum_{l=\nu-i+1}^{(\nu-1)/2} (u \cdot u^{p^l})^{p^{i-1}}$$

and therefore

$$\sum_{1 \le i < k \le \nu} u^{p^{i-1}} u^{p^{k-1}} = \sum_{i-1}^{(\nu-1)/2} \sum_{l=1}^{(\nu-1)/2} (u \cdot u^{p^l})^{p^{i-1}}$$

$$+ \sum_{i=(\nu-1)/2}^{\nu-1} \sum_{l=1}^{\nu-i} (u \cdot u^{p^l})^{p^{i-1}} + \sum_{i=(\nu+3)/2}^{\nu} \sum_{l=\nu-i+1}^{(\nu-1)/2} (u \cdot u^{p^l})^{p^{i-1}}$$

$$= \operatorname{tr} \left(\sum_{i=1}^{(\nu-1)/2} u^{p^{l+1}} \right).$$

Put

$$g(u) = u^2 + 2 \sum_{l=1}^{(\nu-1)/2} u^{p^{l+1}}$$

For every $u \in F_q$ we have

$$z_j^2 = \operatorname{tr} g(u)$$

and, consequently,

$$\sum_{u \in F_q} \psi'(g(u)) = \sum_{z_1, \ldots, z_\nu \in F_p} \psi(z_j^2).$$

For the Gauss sum

$$\sum_{z_j \in F_p} \psi(z_j^2)$$

the equality

$$\left|\sum_{z_j\in F_p}\psi(z_j^2)\right| = p^{1/2}$$

is valid and hence

$$|T_q(g)| = \left|\sum_{u\in F_q}\psi(g(u))\right| = p^{\nu-1/2}.$$

This proves the statement for $\nu \equiv 1 \pmod 2$.

Let now $\nu > 1$ be an even number. In this case we have

$$\begin{aligned}
z_j^2 &= \operatorname{tr} u^2 + 2\sum_{1\le i<k\le\nu} u^{p^{i-1}}u^{p^{k-1}} \qquad = \operatorname{tr} u^2 + 2\sum_{i=1}^{\nu-1}\sum_{l=1}^{\nu-i}(u\cdot u^{p^l})^{p^{i-1}}\\
&= \operatorname{tr} u^2 + 2\sum_{i=1}^{\nu/2}\sum_{l=1}^{(\nu-2)/2}(u\cdot u^{p^l})^{p^{i-1}} \quad + \quad 2\sum_{i=(\nu+2)/2}^{\nu-1}\sum_{l=1}^{\nu-i}(u\cdot u^{p^l})^{p^{i-1}}\\
&\quad + \quad 2\sum_{i=1}^{(\nu-2)/2}\sum_{l=(\nu+2)/2}(u\cdot u^{p^l})^{p^{i-1}} \quad + \quad 2\sum_{i=1}^{\nu/2}(u\cdot u^{p^{\nu/2}})^{p^{i-1}},
\end{aligned}$$

and since

$$\begin{aligned}
\sum_{i=1}^{(\nu-2)/2}\sum_{l=(\nu+2)/2}^{\nu-i}(u\cdot u^{p^l})^{p^{i-1}} &= \sum_{i=1}^{(\nu-2)/2}\sum_{l=(\nu+2)/2}^{\nu-i} u^{p^{i-1}}u^{p^{l+i-1}}\\
= \sum_{k=(\nu+4)/2}^{\nu}\sum_{l=(\nu+2)/2}^{k-1} u^{p^{\nu-l+k-1}}u^{p^{k-1}} &= \sum_{i=(\nu+4)/2}^{\nu}\sum_{l=\nu-i+1}^{(\nu-2)/2}(u\cdot u^{p^l})^{p^{i-1}}
\end{aligned}$$

and

$$2\sum_{i=1}^{\nu/2}(u\cdot u^{p^{\nu/2}})^{p^{i-1}} = \sum_{i=1}^{\nu}(u\cdot u^{p^{\nu/2}})^{p^{i-1}}$$

for any $u \in F_q$, we find that

$$\begin{aligned} z_j^2 &= \operatorname{tr} u^2 + 2\sum_{i=1}^{\nu}\left(\sum_{l-1}^{(\nu-2)/2} u \cdot u^{p^l}\right)^{p^{i-1}} + \sum_{i=1}^{\nu}(u \cdot u^{p^{\nu/2}})^{p^{i-1}} \\ &= \operatorname{tr}\left(u^2 + 2\sum_{l=1}^{(\nu-2)/2} u^{p^l+1} + u^{p^{\nu/2}+1}\right). \end{aligned}$$

Thus, if

$$g(u) = u^2 + 2\sum_{l=1}^{(\nu-2)/2} u^{p^l+1} + u^{p^{\nu/2}+1},$$

then for every $u \in F_q$ we have

$$z_j^2 = \operatorname{tr} g(u)$$

and therefore

$$|T_q(g)| = \left|\sum_{u \in F_q} \psi'(g(u))\right| = p^{\nu-1/2}.$$

This completes the proof.

If char $F_q = 2$ we are able to prove the following result.

Theorem 9. *Let $\nu > 1$ be an odd number, F_q the finite field of characteristic $p = 2$ consisting of $q = p^\nu$ elements and ψ' the character of F_q induced by a non-trivial additive character ψ of the field F_2. There exists a polynomial $g \in F_q[z]$ of the form*

$$g(z) = \alpha_r\alpha_s z^2 + \sum_{l=1}^{(\nu-1)/2} (\alpha_r\alpha_s^{p^l} + \alpha_r^{p^l}\alpha_s) z^{p^l+1}$$

such that

$$T_q(g) = p^{\nu-1}.$$

Proof. Using the same arguments as in the proof of Theorem 8 we find that

$$z_r z_s = \operatorname{tr}\left(\alpha_r\alpha_s z^2 + \sum_{l=1}^{(\nu-1)/2} (\alpha \cdot \alpha_s^{p^l} + \alpha_r^{p^l} \cdot \alpha_s) z^{p^l+1}\right) = \operatorname{tr} g(z)$$

for every $z \in F_q$. Now since

$$\sum_{z_1,\ldots,z_\nu \in F_2} \psi(z_r z_s) = p^{\nu-1}$$

for $r \neq s$, we obtain the equality

$$T_q(g) = p^{\nu-1},$$

which proves the theorem.

4 Applications to Coding Theory

Now we apply the results of previous section to construct rather long geometric Goppa codes with fairly good parameters.

Let F_q be a finite field of characteristic $p > 2$, let $\overline{F}_q$ be an algebraic closure of the field F_q and $\mathbb{A}^{s+1}$ be $(s+1)$-dimensional affine space over $\overline{F}_q$.

Lemma 1. *Let $f_1, \ldots, f_s$ be pairwise coprime square-free monic polynomials in F_q of the same odd degree $l \geq 1$ and Y be the fibre product in $\mathbb{A}^{s+1}$ given by*

$$z_i^2 = f_i(u), \qquad 1 \leq i \leq s. \tag{8}$$

Then the genus $g = g(Y)$ of the curve Y is

$$g = (ls - 3)2^{s-2} + 1.$$

Proof. Let X be a smooth projective model of the curve Y. Denote by v_x the canonical valuation of the function field $\overline{F}_q(X)$, and by $\Omega[X]$ the space of regular differential forms on X. The affine curve Y is easily seen to be smooth. If $\overline{Y}$ is its projective closure, then X is a normalization of $\overline{Y}$ and we have the map $\psi : X \to \overline{Y}$, which is an isomorphism between Y and $\psi^{-1}(Y)$. Hence it follows that $g = g(X)$.

The rational map $(u, z_1, \ldots, z_s) \mapsto u$ of the curve Y into $\mathbb{A}^1$ determines a morphism $\varphi : X \longrightarrow \mathbb{P}^1$ of degree 2^s, so that for $u_0 \in \mathbb{A}^1$ either $\varphi^{-1}(u_0)$ consists of 2^s points of the form $x' = (u, \pm z_1, \ldots, \pm z_s)$ in each of which $v_{x'}(t) = 1$ for a local parameter t at u_0, or else $\varphi^{-1}(u_0)$ consists of 2^{s-1} points of the form $x_i'' = (u, \pm z_1, \ldots, z_{i-1}, 0, \pm z_{i+1}, \ldots, \pm z_s)$, and $v_{x_i''}(t) = 2$.

Let us consider the point at infinity $u_\infty \in \mathbb{P}^1$. If the coordinate on $\mathbb{A}^1$ is denoted by u, then $t = u^{-1}$ is a local parameter at u_∞. If $\varphi^{-1}(u_\infty)$ were to consist of 2^s points $x_\infty^{(\tau)}$, then at each $x_\infty = x_\infty^{(\tau)}$ the function t would be a local parameter. Hence it would follow that $v_{x_\infty}(t) = 1$ and $v_{x_\infty}(f_i(t)) = -l$. But since l is odd, this contradicts the condition that $v_{x_\infty}(f_i(t)) = 2v_{x_\infty}(z_i)$. Thus $\varphi^{-1}(u_\infty)$ consists of $r = 2^{s-1}$ points $x_\infty^{(\tau)}, 1 \leq \tau \leq r$, with projective coordinates $x_\infty^{(\tau)} = (0, 1, \pm 1, \ldots, \pm 1, 0)$. It follows that $X = Y \cup \{x_\infty^{(1)}\} \cup \ldots \cup \{x_\infty^{(r)}\}$. At any such point $x_\infty = x_\infty^{(\tau)}$ we have $v_{x_\infty}(u) = -2$ and $v_{x_\infty}(z_i) = -l$.

Now we shall find a basis of the space $\Omega[X]$ over the field $\overline{F}_q$. Any element $\omega \in \Omega[Y]$ can be written as a $\overline{F}_q$-linear combination of the differential forms $\omega_0 = P_0(u)du$ and

$$\omega_{i_1,\ldots,i_\sigma} = \frac{P_{i_1,\ldots,i_\sigma}(u)du}{z_{i_1}\cdots z_{i_\sigma}},$$

where $i_1, \ldots, i_\sigma$ are integers satisfying the condition $1 \leq i_1 < \cdots < i_\sigma \leq s$ and P_0, $P_{i_1,\ldots,i_\sigma}$ are polynomials in $\overline{F}_q[u]$. Indeed, the differential form

$$\omega_{i_1,\ldots,i_\sigma} = \frac{du}{z_{i_1}\cdots z_{i_\sigma}}$$

is regular at any point $u_0 \in \mathbb{A}^1$ with the condition $z_i(u_0) \neq 0$ for $i \in \{i_1, \ldots, i_\sigma\}$. Now if $z_i(u_0) = 0$ for a unique $i \in \{i_1, \ldots, i_\sigma\}$, then z_i is a local parameter at $x_i'' = (u_0, \pm z_1, \ldots, \pm z_{i-1}, 0, \pm z_{i+1}, \ldots, \pm z_s)$, so that $v_{x_i''}(z_i) = 1$ and $v_{x_i''}(u - u_0) = 2$. Therefore, $v_{x_i''}(du) = 1$ and again $\omega_{i_1,\ldots,i_\sigma}$ is regular at u_0. The form $\omega_0 = du$ is also regular at any point $u_0 \in \mathbb{A}^1$. Thus, the differential forms $\omega_0 = du$ and $\omega_{i_1,\ldots,i_\sigma}$ form a basis of the $\overline{F}_q[u]$-module $\Omega[Y]$.

It remains to clarify which of the forms ω_0 and $\omega_{i_1,\dots,i_\sigma}$ are regular at the points $x_\infty^{(1)},\dots x_\infty^{(r)}$. Let x_∞ be one of these points. If t is a local parameter at x_∞ then $u = t^{-2}u'$, $z_i = t^{-l}z_i'$, where u' and z_i' are units in the local ring $\mathcal{O}_{x_\infty}$. Therefore $\omega'_{i_1,\dots,i_\sigma} = t^{l\sigma-3}\eta_{i_1,\dots,i_\sigma}dt$, with $\eta_{i_1,\dots,i_\sigma}$ a unit in $\mathcal{O}_{x_\infty}$, hence $(\omega'_{i_1,\dots,i_\sigma}) = (l\sigma - 3)\cdot x_\infty$. Thus, the differential form

$$\omega_{i_1,\dots,i_\sigma} = \frac{P_{i_1,\dots,i_\sigma}(u)du}{z_{i_1}\cdots z_{i_\sigma}}$$

is regular at x_∞ if and only if

$$v_{x_\infty}(P_{i_1,\dots,i_\sigma}(u)du) \geq -(l\sigma - 3).$$

This means that $\deg P_{i_1,\dots,i_\sigma}(u) \leq (l\sigma-3)/2$ and hence

$$\deg P_{i_1,\dots,i_\sigma}(u) \leq \begin{cases} \frac{l\sigma-4}{2} & \text{if } \sigma \equiv 0 \pmod 2 \\ \frac{l\sigma-3}{2} & \text{if } \sigma \equiv 1 \pmod 2 \end{cases}$$

The differential form $\omega_0 = P_0 du$ is not regular at x_∞ for any non-zero polynomial $P_0 \in \overline{F}_q[u]$, so the regular differential forms

$$\omega'_{i_1,\dots,i_\sigma}, u\omega'_{i_1,\dots,i_\sigma}, \dots, u^n\omega'_{i_1,\dots,i_\sigma},$$

where $1 \leq i_1 < \cdots < i_\sigma \leq s$ and

$$n = \begin{cases} \frac{l\sigma-4}{2} & \text{if } \sigma \equiv 0 \pmod 2 \\ \frac{l\sigma-3}{2} & \text{if } \sigma \equiv 1 \pmod 2 \end{cases},$$

form a basis of the space $\Omega[X]$ over $\overline{F}$. Therefore

$$\begin{aligned}
\dim_{\overline{F}} \Omega[X] &= \frac{1}{2} \sum_{\substack{\sigma=1 \\ \sigma\equiv 0 \pmod 2}}^{s} \sum_{1\le i_1<\cdots<i_\sigma\le s} (l\sigma - 2) \\
&+ \frac{1}{2} \sum_{\substack{\sigma=1 \\ \sigma\equiv 1 \pmod 2}}^{s} \sum_{1\le i_1<\cdots<i_\sigma\le s} (l\sigma - 1) \\
&= \frac{l}{2}\sum_{\sigma=1}^{s} \sigma\binom{s}{\sigma} - \sum_{\substack{\sigma=1 \\ \sigma\equiv 0 \pmod 2}}^{s} \binom{s}{\sigma} - \frac{1}{2} \sum_{\substack{\sigma=1 \\ \sigma\equiv 1 \pmod 2}}^{s} \binom{s}{\sigma} \\
&= \frac{1}{2}(ls2^{s-1} - 2^{s} - 2^{s-1} + 2)
\end{aligned}$$

and hence

$$g = g(X) = \dim_{\overline{F}_q} \Omega[X] = (ls - 3)2^{s-2} + 1.$$

This completes the proof.

Lemma 2. *Let F_p be a prime finite field of characteristic $p > 2$, let F_q be an extension of F_p of even degree $\nu > 1$ and A be the set of roots in F_q of the polynomial*

$$f(u) = u + u^{p^{\nu/2}}.$$

Then

(i) A is a subgroup of the additive group F_q^+ of the field F_q;

(ii) if $\{A_1 = A, A_2, \ldots, A_r\}$ is the set of all cosets in F_q^+/A and $\{\alpha_1, \ldots, \alpha_r\}$ are distinct representatives of the cosets, the polynomials

$$f_i(u) = (u + \alpha_i) + (u + \alpha_i)^{p^{\nu/2}}, \qquad 1 \le i \le r, \tag{9}$$

are pairwise coprime in $F_q[u]$;

$(iii)\ r = |F_q^+/A| = p^{\nu/2}$.

Proof. The main point is (i). First of all we note that $f(0) = 0$. Now, if α and β are roots of $f(u)$, then

$$\begin{aligned} f(\alpha+\beta) &= (\alpha+\beta)+(\alpha+\beta)^{p^{\nu/2}} = (\alpha+\alpha^{p^{\nu/2}})+(\beta+\beta^{p^{\nu/2}}) \\ &= f(\alpha)+f(\beta)=0, \end{aligned}$$

so that $\alpha+\beta$ is also a root of the polynomial $f(u)$. Thus A is a subgroup of F_q^+.

To prove (ii) let us suppose that $f_i(u)$ and $f_j(u)$ for $i \neq j$ have a common root in F_q, say $u = \theta$. In that case

$$\theta + \alpha_i + (\theta + \alpha_j)^{p^{\nu/2}}$$

and therefore

$$\theta + \alpha_i + \theta^{p^{\nu/2}} + \alpha_i^{p^{\nu/2}} = \theta + \alpha_j + \theta^{p^{\nu/2}} + \alpha_j^{p^{\nu/2}}.$$

This yields

$$\alpha_i - \alpha_j + (\alpha_i - \alpha_j)^{p^{\nu/2}},$$

and we find that $\alpha_i - \alpha_j$ is a root of $f(u)$; hence $\alpha_i - \alpha_j \in A$. But $\alpha_i - \alpha_j \notin A$ by the choice of $\alpha_1, \ldots, \alpha_r$, and we arrive at a contradiction.

Finally, since $|A| = p^{\nu/2}$ we find that

$$r = |F_q^+/A| = p^\nu/p^{\nu/2} = p^{\nu/2}.$$

This finishes the proof.

Lemma 3. *Let F_p be a prime finite field of characteristic $p > 2$, let F_q be an extension of F_p of even degree $\nu > 1$ and $s \leq p^{\nu/2}$ be a positive integer. Let N_q be the number of F_q-rational points of the curve Y given by equations (8) with polynomials*

$$f_i(u) = (u + \alpha_i) + (u + \alpha_i)^{p^{\nu/2}}, \qquad 1 \leq i \leq s,$$

defined by (9). *Then*

$$N_q = (2q^{1/2} - s)q^{1/2}2^{s-1}.$$

Proof. We have

$$\begin{aligned} N_q &= \sum_{u \in F_q} (1 + \chi'(f_1(u))) \cdots (1 + \chi'(f_s(u))) \\ &= \sum_{u \in F_q} \left(1 + \sum_{\sigma=1}^{s} \sum_{1 \le i_1 < \cdots < i_\sigma \le s} \chi'(f_{i_1}(u)) \cdots \chi'(f_{i_\sigma}(u)) \right) \end{aligned}$$

and hence

$$N_q = p^\nu + \sum_{\sigma=1}^{s} \sum_{1 \le i_1 < \cdots < i_\sigma \le s} \chi'(f_{i_1}(u)) \cdots \chi'(f_{i_\sigma}(u)).$$

It follows from Theorem 5 and Lemma 2 that

$$\chi'(f_i(u)) = \begin{cases} 0 & \text{if } u \in A_i\ , \\ 1 & \text{if } u \in F_q \setminus A_i, \end{cases}$$

and since any two distinct sets A_i and A_j have no common element we obtain

$$\begin{aligned} N_q &= p^\nu + \sum_{\sigma=1}^{s} \binom{s}{\sigma} (p^\nu - \sigma p^{\nu/2}) = p_\nu + (2^s - 1)p^\nu - s2^{s-1}p^{\nu/2} \\ &= (2p^{\nu/2} - s)p^{\nu/2}2^{s-1} = (2q^{1/2} - s)q^{1/2}2^{s-1}. \end{aligned}$$

This proves the lemma.

Now we are able to prove the following result (Stepanov [36]).

Theorem 10. *Let* $p > 2$ *be a prime number,* $\nu > 1$ *be an even integer, and* F_q *the a finite field of characteristic* p *consisting of* $q = p^\nu$ *elements. For any positive integers* $s \le q^{1/2}$ *and* $r > (sq^{1/2} -$

$3)2^{s-2}$ *there exists a geometric Goppa* $[n,k,d]_q$*-code* $C = C(D_0, D)$ *with*

$$\begin{aligned} r &< n \leq (2q^{1/2} - s)q^{1/2}2^{s-1}, \\ k &\geq r - (sq^{1/2} - 3)2^{s-2}, \\ d &\geq n - r. \end{aligned}$$

Proof. Let $f_1, \dots, f_s$ be pairwise coprime polynomials in $F_q[u]$ of the same odd degree $l = q^{1/2}$ defined by (9), and $Y \subset \mathbb{A}^{s+1}$ be the affine curve defined over F_q by equations (8). Let $\overline{Y} \subset \mathbb{P}^{s+1}$ be the projective closure of Y and X be a non-singular projective model of $\overline{Y}$ over an algebraic closure $\overline{F}_q$ of the field F_q.

Since the curves $\overline{Y}$ and X are birationally isomorphic, we have $g = g(Y) = g(X)$, and by Lemma 1

$$g = (sq^{1/2} - 3)2^{s-2} + 1.$$

The number $N_q(X)$ of F_q-rational points of the curve X satisfies by Lemma 3 the inequality

$$N_q(X) \geq N_q + 1 = (2q^{1/2} - s)q^{1/2}2^{s-1} + 1.$$

Let $n \leq N_q$ be a positive integer, let $x_1, \dots, x_n$ be F_q-rational points of X at the finite part of X, and x_∞ be a point of X at infinity. Set

$$D_0 = x_1 + \cdots + x_n \quad \text{and} \quad D = r \cdot x_\infty.$$

Applying to X the Goppa construction for $r > (sq^{1/2} - 3)2^{s-1}$, $n > r$, and taking into account (5), (6), we obtain a geometric Goppa $[n,k,d]_q$-code $C = C(D_0, D)$ with

$$\begin{aligned} r &< n \leq (2q^{1/2} - s)q^{1/2}2^{s-1}, \\ k &\geq r - g + 1 = r - (sq^{1/2} - 3)2^{s-2}, \\ d &\geq n - r. \end{aligned}$$

This completes the proof.

Now using (7) we obtain the following result.

Corollary 11. *The relative parameters $R = k/n$ and $\delta = d/n$ of the code $C = C(D_0, D)$ satisfy*

$$R \geq 1 - \delta - \frac{(sq^{1/2} - 3)2^{s-2}}{n}.$$

In particular, for $n = N_q = (2q^{1/2} - s)q^{1/2}2^{s-1}$ we have

$$R \geq 1 - \delta - \frac{sq^{1/2} - 3}{2(2q^{1/2} - s)q^{1/2}}.$$

Similar result is valid in the case of extension of the field F_p of an odd degree $\nu > 1$ (Stepanov, Özbudak [38]).

Theorem 12. *Let $p > 2$ be a prime number, $\nu > 1$ an odd integer and F_q the finite field of characteristic p consisting of $q = p^\nu$ elements. For any positive integers r, s satisfying*

$$s \leq \frac{2q + 4}{p^{(\nu-1)/2}(p+1) - 2}$$

and

$$2^{s-2}((p^{(\nu-1)/2}(p+1) - 2)s - 4) < r < 2^s q$$

there exists a geometric Goppa $[n, k, d]_q$-code with parameters

$$\begin{aligned} r &< n \leq 2^s q, \\ k &\geq r - 2^{s-2}((p^{(\nu-1)/2}(p+1) - 2)s - 4), \\ d &\geq n - r. \end{aligned}$$

Recently these results were extended to the case of codes on fibre product of superelliptic curves (Özbudak [27], Stepanov, Özbudak [39]).

For other applications of character sums to coding theory see Barg [2], Helleseth [15], Lachaud [19], Lachaud, Wolfmann [20], C.J. Moreno, O. Moreno [24], [25], Rodier [29], Tietäväinen [42] and Wolfmann [46].

References

[1] Aaltonen M. J., Notes on the asymptotic behavior of the information rate of block codes, *IEEE Trans. Info. Theory* , 1984, **IT-30**, p. 84-85.

[2] Barg A., Exponential sums and constrained error-correcting codes, *Lect. Notes in Comp. Sciences* , **573**, Springer, 1991, p. 16-22.

[3] Bassalygo L. A., Zinov'ev V. A., Litsyn S. N., A lower estimate of complete trigonometric sums in terms of multiple sums, *Soviet Math. Dokl.*, 1988, **37**, p. 756-759.

[4] Burgess D.A., On character sums and primitive roots, *Proc. London Math. Soc.* (3), 1962, **12**, p. 179-192.

[5] Burgess D. A., On Dirichlet characters of polynomials, *Proc. London Math. Soc.* (3), 1963, **13**, p. 537-548.

[6] Davenport H., Lewis D.J., Character sums and primitive roots in finite fields, *Rend. Cird. Mat. Palermo* (2), 1963, **12**, p. 129-136.

[7] Drinfeld V. G., Vladut S. G., The number of points on an algebraic curve, *Funct. Anal. and Appl.* , 1983, **17**, p. 53-54.

[8] Elliot P.D.T.A., The distribution of primitive roots, *Can. J. Math.*, 1969, **14**, p. 822-841.

[9] Garcia A., Stichtenoth H., A tower of Artin-Schreier extensions of function fields attaining the Drinfeld-Vladut bound, *Invent. Math.*, 1995, **121**, p. 211-222.

[10] van der Geer G., van der Vlugt N., Fibre products of Artin-Schreier curves and generalized Hamming weights of codes, *J. Comb. Theory* , 1995, **70**, no. **2**, p. 337-348.

[11] Gluhov M.M., Lower bounds for character sums over finite fields, *Diskr. Mat.*, 1994, **6**, no. **3**, p. 136-142 (*in Russian*).

[12] Gluhov M.M., On lower bounds for character sums over finite fields, *Preprint* , 1995.

[13] Gluhov M.M., Özbudak F., Codes on superelliptic curves, *Preprint* , 1995.

[14] Goppa V.G., Codes on algebraic curves, *Soviet Math. Dokl.*, 1981, **24**, p. 170-172.

[15] Helleseth T., On the covering radius of cyclic linear codes and arithmetic codes, *Discrete Appl. Math.* , 1985, **11**, p. 157-173.

[16] Ihara Y., Some remarks on the maximum of rational points of algebraic curves over finite fields, *J. Fac. Sci. Tokyo* , 1981, **28**, p. 721-724.

[17] Korobov N.M., Double trigonometric sums and their applications to the estimation of rational sums, *Math. Notes* , 1969, **6**, p. 472-478.

[18] Korobov N.M., Estimate of a sum of Legendre symbols, *Soviet Math. Dokl.*, 1971, **12**, p. 241-245.

[19] Lachaud G., Artin-Schreier curves, exponential sums and the Carlitz-Uchiyama bound for geometric codes, *J. Number Theory* 1991, **39**, p. 18-40.

[20] Lachaud G., Wolfmann J., Sommes de Kloosterman, courbes elliptique et codes cycliques en caracteristique 2, *C. R. Acad. Sci. Paris* , Ser. 1, 1987, **305**, p. 881-883.

[21] Levenshtein V. I., Bounds for packings of metric spaces and some their applications, *Probl. Cybern.*, **40**, Nauka, Moscow, 1983, p. 43-110 (*in Russian*).

[22] Manin Yu. I., What is the maximum of points on a curve over F_2?, *J. Fac. Sci. Tokyo* , 1A, 1981, **28**, p. 721-724.

[23] Mit'kin D.A., Estimation of the sum of Legendre symbols of polynomials of even degree, *Math. Notes* , 1973, **14**, p. 597-602.

[24] Moreno C.J., Moreno O., Exponential sums and Goppa codes I, *Proc. Amer. Math. Soc.* , 1991, **111**, p. 523-531; II, *IEEE Trans. Info. Theory* , 1992, **38**, p. 1222-1229.

[25] Moreno C.J., Moreno O., An improved Bombieri-Weil bound and application to coding theory, *J. Number Theory* , 1992, **38**, p. 32-46.

[26] Özbudak F., On lower bounds for incomplete character sums over finite fields, *Finite Fields and Applications* , 1996, **2**, p. 173-191.

[27] Özbudak F., Codes on fibre products of superelliptic curves, *Preprint* , 1996.

[28] Perret M., Tours ramifiées infininies de corps de classes, *J. Number Theory* , 1991, **38**, p. 300-322.

[29] Rodier F., Minoration de certaines sommes exponentielles binaires, *Lect. Notes in Math.* , **1518**, Springer, p. 199-209.

[30] Serre J.P., Sur le nombre des points rationnels d'une courbe algébrique sur un corps fini, *C.R. Acad. Sci. Paris* , Ser. 1, 1983, **296**, p. 397-402.

[31] Shafarevich I.R., *Basic Algebraic Geometry* I, Second Edition, Springer Verlag, Berlin, 1994.

[32] Stark H.M., On the Riemann hypothesis in hyperelliptic function fields, *Proc. Symp. Pure Math.*, **24**, p. 285-302, Amer. Math. Soc., Providence, R.I., 1973,

[33] Stepanov S.A., *Arithmetic of Algebraic Curves* , Plenum, New York, 1994.

[34] Stepanov S.A., On lower bound for incomplete character sums with polynomials, *Trudy Mat. Inst. Akad. Nauk SSSR* , 1977, **143**, p.175-177 (in Russian).

[35] Stepanov S.A., On lower bounds for character sums over finite fields, *Discr. Math. Appl.*, 1992, **2**, no. **5**, p. 523-532.

[36] Stepanov S.A., Codes on fibre products of hyperelliptic curves, *Discr. Math. Appl.* , to appear.

[37] Stepanov S.A., *Codes on Algebraic Curves* , to be published.

[38] Stepanov S.A., Özbudak F., Fibre products of hyperelliptic curves and geometric Goppa codes, *Discr. Math. Appl.* , 1995.

[39] Stepanov S. A., Özbudak F., Fibre products of superelliptic codes and codes therefrom, *Preprint*, 1996.

[40] Stichtenoth H., *Algebraic Function Fields and Codes* , Springer-Verlag, Berlin, 1993.

[41] Stöhr K.O, Voloch J.F., Weierstrass points and curves over finite fields, *Proc. London Math. Soc.* (3), 1986, **52**, p. 1-19.

[42] Tietäväinen A., Covering radius and dual distance, *Designs, Codes and Crypt.* , 1991, **1**, p. 31-36.

[43] Tsfasman M. A., Vladut S. G., *Algebraic-Geometric Codes* , Kluwer Acad. Publ, Dordrecht, 1991.

[44] Tsfasman M. A., Vladut S. G., Zink Th., Modular curves, Shimura curves, and Goppa codes, *Math. Nachr.* , 1982, **109**, p. 21-28.

[45] Weil A., On some exponential sums, *Proc. Nat. Acad. Sci. U.S.A* , 1948, **34**, p. 204-207.

[46] Wolfmann J., The number of points on certain algebraic curves over finite fields, *Comm. Algebra* , 1989, **17** (8), p. 2055-2060.

[47] Zink Th., Degeneration of Shimura surfaces and a problem in coding theory, *Lect. Notes on Comp. Sci.* , 1986, **199**, p. 503-511.

The Classical Geometry of Vector Bundles

Andrei Tyurin

Algebra Section,
Steklov Mathematical Institute,
Ul. Vavilova 42,
Moscow GSP-I, 117966 Russia

e-mail: `Tyurin@class.mian.su`
or `Tyurin@tyurin.mian.su`
or `Tyurin@Maths.Warwick.Ac.UK`

1 Introduction

One of the expected products of this summer school is an answer to the following question:

What is ALGEBRAIC GEOMETRY?

(We write AG for short.) Actually this is a hard task, because everybody already has the fixed conviction that the objects of AG are algebraic varieties.

An irreducible algebraic variety X has a dimension $\dim X$, and this number is usually a rough indication of the level of completeness of the geometric theory describing it. Algebraic varieties of small dimension carry special names:

$\dim X = 0$:	set of points
$\dim X = 1$:	curves (Xavier Gomez-Mont's lectures)
$\dim X = 2$:	surfaces (Rick Miranda's lectures)
$\dim X = 3$:	3-folds (Miles Reid's lectures)
etc.	

To explain how my lectures fit into this list, I would like to remark that two algebraic varieties of different dimension can be *geometrically identical*. To see this, consider the following chain of examples:

$\dim = 0$	:	a set of 6 distinct points on $\mathbb{P}^1$ up to $\mathrm{PGL}(2, \mathbb{C})$ action;
$\dim = 1$	:	a curve of genus 2;
$\dim = 2$	:	a cubic surface in $\mathbb{P}^3$ with one ordinary double point;
$\dim = 3$	:	a nonsingular intersection of two quadrics in $\mathbb{P}^5$.

The identifications between the objects in $\dim < 3$ are absolutely obvious: the canonical map of a curve (see Rick Miranda's lectures [M] in this volume) of genus 2 is a double cover of $\mathbb{P}^1$ ramified in 6 points; considering $\mathbb{P}^1$ as a conic in $\mathbb{CP}^2$, blowing up 6 points on this conic and constructing the anticanonical map of the resulting surface, we get a cubic in $\mathbb{P}^3$ with an ordinary double point, see [M].

The threefold in our list carries the imposing full name of *Fano threefold of index* 2 *and degree* 4; its halfanticanonical map [M] displays it as the base locus of a pencil of quadrics in $\mathbb{P}^5$. The six singular quadrics of this pencil take us back to a set of 6 distinct points on $\mathbb{P}^1$.

This example of a chain of identifications is of course very classical and simple. A more recent example is Mukai's construction [Mu] of an identification of a plane quartic with a Fano variety V_{22}.

Slogan: *An algebraic geometer is skillful enough if he or she can* *recognize the geometric person under many guises of different dimensions.*

My first aim is to give you some experience in this direction. But my task is a little more complicated, because there is some new person in our game:

algebraic vector bundle.

In some sense this geometric object doesn't have any dimension (or, if you prefer, is infinite dimensional). But in any case, we can't avoid it. Even in our simplest chain of identifications, the intersection of two quadrics in $\mathbb{P}^5$ is a moduli space of stable vector bundles on the corresponding curve ot genus 2.

So my second aim is to construct a simple but a new chain of geometric identifications including a vector bundle as a geometric object.

This new chain isn't quite as simple as the previous one, but it is perhaps the simplest illustration of new geometric observations showing that CLASSICAL AG is a slice of much more general GEOMETRY. Namely, some time ago Gromov observed that many results of ENUMERATIVE AG are true in SYMPLECTIC GEOMETRY. But a recent observation due to Donaldson is much more unexpected: *many constants of ENUMERATIVE AG are invariants of the underlying smooth structure of algebraic surfaces.*

Thus my third aim is to explain these relations between AG and differential geometry.

2 Clebsch and Darboux curves

Let $\mathbb{CP}^2$ be the complex projective plane:

$$\mathbb{CP}^2 = \mathbb{P}T, \quad \text{where } T = \mathbb{C}^3, \quad \text{so that } \mathrm{Pic}\mathbb{CP}^2 = \mathbb{Z} \cdot l,$$

where l is a line. Then $|d\cdot l| = \mathbb{P}S^dT^*$ is the complete linear system of curves of degree d in $\mathbb{CP}^2$. So a homogeneous polynomial $\phi_C \in S^dT^*$ of degree d is the equation of a curve

$$C = \{\phi_C = 0\} \subset \mathbb{CP}^2, \quad \text{that is,} \quad C \in |d \cdot l|.$$

It is a classical enumerative problem in invariant theory to compute the degree $\deg V$ of some $\mathrm{PGL}(3,\mathbb{C})$ invariant subvariety $V \subset |d \cdot l|$.

Example 0: The discriminant hypersurface in $|d \cdot l|$:

$$V_{\text{sing}} = \{C \in |d \cdot l| \mid \mathrm{Sing} C \neq \emptyset\}.$$

This is obviously a subvariety of $|d \cdot l|$ invariant under $\mathrm{PGL}(3,\mathbb{C})$, and an easy calculation shows that

$$\deg V_{\text{sing}} = 3 \cdot (d-1)^2.$$

Now a curve in $\mathbb{CP}^2$ which splits completely as a union of lines

$$\Delta_r = \bigcup_{i=1}^{r} l_i \tag{2.1}$$

is called a *polygon* or an *r-gon*. In §4 we will define an r-gon to be *regular* if all its sides l_i and all its vertices $l_i \cap l_j$ are distinct. Let $P_r \subset |d \cdot l|$ be the subvariety of all r-gons.

Useful exercise: What is $\deg P_r$?

Definition 2.1: *We say that a curve C circumscribes a regular r-gon Δ_r if for every pair (i,j) the vertex (= intersection of sides)*

$$l_i \cap l_j \in C.$$

Let

$$\mathrm{MP}_r^d = \{\Delta_r, C_d\} \subset P_r \times |d \cdot l| \tag{2.2}$$

be the closure of the incidence variety of pairs consisting of a regular r-gon Δ_r and a curve C_d of degree d circumscribing it. We have two projection maps:

$$\begin{array}{ccccc} & & \mathrm{MP}_r^d & & \\ & {}^{p_\Delta}\swarrow & & \searrow^{p_c} & \\ P_r & & & & |d \cdot l| \end{array} \tag{2.3}$$

Thus the subvariety

$$p_c(\mathrm{MP}_r^d) \subset |d \cdot l| \tag{2.4}$$

of curves of degree d circumscribing some r-gon is invariant under $\mathrm{PGL}(3, \mathbb{C})$.

Problem: What is $\deg p_c(\mathrm{MP}_r^d)$? More precisely, what is

$$s_r(d) = \deg p_c \cdot \deg p_c(\mathrm{MP}_r^d)? \tag{2.5}$$

In terms of the defining equations, it is easy to see that

$$(\Delta, C) \in \mathrm{MP}_r^{r-1} \iff \phi_C = \sum_{i=1}^{r} (\phi_\Delta / \phi_{l_i}) \tag{2.6}$$

where $\Delta = \bigcup_{i=1}^{r} l_i$.

Historically, the problem (2.5) is closely related to the following problem:

Definition 2.2: *We say that a polygon $\Delta = \bigcup_{i=1}^{r} l_i$ is apolar to a curve C if*

$$\phi_C = \sum_{i=1}^{r} (\phi_{l_i})^d. \tag{2.7}$$

Let

$$\mathrm{MPA}_r^d = \{\Delta_r, C_d\} \subset P_r \times |d \cdot l| \tag{2.8}$$

be the space of apolar pairs of polygons and curves. We again have two projection maps

$$\begin{array}{ccc} & \mathrm{MPA}_r^d & \\ p_\Delta \swarrow & & \searrow^{p_c} \\ P_r & & |d \cdot l| \end{array} \tag{2.9}$$

Thus the subvariety

$$p_c(\mathrm{MPA}_r^d) \subset |d \cdot l| \tag{2.10}$$

is also invariant under $\mathrm{PGL}(3, \mathbb{C})$.

Problem: What is

$$c_r(d) = \deg p_c \cdot \deg p_c(\mathrm{MPA}_r^d)? \tag{2.11}$$

These problems were solved recently by Geir Ellingsrud and Stein Strømme. Using Bott's formula, they computed the constants $c_r(d)$ for $r < 9$ and $s_r(r-1)$ for $r = 6, 7, 8, 9, 10$.

For example, they find

$$\begin{aligned} 5! \cdot c_5(d) &= d^{10} - 100d^8 + 150d^7 + 3680d^6 - 10260d^5 \\ &\quad -52985d^4 + 224130d^3 + 127344d^2 \\ &\quad -1500480d + 1664640. \end{aligned} \tag{2.12}$$

The following particular cases of the general enumerative problem will be important for us:

Definition 3:

1. *A curve*

$$C \in p_c(\mathrm{MP}_{d+1}^d)$$

is called a Darboux curve.

2. *A curve*

$$C \in p_c(\mathrm{MPA}_{d+1}^d)$$

is called a Clebsch curve.

For special reasons, Darboux curves of degree 4 are called *Lüroth quartics*. These names have a historical explanation. Namely it is easy to see that the virtual (expected) dimension

$$\text{v.dimMP}^d_{d+1} = \text{v.dimMPA}^d_{d+1} = 3d + 2. \tag{2.13}$$

Remark: This dimension is one more than the dimension of the subvariety of rational curves of degree d.

But in 1865, Clebsch observed the following:

Clebsch's Theorem:

1. *The image $p_c(\text{MPA}^4_5)$ is a hypersurface in $|4 \cdot l| = \mathbb{P}^{14}$. (This is in spite of the fact that*

$$\text{v.dimMPA}^4_5 = \dim |4 \cdot l| = 14.)$$

2. *If C is nonsingular then*

$$C \in p_c(\text{MPA}^4_5) \iff p_c^{-1}(C) = \mathbb{P}^1.$$

3.

$$\deg p_c(\text{MPA}^4_5) = 6.$$

Exactly the same facts hold for MP^4_5, that is, for Lüroth quartics, as Lüroth observed in 1868. But the degree of the hypersurface of Lüroth quartics was only computed in 1918 by F. Morley [Mo]:

$$\deg p_c(\text{MP}^4_5) = 54. \tag{2.14}$$

This constant was reproduced in modern investigation (Tyurin, Le Potier, Ellingsrud and Strømme) under absolutely new motivations related to PDEs.

Remarks:

1. It follows from Clebsch's Theorem, that the polynomial (2.12) satisfies
$$c_5(4) = 0,$$
that is, 4 is a root of $s_5(d)$. Can you see this from the display of this polynomial (2.12)?

2. The fibres of the projection p_c of the diagram (2.9) were used by S. Mukai to describe special Fano varieties: let
$$p_c\, \mathrm{MP}_6^4 \to |4 \cdot l|$$
be the right side of the diagram (2.3). Then

(i) for general C, the inverse image $P_c^{-1}(C)$ is a Fano threefold;

(ii) if $C = 2q$ is a double nonsingular conic, then $P_c^{-1}(2q)$ is a compactification of $\mathbb{C}^3$, [Mu].

(iii) The exact formulas of Ellingsrud and Strømme also work when the degree of curve is not small with respect to the number of sides of polygons. More precisely, if $d \geq r - 1$, S. Mukai proved that $p_c(\mathrm{MPA}_7^5) = |5 \cdot l|$, and that the map p_c is birational. But you can see that the constant $c_7(5)$ is negative.

3 Vector Bundles on an Algebraic Surface and Their Sections

Let me recall briefly the main constructions of sheaf theory on algebraic surfaces. The starting point is the structure sheaf $\mathcal{O}_S = \mathcal{O}$ of an algebraic surface S. For a first approach, it is enough to consider a nonsingular surface. Thus the stalk $\mathcal{O}_P$ of $\mathcal{O}$ at a point $p \in S$ is a 2-dimensional regular local ring. Every coherent sheaf F has a stalk F_P at each point $P \in S$, which is a module of finite type over $\mathcal{O}_P$; moreover, in a neighbourhood U of any point, there is a resolution
$$\mathcal{O}_U^p \to \mathcal{O}_U^q \to F_{|U} \to 0.$$

Thus each sheaf F on S defines a filtration

$$S_2 \subset S_1 \subset S$$

by the homological dimension of the stalk. If this filtration is trivial then F is called a *vector bundle* and we will note it as E.

For a sheaf F on S the canonical homomorphism

$$\mathrm{can} : F \to \mathrm{Hom}(\mathrm{Hom}(F, \mathcal{O}), \mathcal{O}) = F^{**} \tag{3.1}$$

can be completed to a 4-term exact sequence

$$0 \to T(F) \to F \xrightarrow{\mathrm{can}} F^{**} \to C(F) \to 0 \tag{3.2}$$

We say that F is a *torsion sheaf* if $F = T(F)$, a *torsion free sheaf* if $T(F) = 0$ and a *reflexive sheaf* if $F = F^{**}$.

It is easy to see that on a surface, *a reflexive sheaf is a vector bundle.* Moreover, for a torsion free sheaf F, we have $\dim \mathrm{Supp} C(F) = 0$; that is, in this case $C(F)$ is an *Artinian sheaf.*

A pair of sheaves F_1 and F_2 defines three vector spaces $\mathrm{Ext}^i(F_1, F_2)$, for $i = 0, 1, 2$ with the usual functorial properties.

In the short exact sequence of sheaves

$$0 \to F_2 \to F \to F_1 \to 0, \tag{*}$$

the sheaf F is called an *extension* of F_1 by F_2; such an extension is given by an element $e(F)$ in the vector space $\mathrm{Ext}^1(F_1, F_2)$, so the set of classes of such extensions has the structure of the vector space $\mathrm{Ext}^1(F_1, F_2)$. For the zero class $0 \in \mathrm{Ext}^1(F_1, F_2)$ we have $F = F_1 \oplus F_2$.

Exercises:

1. Prove that on an algebraic curve C, every coherent sheaf F is a direct sum
$$F = T(F) \oplus F^{**}.$$

2. Suppose that we have two extensions F and F' of F_1 by F_2 and F_1' by F_2, together with a homomorphism $\varphi\ F_1' \to F_1$. Then the identity map $F_2 = F_2$ and the given map φ extend to a homomorphism $F' \to F$ if and only if the homomorphism

$$\tilde{\varphi}\ \mathrm{Ext}^1(F_1, F_2) \to \mathrm{Ext}^1(F_1', F_2)$$

induced by φ satisfies

$$\tilde{\varphi}(e(F)) = e(F'). \tag{3.3}$$

Of course, we would prefer to work only with vector bundles, which is enough for working over algebraic curves. But over algebraic surfaces, it is absolutely necessary to use torsion free sheaves.

Any rank 1 torsion free sheaf J on an algebraic surface S admits an exact sequence of the form (3.2):

$$0 \to J \to \mathcal{O}_S(D) = J^{**} \to C(J) \to 0, \tag{3.4}$$

where D is some divisor on S, and we can untwist this sequence by tensoring with $\mathcal{O}_S(-D)$:

$$0 \to J(-D) \to \mathcal{O}_S \to \mathcal{O}_\xi \to 0. \tag{3.5}$$

The last sheaf is the structure sheaf of 0-dimensional subscheme (a *cluster*, or a "fine 0-cycle") ξ of S, and $J(-D) = \mathcal{I}_\xi \subset \mathcal{O}_S$ is the ideal sheaf of this subscheme. A cluster ξ defines a cycle of points

$$[\xi] = \sum \deg(\xi, p_i) \cdot p_i.$$

We say that ξ is *reduced* if we have

$$\deg(\xi, p_i) = 1 \quad (\text{or } 0) \quad \text{for every } i.$$

In this case $\xi = [\xi]$, and the cluster is a configuration of distinct points on S.

Thus a rank 1 torsion free sheaf admits two invariants: $c_1(J) = c_1(J^{**})$ and $c_2(J) = \deg \xi = h^0(\mathcal{O}_\xi)$.

Now let $s\ \mathcal{O}_S \to E$ be a section of a vector bundle E of rank 2. We say that a section is *regular* if its zero set is a 0-dimensional subscheme:

$$(s)_0 = \xi.$$

In this case, by definition, $\deg \xi = c_2(E)$.

Remark: If the zero set of a section contains an effective curve C, we can untwist it by $-C$ to obtain a regular section $s\,\mathcal{O}_S \to E(-C)$. In this case

$$\deg \xi = c_2(E) - C \cdot (c_1(E) - C).$$

For a rank 2 vector bundle E, the dual map to a regular section s can be extended to the Koszul resolution (as in David Eisenbud's lectures)

$$0 \to \bigwedge^2 E^* \xrightarrow{\wedge S^*} E^* \xrightarrow{s^*} \mathcal{O}_S \xrightarrow{\text{can}} \mathcal{O}_\xi \to 0 \tag{3.6}$$

of the zero set $(s)_0 = \xi$ of s.

The kernel of can is just the ideal sheaf of ξ from (3.5) and from the first part of the sequence (3.6) we get the exact sequence

$$0 \to \bigwedge^2 E^* \to E^* \to \mathcal{I}_\xi \to 0.$$

Tensoring this sequence by the invertible sheaf $\bigwedge^2 E = \det E$ we get finally the *short exact sequence of a regular section*:

$$0 \to \mathcal{O}_S \xrightarrow{s} E \to \mathcal{I}_\xi(c_1(E)) \to 0. \tag{3.7}$$

As we know, an extension of this type is given by an element $e \in \operatorname{Ext}^1(\mathcal{I}_\xi(c_1(E))$. For the last space, by Serre duality we have

$$\operatorname{Ext}^1(\mathcal{I}_\xi(D), \mathcal{O}_S) = \operatorname{Ext}^1(\mathcal{O}_S, \mathcal{I}_\xi(D+K_S))^* = H^1(\mathcal{I}_\xi(D+K_S))^* \tag{3.8}$$

where K_S is the canonical class of S.

Thus a pair (s, E) consisting of a vector bundle and a section is given by a cluster $(s)_0 = \xi$ and a hyperplane $p \subset H^1(\mathcal{I}_\xi(c_1(E) + K_S))$.

4 The First Interpretation – moduli spaces of stable pairs

Now we will consider the space MP^d_{d+1} of pairs (2.2) only. A polygon $\Delta = \sum l_i$ is called *regular* if $i \neq j \Rightarrow l_i \neq l_j$, and $l_i \cap l_j = l_k \cap l_n \Rightarrow (i,j) = (k,n)$. That is, all the sides of Δ are distinct, and all the vertices of Δ (= intersections of sides) are different too.

A pair (Δ, C) is called *regular* if Δ is regular and C is nonsingular.

Let P^0_r be the open subset of regular polygons. Then we have the open subset

$$M_0P^d_{d+1} = p_\Delta(P^0_r)^{-1} \cap p_c^{-1}\Big(p_c(\mathrm{MP}^d_{d+1}) \setminus V_{\text{sing}} \cap p_c(\mathrm{MP}^d_{d+1})\Big) \quad (4.1)$$

of regular pairs.

Every regular polygon $\Delta = \bigcup_{i=1}^r l_i$ defines a cycle of points

$$\Delta^* = l_1^* + \cdots + l_r^* \quad (4.2)$$

on the *dual plane* $\mathbb{P}^{2*}$. It is a fine cycle, and we want to consider it as a cluster (0-dimensional subscheme) of the dual plane.

On the other hand, Δ also determines the cycle of vertices

$$\mathrm{Ver}\Delta = \bigcup_{i,j} l_i \cap l_j \quad (4.3)$$

on the plane $\mathbb{CP}^2$ itself. The cluster Δ^* defines an ideal sheaf $\mathcal{I}_{\Delta^*}$, and the family of extensions

$$0 \to \mathcal{O}_{\mathbb{P}^{2*}} \to E \to \mathcal{I}_{\Delta^*}(2) \to 0 \quad (4.4)$$

parametrized by the space $\mathbb{P}H^1(\mathcal{I}_{\Delta^*}(-1))^*$ (see the end of the previous section).

From the exact sequence

$$0 \to \mathcal{I}_{\Delta^*}(-1) \to \mathcal{O}_{(\mathbb{CP}^2)^*}(-1) \to \mathcal{O}_{\Delta^*}(-1) \to 0$$

we get the isomorphism

$$H^1(\mathcal{I}_{\Delta^*}(-1) = H^0(\mathcal{O}_{\Delta^*}(-1)), \tag{4.4'}$$

so the extension (4.4) is given by a hyperplane in $H^0(\mathcal{O}_{\Delta^*}(-1))$.

On the other hand, the space of curves of degree d circumscribing Δ is the following:

$$|d \cdot l - \mathrm{Ver}\Delta| = \mathbb{P}H^0(\mathcal{I}_{\mathrm{Ver}_\Delta}(d)). \tag{4.5}$$

It's easy to see that ranks of the spaces (4.4') and (4.5) are equal. We would like to prove that

$$H^1(\mathcal{I}_{\Delta^*}(-1) = H^0(\mathcal{I}_{\mathrm{Ver}_\Delta}(d))^*. \tag{4.6}$$

Let me emphasize again that on the left-hand side we have a sheaf on $\mathbb{CP}^2$ but on the right-hand side we have a sheaf on the *dual* plane $(\mathbb{CP}^2)^*$.

Remark: Actually the proof of this equality is a very good exercise for David Eisenbud's lectures.

Here is the heart of our lectures: for a geometric object (Δ, C) on the plane $\mathbb{CP}^2$ we get a new interpretation as a pair (s, E) (see the end of the §3) on the dual plane $(\mathbb{CP}^2)^*$.

Let $\mathbb{CP}^2 = \mathbb{P}T$, where $T = \mathbb{C}^3$, and $(\mathbb{CP}^2)^* = \mathbb{P}T^*$. Let

$$H^1(\mathcal{I}_{\Delta^*}(k)) = V_k. \tag{4.7}$$

Then every line l on $(\mathbb{CP}^2)^*$ defines a homomorphism

$$H^1(\mathcal{I}_{\Delta^*}(-1)) = V_{-1} \to H^1(\mathcal{I}_{\Delta^*}) = V_0 \tag{4.8}$$

given by multiplication by ϕ_l.

When l sweeps out $\mathbb{P}T = ((\mathbb{CP}^2)^*)^*$, we get a homomorphism

$$T \otimes V_{-1} \to V_0, \tag{4.9}$$

which we can consider as a homomorphism of vector bundles on $\mathbb{CP}^2$!

$$V_{-1} \otimes \mathcal{O}_{\mathbb{CP}^2}(-1) \xrightarrow{\varphi} V_0 \otimes \mathcal{O}_{\mathbb{CP}^2}. \tag{4.10}$$

The homomorphism φ is nothing other than a $(d+1) \times d$ matrix of linear forms on $\mathbb{CP}^2$, which we can extend to the exact sequence

$$0 \to \mathcal{O}_{\mathbb{CP}^2}(-d-1) \to V_{-1} \otimes \mathcal{O}_{\mathbb{CP}^2}(-1) \xrightarrow{\varphi} V_0 \otimes \mathcal{O}_{\mathbb{CP}^2} \to \operatorname{coker} \to 0 \tag{4.11}$$

It is easy to see that $\operatorname{Supp}\operatorname{coker} = \operatorname{Ver}\Delta$.

Now applying the functor $\operatorname{Hom}(*, \mathcal{O}_{\mathbb{CP}^2}(-1))$ to this exact sequence, we get *Eagon–Northcott resolution*

$$V_0^* \otimes \mathcal{O}_{\mathbb{CP}^2}(-1) \to V_{-1}^* \otimes \mathcal{O}_{\mathbb{CP}^2} \to \mathcal{I}_{\operatorname{Ver}\Delta}(d) \to 0 \tag{4.12}$$

of the ideal sheaf $\mathcal{I}_{\operatorname{Ver}\Delta}(d)$.

Remark: Our Eagon-Northcott resolution is a slight generalization of the Koszul complex (see Eisenbud's lectures).

Now the cohomology long exact sequence of (4.11) provides the required equality (4.6) and an embedding

$$M_0 P_{d+1}^d \hookrightarrow \mathcal{MP}(2, 2, d+1) \tag{4.13}$$

to the moduli space $\mathcal{MP}(2,2,d+1)$ of *stable pairs* (s, E) where E is a vector bundle of rank 2 with $c_1 = 2$, $c_2 = d+1$. Here the zero set of s is a simple cluster in the dual plane, that is, a $(d+1)$-gon in $\mathbb{CP}^2$.

Now the left-hand side of (4.13) admits a projection map p_c to $|d \cdot l|$, and the right-hand side admits the projection on the second component – the vector bundle.

To compare these projections and to compute the fibres of p_c, we have to consider a new geometric object, the *noncommutative plane.*

5 Noncommutative Planes

For any pair $(\Delta, C) \in M_0P^d_{d+1}$, the nonsingular curve C contains the effective divisor

$$\mathrm{Ver}\Delta = \bigcup_{i,j}(l_i \cap l_j) \qquad (5.1)$$

(4.3) of degree $\frac{1}{2}d(d+1)$. Let $\mathcal{O}_C(h) = \mathcal{O}_{\mathbb{CP}^2}(1)_{|C}$.

Lemma 5.1: *The divisor class*

$$\mathrm{Ver}\Delta - 2h = \theta \qquad (5.2)$$

is a regular theta characteristic of C. That is, $2\theta = K_C$ is the canonical class of C, and $h^0(\mathcal{O}_C(\theta)) = 0$, in other words, this theta characteristic is ineffective.

Proof: Consider the polygon as a curve of degree $d+1$. Then the support of the intersection

$$\mathrm{Supp}(\Delta \cdot C) = \mathrm{Ver}(\Delta),$$

because for every line l_i the intersection

$$C \cap l_i = l_i \cap (\bigcup_{j \neq i} l_j).$$

(Both sides have degree d and by definition the curve C contains the set in the right-hand side of this equality). Now by definition

$$\mathrm{Ver}\Delta = \mathrm{Sing}\Delta$$

and every singular point of Δ is quadratic. Hence as divisor classes on C, we have

$$2\mathrm{Ver}\Delta = C \cdot \Delta = (d+1)h,$$

and

$$2\mathrm{Ver}\Delta - 4h = (d-3)h = K_C$$

by the adjunction formula.

Now if $\mathrm{Ver}\Delta - 2h = \eta$ is effective then

$$\eta = (d-1)h - \mathrm{Ver}\Delta$$

and there exists a curve C' of degree $(d-1)$ which contains $\mathrm{Ver}\Delta$. But then C' and Δ have a common component, because

$$C' \cdot \Delta \geq 2 \deg \mathrm{Ver}\Delta = d(d+1) > \deg C' \cdot \deg \Delta = (d-1) \cdot (d+1).$$

Thus

$$C' = C_0 + \bigcup_{i=1}^{n} l_i,$$

where C_0 doesn't contain lines. Repeating this arguments for C_0 and Δ_{d+1-n}, we get a contradiction. Q.E.D.

Now the pair (C, θ) defines a net of quadrics. Namely, if θ is an ineffective theta characteristic on C then the complete linear system $|\theta + h|$ is base point free and $h^0(\mathcal{O}_C(\theta + h)) = d$. Consider $\mathcal{O}_C(\theta + h)$ as a $\mathcal{O}_{\mathbb{CP}^2}$-sheaf, and the canonical surjective map

$$H^0(\mathcal{O}_C(\theta + h)) \otimes \mathcal{O}_{\mathbb{CP}^2} \to \mathcal{O}_C(\theta + h) \to 0.$$

We have the exact sequence

$$0 \to \ker \xrightarrow{\alpha} H^0(\mathcal{O}_C(\theta + h)) \otimes \mathcal{O}_{\mathbb{CP}^2} \to \mathcal{O}_C(\theta + h) \to 0,$$

and it is easy to see that

$$\ker = H^0(\mathcal{O}_C(\theta + h))^* \otimes \mathcal{O}_{\mathbb{CP}^2}(-1)$$

and we have the net of correlations

$$\alpha\; H \otimes \mathcal{O}_{\mathbb{CP}^2}(-1) \to H^* \otimes \mathcal{O}_{\mathbb{CP}^2} \tag{5.3}$$

where $H = H^0(\mathcal{O}_C(\theta + h))^*$.

Under any identification $H = H^*$ and a choice of the homogeneous coordinates $(\lambda_0, \lambda_1, \lambda_2)$ of $\mathbb{CP}^2$, we can consider the homomorphism (5.3) as a linear combination of a triple of symmetric $d \times d$ matrices

$$\lambda_0 \cdot A_0 + \lambda_1 \cdot A_1 + \lambda_2 \cdot A_2, \tag{5.4}$$

where the equation of curve C is

$$\varphi_C = \det(\lambda_0 \cdot A_0 + \lambda_1 \cdot A_1 + \lambda_2 \cdot A_2) \tag{5.5}$$

and we can consider the line bundle $\mathcal{O}_C(\theta + h)$ as the family of cokernels of the net of correlations (5.3).

Now the group $\mathrm{GL}(d, \mathbb{C})$ acts on the set of triples in the usual way:

$$g(A_0, A_1, A_2) = (gA_0g^*, gA_1g^*, gA_2g^*) \tag{5.6}$$

Let $\{(A_0, A_1, A_2)\}^{\mathrm{SS}}$ be the set of semistable points with respect to this action. Then the variety

$$\{(A_0, A_1, A_2)\}^{\mathrm{SS}} \;/\; \mathrm{PGL}(d, \mathbb{C}) = \mathcal{P}_d^2 \tag{5.7}$$

is called the *noncommutative plane.*

C.T.C. Wall proved that

$$\begin{gathered}(A_0, A_1, A_2) \in \{(A_0, A_1, A_2)\}^{\mathrm{SS}} \Rightarrow \\ \phi_C = \det(\lambda_0 \cdot A_0 + \lambda_1 \cdot A_1 + \lambda_2 \cdot A_2) \neq 0.\end{gathered} \tag{5.8}$$

Thus we have a regular map

$$p_c\, \mathcal{P}_d^2 \to |d \cdot l| \tag{5.9}$$

sending a triple to ϕ_C (5.5). Thus

$$\deg p_c = 2^{g-1} \cdot (2^g + 1), \quad \text{where } g = \tfrac{1}{2}(d-1)(d-2)$$

is the number of even theta characteristics of a nonsingular plane curve of degree d.

Now assume that a triple (A_0, A_1, A_2) satisfies $\det A_0 \neq 0$, and consider the skew symmetric matrix

$$[A_0, A_1 \wedge A_2] = A_1 \cdot A_0^{-1} \cdot A_2 - A_2 \cdot A_0^{-1} \cdot A_1. \tag{5.10}$$

It easy to see that the rank of this matrix rankα is an invariant of a class of a net of quadrics. Thus we have a filtration

$$\mathcal{P}_d^2(0) \subset \mathcal{P}_d^2(2) \subset \cdots \subset \mathcal{P}_d^2, \tag{5.11}$$

where

$$\mathcal{P}_d^2(2r) = \{(A_0, A_1, A_2) \mid \mathrm{rank}[A_0, A_1 \wedge A_2] \leq 2r\}.$$

Now we have the pseudoclassical

Problem: What is

$$\deg p_c(\mathcal{P}_d^2(2r)) \cdot \deg p_c? \tag{5.12}$$

The relationship between our geometric objects is following

Proposition 5.1:

1. $p_c(\mathrm{MP}_{d+1}^d) \subset p_c(\mathcal{P}_d^2(2))$;
2. $d \leq 5 \Rightarrow p_c(\mathrm{MP}_{d+1}^d) = p_c(\mathcal{P}_d^2(2))$.

To prove these statements, we need a final interpretation of the geometric objects.

Consider the flag diagram

$$\begin{array}{ccc} & F = \{p \in l\} & \\ {}^{p}\swarrow & & \searrow^{q} \\ \mathbb{CP}^2 & & (\mathbb{CP}^2)^* \end{array} \tag{5.13}$$

Then we can apply the functor $q_* \circ p^*$ for any net of quadrics α (5.3). We get the exact sequence of sheaves on the dual plane $(\mathbb{CP}^2)^*$:

$$0 \to H \otimes \mathcal{O}_{(\mathbb{CP}^2)^*}(-1) \overset{q_* \circ p^*(-1)(\alpha(1))}{\longrightarrow} H^* \otimes \Omega(1) \to \mathrm{coker}\alpha \to 0 \tag{5.14}$$

and as second invariant of a net we have the number

$$\mathrm{rank}\mathrm{Hom}(\mathrm{coker}\alpha, \mathcal{O}_{(\mathbb{CP}^2)^*}).$$

But actually it isn't a new invariant, because of the following result:

Barth's theorem (see [B]):

$$\operatorname{rank}\operatorname{Hom}(\operatorname{coker}\alpha, \mathcal{O}_{(\mathbb{CP}^2)^*}) = d - \operatorname{rank}\alpha$$

where $2r$ *is the rank of the net of quadrics* (5.10)–(5.11).

Thus for every net α of rank 2 on $\mathbb{CP}^2$ we have the complex on its dual plane

$$0 \to H \otimes \mathcal{O}_{(\mathbb{CP}^2)^*}(-1) \xrightarrow{q_* \circ p^*(-1)(\alpha(1))} H^* \otimes \Omega(1) \xrightarrow{\text{can}} \mathbb{C}^{d-2} \otimes \mathcal{O}_{(\mathbb{CP}^2)^*} \to 0 \tag{5.15}$$

which is called a *monad*, and the middle cohomology

$$\ker \text{can}/\text{im} q_* \circ p^*(-1)(\alpha(1)) = E \tag{5.16}$$

is a semistable torsion free sheaf of rank 2 with the Chern classes $c_1 = 0$, $c_2 = 2$.

So we have the map

$$\mathcal{P}_d^2(2)) \to \overline{M(2,0,d)}$$

to the Gieseker closure of the moduli space of stable vector bundles on $\mathbb{CP}^2$. The construction of the inverse map is as follows: a point $p \in \mathbb{CP}^2$ gives a line $l_p \in (\mathbb{CP}^2)^*$ in the dual plane. A line $l_p \in (\mathbb{CP}^2)^*$ is called a *jumping line* for E if

$$E_{|l_p} \neq \mathcal{O}_{l_p} \oplus \mathcal{O}_{l_p}, \quad \text{which happens iff} \quad h^0(E(-1)_{|l_p}) \neq 0. \tag{5.17}$$

Thus in the dual plane $(\mathbb{CP}^2)^*$ we have the curve $C(E)$ of jumping lines of E.

Now it is easy to see that $H^1(E(-2)) = H = \mathbb{C}^d$ and the Serre dual space $H^1(E(-1)) = H^*$. Now multiplication by the equation of any line ϕ_{l_p} defines the correlation

$$H^1(E(-2)) = H \to H^1(E(-1)) = H^* \tag{5.18}$$

as an element of the net of correlations (5.4) $C(E) = C(5.5)$.

Now consider a regular pair $(\Delta, C) \in M_0 P^d_{d+1}$ (see (5.1)), where $\Delta = \bigcup l_i$. Then we have the chain of identifications:

$$(\Delta, C) = (\mathbb{C}^* \cdot s, E) \tag{5.19}$$

where s is a regular section of E;

$$E = (C, \theta) = (\alpha) \tag{5.19'}$$

So a pair (Δ, C) is geometrically equivalent to the exact sequence on $(\mathbb{CP}^2)^*$:

$$0 \to \mathcal{O}_{(\mathbb{CP}^2)^*}(-1) \to E \to \mathcal{I}_{\Delta^*}(1) \to 0 \tag{5.20}$$

where $\mathcal{I}_\Delta$ is the ideal sheaf of 0-dimensional cycle Δ^* on $(\mathbb{CP}^2)^*$.

Corollary: *The space of circumscribed $(d+1)$-gons to a Darboux curve C is a rational irreducible variety.*

Indeed, it is birationally equivalent to $\mathbb{P}H^0(E)$!

These geometric identifications were done for "regular" geometric objects. It is reasonable to construct some "natural" nonsingular compactification of the space of regular objects sending the computation of constants of type (5.12), (2.11), (2.5) to the regular procedure of computations of Chern classes of standard vector bundles on "moduli space" our geometric figures.

6 Compactifications

A regular $(d+1)$-gon (2.1) is of course a curve of degree $d+1$ on $\mathbb{CP}^2$, and so the space P_{d+1} of all polygons is a compact irreducible subvariety in the complete linear system $|(d+1) \cdot l|$. Geometrically,

$$P_{d+1} = S^{d+1}(\mathbb{CP}^2)^* \tag{6.1}$$

is the $(d+1)$-st symmetric power of $(\mathbb{CP}^2)^*$, a rather singular algebraic variety. Fortunately for algebraic surfaces, there is a canonical desingularization of it. Let us recall that a regular polygon Δ defines a zero dimensional subscheme Δ^* (4.2). Thus as a compactification of the space of regular polygons on $\mathbb{CP}^2$, we can consider the moduli space of zero dimensional subscheme of $(\mathbb{CP}^2)^*$ of degree d:

$$\mathrm{Hilb}^{d+1} = M(1, 0, d+1) \tag{6.2}$$

which is called the *Hilbert scheme* on $(\mathbb{CP}^2)^*$. The beautiful and very important theory of Hilbert schemes says that for a nonsingular algebraic surface, this scheme is again nonsingular (see [F]). The space of extensions of type (5.20) is given by the projectivization of the vector space $H^1(\mathcal{I}_\xi(-1))$ (see (3.8)), because of $c_1(E(1)) = 2h$, and because the canonical class of the plane is given by $K_{(\mathbb{CP}^2)^*} = -3h$. Thus it is natural to represent the space of all nontrivial extensions (5.20) as a projectivization of a vector bundle on Hilb^{d+1}. Of course this variety is nonsingular.

From now on, all of our geometric objects are defined on the dual projective plane $(\mathbb{CP}^2)^*$, and we omit the star.

First of all, on the Hilbert scheme we have the special divisor class H defined by clusters intersecting a fixed line. On the other hand, the Hilbert scheme defines the universal subscheme $Z \subset \mathbb{CP}^2 \times \mathrm{Hilb}$, and the two projection maps to the direct components define the diagram

$$\begin{array}{ccccc} & & Z_{d+1} & & \\ & {}^{p_{\mathbb{CP}^2}}\swarrow & & \searrow^{p_H} & \\ \mathbb{CP}^2 & & & & \mathrm{Hilb}^{d+1} \end{array} \tag{6.3}$$

For any divisor class $\mathcal{O}_{\mathbb{CP}^2}(k)$, consider the vector bundle

$$\mathcal{E}_k = R^0 p_H(p_S^* \mathcal{O}_{\mathbb{CP}^2}(k)). \tag{6.4}$$

These sheaves are locally free, because the canonical homomorphism is surjective.

In particular, in our case $k = -1$, the fibre of this vector bundle over $\xi \in \mathrm{Hilb}$ is $H^0(\mathcal{O}_\xi(-1))$. Now the cohomology long exact sequence of

$$0 \to \mathcal{I}_\xi(-1) \to \mathcal{O}(-1) \to \mathcal{O}_\xi \to 0 \tag{6.5}$$

gives the isomorphism

$$H^0(\mathcal{O}_\xi(-1)) = H^1(\mathcal{I}_\xi(-1)) \tag{6.6}$$

Thus the space of all extensions of type (5.20) is the projectivization

$$\mathbb{P}\mathcal{E}^*_{-1}. \tag{6.7}$$

The next thing we have to understand is that our constant $s_{d+1}(d)$ (2.5) is the top Segre class of the standard vector bundle on Hilb:

$$s_{d+1}(d) = s_{\text{top}}(\mathcal{E}_{-1}(H)) \tag{6.8}$$

To see this, we have to return to the isomorphism (4.6) and remark that for $s \in H^0(\mathcal{I}_{\text{Ver}_\Delta}(d)$ the condition "a curve $C = (s)_0$ passes through a point $l^* \in (\mathbb{CP}^2)^*$" defines a section of $\mathcal{E}_{-1}(H)$! Thus by the definition of the Segre class, $3d+2$ general points determine a general $(3d+2)$-dimensional subspace W of $H^0(\mathcal{E}_{-1}(H))$. Now the canonical homomorphism

$$W \otimes \mathcal{O}_{\text{Hilb}} \xrightarrow{\text{can}} \mathcal{E}_{-1}(H)$$

is general enough and

$$\deg \operatorname{coker} \text{can} =: s_{\text{top}}(\mathcal{E}_{-1}(H)) = s_{d+1}(d) \tag{6.9}$$

is the number of Darboux curves through $3d+2$ general points.

Using this beautiful interpretation, G. Ellingsrud and S. Strømme computed (2.12) (using Bott's formula for the $\mathbb{C}^*$ action on Hilb, see [E–S]):

$$\begin{aligned}
s_5(4) &= 54 \\
s_6(5) &= 2540 \\
s_7(6) &= 583020 \\
s_8(7) &= 99951390 \\
s_9(8) &= 16059395240 \\
s_{10}(9) &= 2598958192572.
\end{aligned}$$

This list can be extended if your computer is good enough and you have S. A. Strømme as a collaborator. But to understand the nature of these numbers (it's new shape of mathematical questions, isn't it?), you have to use new identifications, proposed below, and the collection of beautiful new results provided by differential topologists such as Fintushel and Stern, Kotschick and Lisca and many others. We will discuss this in the next section.

Thus this story isn't finished yet. The nonsingular compactification (6.7) of the moduli space of pairs (Δ, C) (on $(\mathbb{CP}^2)^*!$) is called the moduli space of stable pairs $(\mathcal{C}^* \cdot s, E)$ (see [T3], Lecture 6):

$$\mathbb{P}\mathcal{E}^*_{-1} = \mathcal{MP}(2,2,d+1). \tag{6.10}$$

Let us consider the general diagram of our identifications:

$$\begin{array}{ccccccc} & \mathbb{P}\mathcal{E}^*_{-1} & = & \mathcal{MP}(2,2,d+1) & & & \\ {}^{p_\Delta}\swarrow & & \searrow^{p_c} & & & \searrow^{p_E} & \\ \mathrm{Hilb}^{d+1} & & |d \cdot l| & & \overleftarrow{p_j} & & M(2,2,d+1) \end{array} \tag{6.11}$$

where $M(2,2,d+1)$ is the moduli space of semistable bundles, the map p_E sends a pair (E,s) to the vector bundle E, and p_j sends the vector bundle E to its curve of jumping lines (5.17). Now twisting sheaves by $\mathcal{O}_{\mathbb{CP}^2}(1)$ gives the isomorphism

$$M(2,0,d) = M(2,2,d+1)$$

and we can use the chain of identifications (5.19).

Remark: The extension of the map p_j to the compactification of moduli spaces is a nontrivial task.

The map p_E is only a rational map, as treated in detail in [T–T]: we described there what has to be blown up and what gets blown down. But in any case, we can describe the image of this rational map: the final moduli space contains the Brill–Noether locus

$$M_1 = \{E \in M(2,2,d+1) \mid h^0(E) \geq 1\}. \tag{6.12}$$

Thus we have

$$M_1 = p_E(\mathcal{MP}(2,2,d+1)) \tag{6.13}$$

This is just what we need. Now by the R–R theorem,

$$d \leq 5 \Rightarrow M_1 = M(2,2,d+1). \tag{6.14}$$

Moreover, if $d = 4$, that is, for the Lüroth quartics we have the following construction: let Q be a nonsingular plane conic and $|\eta|$ be a general linear pencil of divisors of degree 5 on Q. For an element of this pencil $p_1 + \cdots + p_5$, consider the pentagon $\Delta = \bigcup l_i$, where l_i is the tangent line to Q at p_i. When elements sweep out this pencil, the cycles Ver$l(\Delta)$ sweep out a quartic curve C.

So we have the divisor classes

$$\mathcal{O}_{\mathbb{P}\mathcal{E}^*_{-1}}(1) = p_c^*(\mathcal{O}_{|d \cdot l|}(1)) = \mathcal{O}_{\mathcal{MP}(2,2,d+1)}(D) \tag{6.15}$$

$$\mu(l) = p_j^*(\mathcal{O}_{|d \cdot l|}(1)) \tag{6.16}$$

on $\mathcal{MP}(2,2,d+1)$ and M_1, which are related by the birational map p_E. This situation is a beautiful exercise in practical birational geometry. As you know from Miles Reid's lectures, a birational map may well alter the degree of a divisor. But in our case (this beautiful observation is due to Dmitry Orlov) the existence of the regular maps p_c and p_j relating our divisors as in (6.15) and (6.16) gives the equality

$$s_{\text{top}}(\mathcal{E}_{-1}(H)) = c_{d+1}(d) = D^{3d+2} = (\mu(l))^{3d+2} \tag{6.17}$$

The last dearth of this beautiful chain of identifications of geometric objects is the following: the points of each of the moduli spaces $\mathcal{MP}(2,2,d+1)$, $M(2,2,d+1)$ and M_1 describe geometric objects on the same algebraic surface, the plane $\mathbb{CP}^2$. But the construction of the divisor class $\mu(l)$ sends us to geometric objects (jumping curves) on the dual plane $(\mathbb{CP}^2)^*$. From the *algebraic geometric* point of view this is reasonable, but we promised to extend these constructions to objects of *differential geometry.* As a differential topological object, the projective plane doesn't define the dual plane. Avoiding this obstacle, we would like to describe the constants $s_{d+1}(d)$ in terms of $\mathbb{CP}^2$. That is, we want to define μ-class $\mu(l)$ in terms of objects on $\mathbb{CP}^2$ only.

Now on the direct product $\mathbb{CP}^2 \times M(2,2,d+1)$, the universal sheaf $\mathcal{F}$ exists locally only (for technical details see [T2]), but the Pontrjagin class

$$p_1(\mathcal{F}) = 4c_2(\mathcal{F}) - c_1^2(\mathcal{F}) \in \mathrm{Pic}\mathbb{CP}^2 \otimes \mathrm{Pic}M(2,2,d+1) \quad (6.18)$$

is defined correctly in any case. The intersection number on $\mathrm{Pic}\mathbb{CP}^2$ gives the isomorphism $(\mathrm{Pic}\mathbb{CP}^2)^* = \mathrm{Pic}\mathbb{CP}^2$ so we can consider the Pontrjagin class (6.18) as the homomorphism

$$\mu = 1/4p_1(\mathcal{F}) : \mathrm{Pic}\mathbb{CP}^2 \to \mathrm{Pic}M(2,2,d+1). \quad (6.19)$$

Now it's easy to see that the divisor class $\mu(l)_{|M_1}$ is just (6.16).

Now to get our classical enumerative algebraic geometry constants (2.11) in the more general set-up, we have to use the equality

$$c_{d+1}(d) = (\mu(l))^{3d+2}, \quad (6.20)$$

and to extend the definitions of $M_1 \subset M(2,2,d+1)$ and $\mu(l)$ in differential geometric terms. Of course now $M_1 \subset M(2,2,d+1)$ will be compact spaces and $\mu(l) \in H^2(M(2,2,d+1)$ is a 2-cohomology class only but this is quite enough to define the constant (6.20)!

7 Differential Geometry

Algebraic geometry can be considered as a part of differential geometry, namely as Kähler geometry. Then we have to use new notions like connections, differential forms and so on; you can learn this approach from the standard monograph [G–H]. We also strongly recommend the monograph [D–K] as a unique source of new style to use these ideas. But it is very important to understand that classical algebraic geometry is a foundation of almost all the local constructions of Riemannian geometry. We will discuss the special case of the projective plane $\mathbb{CP}^2$, but you can determine the generality quantor yourself.

Let M be the underlying 4-manifold of the complex projective plane $\mathbb{CP}^2$. Any Riemannian metric g on M defines a decomposition of the complexified tangent bundle $TM_{\mathbb{C}}$ as a tensor product

$$TM_{\mathbb{C}} = (W^-)^* \otimes W^+$$

of two rank 2 Hermitian vector bundles $W^{\pm}$ with

$$c_1(W^{\pm}) = -3h,$$

where h is the generator of $H^2(M, \mathbb{Z})$.

Write $*$ for the Hodge star operator on $\Omega^2(M)$ determined by the metric g. Moreover, for any U(2)-bundle E on M of topological type $(c_1 = 2, c_2 = d+1)$ and any Hermitian connection $a \in \mathcal{A}_h$ on E, putting any Hermitian connection ∇_0 on $\bigwedge^2 W^{\pm}$ gives a coupled Dirac operator

$$D_a^{g,\nabla_0}\ \Gamma^\infty(E \otimes W^+) \to \Gamma^\infty(E \otimes W^-). \tag{7.1}$$

Now the orbit space of irreducible connections modulo the gauge group

$$\mathcal{B}(2,2,d+1) = \mathcal{A}_h^*(2,2,d+1)/\mathcal{G}$$

contains the subspace

$$\begin{aligned} \mathcal{M}^g(2,2,d+1) &= \{(a) \in \mathcal{M}^g(2,2,d+1) \mid *F_a = -F_a\} \\ &\subset \mathcal{B}(E) \end{aligned} \tag{7.2}$$

of antiselfdual connections with respect to the Riemannian metric g (here F_a is the curvature form of a connection a).

Now we can consider the subspace of jumping connections:

$$\begin{aligned} \mathcal{M}_1^g(d) &= \{(a) \in \mathcal{M}^g(2,2,d+1) \mid \operatorname{rank}\ker D_a^{g,\nabla_0} \geq 1\} \\ &\subseteq \mathcal{M}^g(2,2,d+1). \end{aligned} \tag{7.3}$$

The virtual codimension of $\mathcal{M}_{(d)}^g$ (that is, the expected codimension determined by the Atiyah–Singer index theorem) is given by

$$\text{v.codim}\mathcal{M}_1^g(d) = 2 - 2\chi = 2(5-d), \tag{7.4}$$

where χ is the index of the coupled Dirac operator (7.1), which depends only on the Chern classes of E.

So you can see that for $d \leq 5$

$$\mathcal{M}_1^g(d) = \mathcal{M}^g(2, 2, d+1) \tag{7.5}$$

Please compare this fact with (6.14)!

For a generic metric g, the moduli spaces $\mathcal{M}^g(2, 2, d+1)$ and $\mathcal{M}_1^g(d)$ (7.3) are smooth manifolds of the expected dimension with regular ends (see [D–K] and [P–T], Chap. 2, §3). Moreover, $\mathcal{M}^g(2, 2, d+1)$ admits a natural orientation (see [D–K]) inducing an orientation on $\mathcal{M}_1^g$, because its normal bundle has a natural complex structure. This orientation is described in detail in [P–T], Chap. 1, §5.

Moreover there exists the so-called Uhlenbeck compactification of our moduli spaces

$$\overline{\mathcal{M}^g(2, 2, d+1)} \supseteq \overline{\mathcal{M}_1^g(d)} \tag{7.6}$$

Now for any element of our filtration the first Pontrjagin class of the universal connection on the direct product $M \times \mathcal{M}^g(2, 2, d+1)$ (by the slant product) defines cohomological correspondences

$$\begin{aligned} \mu_d : H_i(M, \mathbb{Z}) &\rightarrow H^{4-i}(\overline{\mathcal{M}^g(2, 2, d+1)}, \mathbb{Z}), \\ \mu_d^1 : H_i(M, \mathbb{Z}) &\rightarrow H^{4-i}(\overline{\mathcal{M}_1^g(d)}, \mathbb{Z}), \end{aligned}$$

and two collections of numbers

$$D_g(d) = (\mu_d(h))^{4d-3}, \tag{7.7}$$

the so-called *Donaldson numbers* (Donaldson polynomials) of $\mathbb{CP}^2$, and

$$s\gamma_g^d = (\mu_d^1(h))^{3d+2}. \tag{7.8}$$

Now suppose as a special case that our metric g is the Fubini–Study metric g_{F-S}. In this case, by the Donaldson–Uhlenbeck identification theorem, we have

$$\mathcal{M}^{g_{F-S}}(2, 2, d+1) = M(2, 2, d+1),$$

where the right-hand side is the moduli space of holomorphic stable bundles on $\mathbb{CP}^2$ (6.11).

Making this identification $(a) = E$, we have identifications

$$\ker D_a^{g_{F-S}} = H^0(E) \oplus H^2(E) \text{ and } \operatorname{coker} D_a^{g_{F-S}} = H^1(E), \quad (7.9)$$

where $H^i(E)$ denote coherent cohomology groups (see [D–K]).

But by Serre duality $H^2(E) = H^0(E(-2))^* = 0$, by the stability of E. Thus the subspace $\mathcal{M}_1^{g_{F-S}}(d)$ is

$$\mathcal{M}_1^{g_{F-S}}(d) = \{E \in M(2, 2, d+1) \mid h^0(E) \geq 1\}, \quad (7.10)$$

that is (see (6.12)),

$$\mathcal{M}_1^{g_{F-S}}(d) = M_1$$

and our constants (7. 8)

$$s\gamma_{F-S}^d = s_{d+1}(d) \quad (7.11)$$

are constants (2.5) and (6.8).

These integers don't depend on the metric g, because the space of all Riemannian metrics is contractible. (In fact, to be rigorous, we have to use much more sophisticated bordism arguments, similar to those in [D–K], where the same statement was proved for the Donaldson's numbers (7. 7)).

Let us remark that the initial terms of both collections (7.7) and (7.8) are coincidence by (7.5) and (6.14)

$$\begin{aligned} D_g(4) &= s\gamma_g^4 = 54; \\ D_g(5) &= s\gamma_g^5 = 2540; \end{aligned}$$

and the collection of Donaldson's constants was extended to infinity by Ellingsrud and Göttsche (see [E–G]):

$$\begin{aligned} D_g(6) &= 233208; \\ D_g(7) &= 35825553; \\ D_g(8) &= 8365418914; \\ D_g(9) &= 2780195996868; \\ D_g(10) &= 1253558847090600; \\ &\text{etc.} \end{aligned}$$

Let me remark that we have got the following striking fact:

Theorem: *The constants* (7.11) = (2.5) = (6.8) *are invariants of the underlying differentiable structure of* $\mathbb{CP}^2$.

It is well known that the complex structure on M is unique. Recently it was proved that the symplectic structure on M is unique. The next question is the following differentiable version of the Poincaré conjecture for $\mathbb{CP}^2$:

Conjecture (DPC for $\mathbb{CP}^2$): *The complex projective plane* $\mathbb{CP}^2$ *has a unique differentiable structure.*

In particular this statement would imply the following fact

Corollary: *The constants* (7.11) = (2.5) = (6.8) *are invariants of the topological structure of* $\mathbb{CP}^2$.

There is overwhelming direct evidence for this statement, and hence a partial confirmation of Conjecture DPC for $\mathbb{CP}^2$. Namely there are two possible methods to construct our constants using the topological structure of the plane only. The first approach is related to the following fundamental problem:

Hilbert scheme problem: *Can we give a purely topological construction of the Hilbert scheme* Hilb^d *and of the standard vector bundles* (6.3)*?*

Of course this problem is interesting in full generality for all 4-manifolds. If we could realize the scheme (6.3) topologically then, as proposed by G. Ellingsrud, we could use induction over d to prove

Proposition 7.1: *The constants* (6.8) *are topological invariants.*

The second approach to prove the statement of Corollary is related to proving of the same fact for Donaldson's constants (7.7) proposed by Kotschick and Lisca using the Kotschick–Morgan Conjecture from [K–M].

The idea of this program is the following: let us blow up one or two points on $\mathbb{CP}^2$ and for this new surface $\widetilde{\mathbb{CP}^2}$ let us consider the same numbers as (7.7). But now these numbers depend on the Riemannian metric in an essential way. However it can be shown that the dependence of these numbers on the metric can be controlled explicitly! Going from $\mathbb{CP}^2$ to $\widetilde{\mathbb{CP}^2}$, we have to consider the collection $\{\alpha\}_d \subset H^2(\widetilde{\mathbb{CP}^2}, \mathbb{Z})$ of classes such that

$$\alpha = 0 \bmod 2 \quad \text{and} \quad -4d \leq \alpha^2 < 0. \tag{7.12}$$

The intersection of the positive cone in $H^2(\widetilde{\mathbb{CP}^2}, \mathbb{R})$ with the hyperplane $\alpha^{\perp}$ is called a *d-wall* (or the d-wall defined by α). Let Δ_d be the set of open chambers into which the positive cone is divided by all d-walls.

Proposition 7.2:

1. *The constant $D_g(d)$ of (7.7) for $\widetilde{\mathbb{CP}^2}$ depends on the chamber $C \in \Delta_d$ which contains the g-self dual harmonic 2-form.*
2. *If C and C' are chambers then*

$$D_C(d) - D_{C'}(d) = \sum_{\alpha} \delta_d(\alpha) \tag{7.13}$$

where the sum is taken over all d-walls α such that

$$\alpha \cdot C' < 0 < \alpha \cdot C$$

that is over all walls dividing C and C'

Now the following result has been proved:

Proposition 7.3: *The constant $D_g(d)$ of (7.7) for $\widetilde{\mathbb{CP}^2}$ are determined by the difference terms δ_d only.*

The following fact is "almost" proved:

Kotschick–Morgan Conjecture (see [K–M]): *The difference terms δ_d are homotopy invariants.*

So finishing the proof of this conjecture implies the topological definition of the Donaldson constants (7.7).

Remark: Mixing the Hilbert scheme method and the difference terms method is a very fruitful technique. G. Ellingsrud and L. Göttsche [E–G] can use it to compute any Donaldson number exactly, but the real nature of these expressive numbers remains an open question.

Finally, to prove the topological nature of the constants $\sigma\gamma^d_{F-S} = s_{d+1}(d)$ (7.11), we have to mimic these constructions for the moduli spaces of jumping instantons (7.3) or, in full generality, for the spin polynomials (see [T1]) in place of the Donaldson polynomials.

I would like to finish by drawing the following conclusion:

Moral: *Different interpretations of classical algebraic geometric figures provide very fruitful approaches to understanding the nature of results of ENUMERATIVE AG.*

References

[B] W. Barth, *Moduli of vector bundles on the projective plane*, Invent. Math., **42** (1977), 63–91.

[D–K] S. Donaldson and P. Kronheimer, *The Geometry of Four-Manifolds*, Clarendon Press, Oxford, 1990.

[E–G] G. Ellingsrud and L. Göttsche, *Wall change formulas, Bott residue formula and the Donaldson invariants of rational surfaces,* preprint 1995.

[E–S] G. Ellingsrud and S. A. Strømme, *Botts formula and enumerative geometry*, to appear in Journal of the AMS.

[F] J. Fogarty, *Algebraic families on an algebraic surface*, Amer. J. Math., **90** (1968), 511–521.

[G–H] Ph. Griffiths and J. Harris, *The Principles of Algebraic Geometry,* Wiley, New York 1978.

[K–M] D. Kotschick, J. Morgan, *SO3-invariants for 4-manifolds with* $b_2^+ = 1$, *II,* J. Diff. Geom., **39** (1994), 433–456.

[M] R. Miranda, An Overview of Algebraic surfaces, This volume.

[Mo] F. Morley, *On the Lüroth quartic curve,* Amer. J. of Math., **36** (1918).

[Mu] S. Mukai, *Fano 3-folds,* London Math. Soc. Lect. Notes Series, **179** (1992), 255–263.

[P–T] V. Pidstrigach and A. Tyurin, *Invariants of the smooth structures of an algebraic surfaces arising from the Dirac operator,* Izv. AN SSSR, **52:2** (1992), 279–371. (Russian) English transl. in Warwick preprint, **22**(1992).

[T–T] A. Tikhomirov and A. Tyurin, *Application of geometric approximation procedure to computing the Donaldson polynomials for* $\mathbb{CP}^2$, Mathematica Goettingensis, Sonderforschungsbereichs Geometry and Analysis, **Heft 12** (1994), 1–71.

[T1] A. Tyurin, *Spin polynomial invariants of the smooth structures of algebraic surfaces,* Iz. AN SSSR **57:2** (1993), 279–371.

[T2] A. Tyurin, *Canonical and almost canonical spin polynomials of algebraic surfaces,* Proceedings of the Conference "Vector bundles", Durham 1993, **Aspects of Mathematics,** (1993), 250–275.

[T3] A. Tyurin, *Six lectures on four manifolds,* to appear in LNM, (1996), 1–58.

Index